W9-ANN-521

ADVENTURES IN CHEMICAL PHYSICS

A SPECIAL VOLUME OF ADVANCES IN CHEMICAL PHYSICS

VOLUME 132

Adventures in CHEMICAL PHYSICS

ADVANCES IN CHEMICAL PHYSICS VOLUME 132

Edited by

R. STEPHEN BERRY and JOSHUA JORTNER

Series Editor

STUART A. RICE

Department of Chemistry
and
The James Franck Institute
The University of Chicago
Chicago, Illinois

AN INTERSCIENCE PUBLICATION
JOHN WILEY & SONS, INC.

Published by John Wiley & Sons, Inc., Hoboken, New Jersey.
Published simultaneously in Canada.

Library of Congress Catalog Number: 58-9935

ISBN-13 978-0-471-73842-8
ISBN-10 0-471-73842-5

Printed in the United States of America

10 9 8 7 6 5 4 3 2 1

CONTRIBUTORS

R. STEPHEN BERRY, Department of Chemistry, The University of Chicago, Chicago, Illinois 60637, USA

THORSTEN M. BERNHARDT, Institut für Experimentalphysik, Freie Universität Berlin, D-14195 Berlin, Germany

VLASTA BONAČIĆ-KOUTECKÝ, Department of Chemistry, Humboldt-Universität zu Berlin, D-12489 Berlin, Germany

I. P. DAYKOV, Department of Physics, Cornell University, Ithaca, New York, 14850, USA

R. EDWARDS, Department of Mathematics and Statistics, University of Victoria, Victoria, BC, Canada V8W 3P4

GRAHAM R. FLEMING, Department of Chemistry, University of California, Berkeley; and Physical Biosciences Division, Lawrence Berkeley National Laboratory, Berkeley, California 94720-1460, USA

L. GLASS, Centre for Nonlinear Dynamics in Physiology and Medicine, Department of Physiology, McIntyre Medical Sciences Building, McGill University, Montréal, Qúebec, Canada H3G 1Y6

ROBIN M. HOCHSTRASSER, Department of Chemistry, University of Pennsylvania, Philadelphia, Pennsylvania 19104-6323, USA

JOSHUA JORTNER, School of Chemistry, Tel Aviv University, 69978 Tel Aviv, Israel

ROLAND MITRIĆ, Department of Chemistry, Humboldt-Universität zu Berlin, D-12489 Berlin, Germany

A. PROYKOVA, University of Sofia, Atomic Physics, Sofia 1126, Bulgaria

MICHAEL ROSENBLIT, School of Chemistry, Tel Aviv University, Tel Aviv 69978, Israel. *Present address*: Ilse Katz Center for Meso- and Nanoscale Science and Technology, Ben Gurion University of the Negev, Beer Sheva 84105, Israel

GREGORY D. SCHOLES, Lash Miller Chemical Laboratories, University of Toronto, Toronto, Canada M5S 3H6

LUDGER WÖSTE, Institut für Experimentalphysik, Freie Universität Berlin, D-14195 Berlin, Germany

INTRODUCTION

Few of us can any longer keep up with the flood of scientific literature, even in specialized subfields. Any attempt to do more and be broadly educated with respect to a large domain of science has the appearance of tilting at windmills. Yet the synthesis of ideas drawn from different subjects into new, powerful, general concepts is as valuable as ever, and the desire to remain educated persists in all scientists. This series, *Advances in Chemical Physics*, is devoted to helping the reader obtain general information about a wide variety of topics in chemical physics, a field that we interpret very broadly. Our intent is to have experts present comprehensive analyses of subjects of interest and to encourage the expression of individual points of view. We hope that this approach to the presentation of an overview of a subject will both stimulate new research and serve as a personalized learning text for beginners in a field.

Stuart A. Rice

STUART ALAN RICE: SCIENTIST WITHOUT BOUNDS

We, the editors of this volume, have had the remarkably good fortune to be close friends and scientific associates of Stuart Rice almost "from the days when dinosaurs roamed the earth." One of us met Stuart in 1952 on the first day of graduate school at Harvard. The other began working with Stuart ten years later in 1962, five years after Stuart had joined the faculty of the University of Chicago. Stuart came to Harvard from Bronx Science High School and then Brooklyn College. In 1952, when he arrived as a graduate student at Harvard, he joined the research group of Paul Doty, working on biopolymers, including the unwinding of the strands of the DNA helix. Characteristically, that work did not keep him completely occupied, so he simultaneously worked on several other problems such as polyelectrolytes and the folding of proteins, among other things. He became a Junior Fellow at Harvard and switched to doing high-temperature infrared spectroscopy of gaseous molecules, in collaboration with William Klemperer. After a year he moved to Yale, where he spent his second year of the Fellowship in the group of John Kirkwood.

Stuart came to the University of Chicago in 1957, and has been a major figure in physical chemistry and other areas of science for almost 50 years. He made central contributions to the intellectual environment, to the highest-quality scientific endeavor, to the remarkable interdisciplinary scientific collaboration, and to the intense focus on excellence at the University of Chicago. In an admirable way he promoted and perpetuated the unique culture of the University. In 2005, the University of Chicago Alumni Association awarded him its Faculty Achievement Medal, in recognition of what his teaching and guidance meant to many, many students over the years.

Stuart's contributions to science have ranged across virtually the entire domain of modern physical chemistry. His research uses state-of-the-art experimental methods and fundamental theoretical approaches, spanning from isolated molecules to the condensed phase. His work has consistently been pioneering, often constituting the first attack on a new subject and, most characteristically, always addressing an important research area. Stuart as a scientist is universal, deep and demanding of the highest intellectual standards. His work has had, and continues to have, great influence on the development of chemistry and other related areas of science.

Stuart has taught us new fundamental concepts in fields ranging from the nature of liquids through the puzzling subject of liquid metal surfaces to coherent quantum control of chemical reactions. His studies of active control of molecular dynamics can be seen as evolving through a long series of contributions to predecessor forefront areas of physical chemistry. He made seminal advances to the theory of electronic states of molecular solids, polymers, and liquids, including theoretical studies of singlet and triplet exciton band structure, exciton–exciton annihilation reactions, hole and electron mobility and band structure, and exciton states in liquids. Although now 40 years old, these calculations have not been superseded, and recent experimental data testify to their accuracy. The studies of condensed matter electronic structure led to theoretical and experimental studies of radiationless transitions, including landmark experimental and theoretical studies of the vibrational state dependence of the decay of optically excited molecules under collision-free conditions and the generalization of the theory to describe unimolecular reactions. In turn, these studies led to the examination of vibrational energy flow in polyatomic molecules. He was a pioneer in the study of the influence of deterministic classical mechanical chaos on the classical theory of the unimolecular reaction rate, and he has published seminal studies of quantum chaos and its relevance to chemical reactions. These diverse studies laid the foundation for Stuart's development of the theory of optical control of molecular dynamics as applied to controlling product selection in a chemical reaction, introducing the concepts of multiple pulse timing control. Stuart studied the conditions for the existence of the optimal control field for a system with a spectrum that is typical of a reacting molecule. Recently, Stuart has focused his attention on adiabatic transfer processes that can be used to control molecular dynamics and has started to develop the theory of control of molecular dynamics in a liquid, the medium in which the vast majority of chemical reactions take place.

Stuart's studies of the structure of the liquid–vapor interfaces of metals and alloys can also be related to his previous research. He developed the first theory of transport in dense simple fluids that explicitly recognizes, and accounts for, the different dynamics associated with short-range repulsion and longer-ranged attraction. He has contributed to the theory of the three-molecule distribution function in a liquid and the theory of melting, and he developed the Random Network Model of water and the first consistent description of the amorphous solid phase of water. His work provided the first evidence for the existence of a high-density form of amorphous solid water, opening the study of polymorphism in disordered phases. Concurrently, he initiated studies of the structures of the liquid–vapor interfaces of metals and alloys. This work showed that the character of the interaction between the atoms in a liquid metal (e.g., a "pool of mercury") is fundamentally different from that between the atoms in a dielectric liquid. Consequently, it is reasonable to expect that their respective interfaces have different structures. The theoretical challenge is that in the inhomogeneous region the various length scales associated with the width, the depth over which the excess concentration of a segregated component is distributed, the range of positional correlations, the range of the effective ion–ion interactions, and the distances over which the electrons undergo a transition from delocalized states (in the bulk liquid) to localized states (e.g., in the vapor) are all of comparable magnitude. He developed the modern theoretical description of the liquid–vapor interfaces of pure metals and alloys that correctly accounts for the electronic structure of the metal and its dependence on the atomic distribution and composition across the interface, and he advanced the first prediction that the liquid–vapor interface of a metal is stratified and that in a dilute alloy the solute segregates to form a complete monolayer at the interface. These predictions have now been multiply verified in experimental studies. Some recent, fascinating experimental work led to the discovery that some solutes can form a crystalline monolayer in the liquid–vapor interface, a finding that was not anticipated by theory and which does not yet have a theoretical interpretation.

Stuart has held many responsible administrative posts at the University of Chicago, ranging from the Director of the James Franck Institute to the Chairman of the Chemistry Department. Subsequently he became the longest-serving Dean of Physical Sciences in the Division's history. Through all that, Stuart's research continued at full speed and productivity. He served on the National Science Board and received the Presidential Medal of Science of the United States. And he still continues his research program, working closely with his students and postdoctorals, even when he changed roles to Professor Emeritus at Chicago and became Special Advisor to the Director at Argonne National Laboratory.

Stuart has been a major contributor to building and maintaining the strength of the Chemical Sciences in Chicago, in the United States, and throughout the world. The more than 100 Ph.D. research students and many postdoctoral fellows who worked with him have become important figures in the field of physical chemistry and in other areas of science. Stuart has been an advocate for chemistry and for science generally, both nationally and internationally, and has helped to shape the direction of science for the future.

We both want to take this opportunity to express the great pleasure and honor it has been and still is to be Stuart's colleagues and collaborators, and we look forward to many new explorations with him into the mysteries that science can unravel.

R. Stephen Berry
Joshua Jortner

CONTENTS

DYNAMICAL MODELS FOR TWO-DIMENSIONAL INFRARED SPECTROSCOPY OF PEPTIDES†

ROBIN M. HOCHSTRASSER

Department of Chemistry, University of Pennsylvania, Philadelphia, Pennsylvania 19104-6323, USA

CONTENTS

†In recognition of his amazing impact on chemical physics, this survey is dedicated to Stuart Rice on the occasion of his 70th birthday.

Adventures in Chemical Physics: A Special Volume in Advances in Chemical Physics, Volume 132, edited by R. Stephen Berry and Joshua Jortner. Series editor Stuart A. Rice

I. INTRODUCTION

The development of methods that can determine the time dependence of structural changes in complex systems, particularly biological systems, represents an exciting challenge for chemical physics. The new multidimensional infrared spectroscopies, 2D- and 3D-IR [1–19], which have essentially unlimited time resolution on the scale of large structural changes, can be expected to contribute significantly to this goal. Such approaches are expected to complement the vast knowledge of average structures obtained by the established methods of structural biology and their time dependent variants.

Decades of theoretical and experimental research on nonlinear optical and infrared spectroscopy have established the concepts underlying the operational aspects of multidimensional infrared experiments. However, the principles now used for the manipulation of the multidimensional IR data sets in time or frequency domains, phase manipulation, properties of multidimensional Fourier transforms, and many other procedures, often of significant complexity, are closely related to textbook material in NMR [20] even though the practical aspects of the two types of experiment are quite different. Analogous to NMR, the signal generation in 2D-IR is based on the interaction of successive phase-locked pulses with a sample followed by detection of the field generated after the last pulse. For experiments in the IR or the optical regime, which are at much higher frequencies than most detectors can respond, heterodyne methods must be used to obtain the generated *field*. Heterodyning, which permits the measurement of optical electric fields by mixing on a slow square law detector, has been employed since the earliest days of optical nonlinear spectroscopy [21]. In the higher-frequency regimes of optical and IR fields, in contrast to radio and microwaves, the detected field is generally in the weak signal limit and therefore chosen from a particular order of nonlinearity. On the other hand, NMR and EPR are generally conducted near the saturation limit. The optical and IR approaches have mainly been third-order susceptibility measurements with the exception of some recent Raman spectroscopy experiments on liquids which were in the fifth order [22].

The earliest 2D-IR experiments used a versatile pump-probe technique [1], but the first 2D-IR photon-echo results were reported soon after that [3]. The spectral line narrowing or optimization aspect of 2D-IR arises from the contribution of the *photon echo* to the signal. These echoes are part of the pump-probe signal also, but they can be examined free from other influences by means of photon-echo spectroscopic methods. Since the announcement of the first photon-echo experiment with two incident visible light pulses on ruby crystals [23], it has been well known that the echo signal separates inhomogeneous and homogeneous contributions to the spectral line width in the optical spectrum in analogy with what already had been clear for radio-frequency and microwave spin echoes of two-level systems. However, the work of Mukamel and co-workers [24] has shown that the dynamics of optically prepared states cannot generally be considered in terms of Bloch parameters, so the echo responses in the optical and IR spectral regions are indeed considerably different and often more difficult to model than those in NMR. The first high-frequency two-pulse photon echoes on individual molecules, rather than solid-state materials, were actually carried out in the *infrared* around 10.6 μm on SF_6 by Patel and co-workers [25] and at 3 μm by Brewer and Shoemaker [26] who demonstrated that most of the pulsed RF responses could be reproduced in the infrared with vibrational modes of methyl fluoride acting as the two-level systems. This work was a landmark achievement in quantum optics and led to many other infrared photon echo studies of the dephasing of vibrational transitions. In 1974 Wiersma and Aartsma reported two-pulse photon echoes of two-level electronic transitions of molecules in mixed molecular crystals at low temperatures on nanosecond time scales, in work that gave birth to a new dimension in the field of time-dependent spectroscopy of molecular solids [27]. With the advent of reliable, shorter, laser pulses these optical measurements were naturally extended to the available pulse time scales and to a range of media such as liquids and glasses in which the motions were faster, matching the available time resolution. For example, in 1991, Shank and co-workers [28] reported two-pulse photon echoes in the optical spectrum with 6-fs time resolution, the then shortest available pulses. Many variants of the photon echo at a variety of time scales including heterodyning, gating, three-pulse methods (see, for example, Ref. 29), and more recent two-dimensional techniques [30] were developed for optical pulse experiments along with methods for deducing the time correlation functions of the frequency fluctuations [31]. On the theoretical side, predictions of the form and possible importance of multidimensional optical spectroscopies had been predicted already in 1993 [32]. Although various femtosecond-time-scale IR experiments had been carried out on a variety of proteins and aqueous systems in this early period [33–38], suitably short, sufficiently stable, and tunable pulses were not so readily available in the infrared region until the titanium sapphire laser and modern nonlinear optical materials for infrared frequency generation became more established. Nevertheless, the two-pulse echo technique with infrared radiation was extended to the picosecond regime in experiments of Fayer and co-workers, who used an infrared free electron laser source to determine the two-level system dynamics of vibrators in solutions and glasses [39]. Femtosecond-time-scale three-pulse echoes of vibrations in the infrared were first accomplished in 1998 for ions in liquid water [40] and later for peptides [41] and proteins [41, 42]. In 2000, three-pulse phase-locked echo experiments with heterodyne detection on peptides [3] finally enabled the assembly of multidimensional vibrational spectra in the mid-infrared. The 2D-IR spectra had also been constructed from pump-probe experiments on peptides and proteins [1]: the spectra obtained in this approach are closely related to the real part of the heterodyned 2D-IR experiment [43]. The theory of two-dimensional vibrational spectroscopy is also in place [44]. The field of multidimensional IR spectroscopy of vibrators is now very active and is replete with recent important technical and scientific advances from many different laboratories and diverse areas of application [43, 45, 46], including liquids [47, 48], which attests to the outstanding potential of such methods for the study of structure and molecular dynamics in liquids, glasses, and biological systems.

The backbones of protein structures are the polypeptides whose amide units, -NHCH(R)CO-, have infrared spectra that are ultrasensitive to the details of the many possible secondary structures that exist in proteins. The 2D-IR method exposes much more information regarding the potential surfaces of polypetides than conventional FTIR spectroscopy because it accesses anharmonic contributions directly, but the interpretation of the results depends on having a deeper understanding of the dynamics of vibrational states than can be obtained from pump-probe experiments. Already a few nonlinear IR spectroscopic investigations have been carried out on the amide-I and amide-II transitions of polypeptides and peptides: These transitions involve mainly the carbonyl stretching mode and in-plane CNH bending modes. Another structure sensitive vibration of the peptide group is the amide-A mode that is mainly the N–H stretching motion. Recently the first series of experiments using dual frequencies in 2D-IR were used to examine the coupling between the amide-I and amide-A modes [18].

Nonlinear infrared spectroscopy can in principle provide knowledge of all the relaxation processes of oscillators, including those that do not manifest themselves in the linear spectral line shapes. The $v = 0 \rightarrow v = 1$ transition line shape is determined by the overall rotation of the molecule, population relaxation time T_1 and by the vibrational frequency correlation function. The experimental line-shape is not a very useful determinant of this correlation function [40, 49] because it provides experimental data only along one axis, either frequency or time, and the line-shape function is usually too complex to be described by a few parameters. The third-order nonlinear IR experiments provide data along three axes in principle, and even 2D-IR obtains a square grid of data points. These factors result in the correlation function being much better determined than by linear methods simply because of the increase in the number of observables dependent on the same set of parameters. The nonlinear experiments probe levels beyond $v = 0$ and $v = 1$ and so generate relaxation properties that are not part of the IR line shape. Furthermore, by judicious choice of phase-matching conditions and pulse sequences, the nonlinear signal can be chosen to emphasize different characteristics of the dynamics and of the correlation function by means of the pump-probe, transient grating and two- or three-pulse photon echo experimental arrangements [6, 40, 42, 47, 50–54]. Furthermore, the methods allow the determination of key parameters of the anharmonic potential surfaces of peptides and hence provide important tests of theoretical calculations of molecular structure and dynamics. These coherent nonlinear infrared techniques permit experimental determination of the coupling and angular relations of vibrators using experimental protocols that are analogous to those developed for NMR. The first such experiments concerned the amide-I modes of peptides, which are mainly C=O vibrators. In that case all the relevant frequencies of an interacting ensemble of modes could readily be bracketed by the spectral bandwidth of 120-fs infrared laser pulses. The response of such a system to sequences of three pulses, each with the same center frequency in the amide-I region, gave rise to coherent signals whose two- and three-dimensional correlation spectra yielded the relevant structural and dynamical information. We have recently carried out dual-frequency phase-locked 2D-IR experiments in which the coupling of different modes can be examined, free from the contributions of the fundamentals themselves [15, 16, 18].

In the present chapter we discuss the signal processing of heterodyned three-pulse echo experiments in the infrared using single and dual frequencies. The basic approaches to understanding these experiments have long been part of nonlinear spectroscopy on which subject there have been many reviews [8, 11, 43, 55–62] and textbooks [21]; the underlying theory of nonlinear spectroscopic experiments with special focus on pulsed laser responses is unified in the recent book by Mukamel [29]. An important part of all nonlinear experiments, including 2D-IR, is the processing and engineering of the signals. All of the procedures used are common in other fields such as radio-frequency communications, acoustics, and nuclear magnetic resonance. However, until recently, such approaches have not been widely used for high-frequency signals as in the optical and mid-IR regimes. Therefore a very brief review is given of elementary properties of electromagnetic fields and the way they enter into nonlinear experiments. Different types of interferometry are then briefly introduced with reference to model pulses. There are a number of recent, useful accounts of the technical aspects of nonlinear spectroscopy using short pulses that focus on multidimensional methods. But the current activity in multidimensional methods derives from basic nonlinear optical spectroscopy developed for molecules mainly in the 1970s and 1980s, the vast literature on signal processing and spectral analysis (see, for example, the Prentice-Hall Signal Processing Series), and gradual enlightenments on the relationships between nonlinear spectroscopy and NMR [63].

In the 2D-IR experiments there exists a useful simplification of the description of the spectra when the vibrational dynamics is in the separation of time-scales limit of the so-called Bloch dynamics. Then the correlations of the fluctuations of the various quasi-degenerate amide modes dominate the signals and the interpretations are quite straightforward and analytic. On the basis of experimental determinations of the correlation functions, we explore some of the sensitivities of the 2D-IR signals to the dynamic approximations. In addition, we discuss some of the important possible manipulations of the 2D-IR spectra that permit the display of essentially all possible third-order nonlinear responses from a single data set.

II. RELEVANT ASPECTS OF THE GENERATED FIELDS AND LINEAR RESPONSE

In order to introduce some notation, we first recall a few of the well-known properties of the interaction of light pulses with molecules in the linear approximation. Frequently, the signals in nonlinear optical experiments are expressed in terms of the polarization induced in the medium by the incident pulses. The complex linear polarization $P(t)$ vector for a distribution of identical two-level systems is obtained from an elementary calculation of the density matrix using the Liouville equation of a system perturbed by an electric field and proceeding as follows:

$$P(t) = \mu_{10}\rho_{01}(t) = -\left(\frac{i}{2\hbar}\right)\mu_{10}\mu_{01}\cdot\hat{\varepsilon}\int_0^\infty dTE(t-T)e^{-(i\omega_{01}+\gamma)T} \tag{1}$$

where $\hbar\omega_{01} = \hbar(\omega_0 - \omega_1)$, $\rho_{01}(t)$ is the coherence in the two-level system, $E(t)$ is the applied field in the rotating wave approximation (i.e., the envelope times exp(-iωt)), μ_{10} is the $1\rightarrow 0$ transition moment dipole vector, $\hat{\varepsilon}$ is the field polarization vector, and γ is the relaxation rate of the coherent state. More generally, the dynamics is not representable by a distribution of homogeneously broadened transitions but requires more elaborate types of frequency correlation functions. However, this so-called Bloch model is useful to demonstrate the character of the interaction of light and molecules.

A delta-pulse field of unit area and angular frequency ω is obtained from a Gaussian pulse by tending its time width to 0:

$$E(t-T) = \mathrm{Lim}_{\sigma->0}\left(\frac{1}{\sigma\sqrt{\pi}}\right)e^{-(t-T)^2/\sigma^2+i\omega(t-T)} = \delta(t-T)e^{i\omega(t-T)} \quad (2)$$

The complex polarization [Eq. (1)] becomes

$$P_\delta(t) = -\left(\frac{i}{2\hbar}\right)\mu_{10}\mu_{01}\cdot\hat{\varepsilon}e^{-(i\omega_{01}+\gamma)t} = R(\omega_{01},t) \quad (3)$$

which defines a linear response function for a single oscillator from a distribution of oscillators undergoing spontaneous decay with rate γ. In a very weakly absorbing medium where the probability distribution of frequencies is $G(\omega_{01})$, the corresponding ensemble polarization is obtained as

$$\langle R(\omega_{01},t)\rangle = \int R(\omega_{01},t)G(\omega_{01})\,d\omega_{01} \quad (4)$$

If the deviation from the mean frequency, $\bar{\omega}_{01}$, is Gaussian with standard deviation σ, the complex polarization response to a delta function excitation becomes

$$-\left(\frac{i}{2\hbar}\right)\mu_{10}\mu_{01}\cdot\hat{\varepsilon}e^{-(i\bar{\omega}_{01}+\gamma)t-\sigma^2t^2/2} \quad (5)$$

This polarization generates the so-called free decay field of the sample which, when dominated by the inhomogeneous contribution, exhibits a Gaussian decay of the oscillations at $\bar{\omega}_{01}$. This emission trails behind the excitation pulse and its peak amplitude is related to the absorption coefficient of the sample. The Fourier transform of this signal is the Voigt profile. In a conventional linear experiment, this free induction decay (FID) of the sample is collinear with the driving field, as specified by Maxwell's equations. In the next paragraph we imagine that the FID is measured independently of the driving field, which can be arranged in a variety of different experimental arrangements, one of which is by combining the signal on the detector with a variably delayed ultra-short pulse excitation.

In any experiment the generated signal *after the sample* is actually a real field that is generated by the oscillating polarization *in the sample* over the path length l. This complex electric field is $(2\pi i\omega l/c)P(t)$, so that the envelope of the cosine part of the polarization is the envelope of the sine part of the electric field, and vice versa. If we carry out a heterodyne measurement on this field with a very short pulse, we measure a real signal $S(t)$ which is proportional to $\mathrm{Re}\{iP(t)\}$, so that for the homogeneous system the signal is

$S(t) = e^{-\gamma t} \cos \omega_{01} t$. The half-Fourier transform (HFT) of this signal is the complex spectrum. We take the HFT because there is no signal prior to the excitation time $t = 0$:

$$\mathbf{S}(\omega) = \int_0^\infty dt S(t) e^{-i\omega t} = \frac{(\gamma + i\omega)}{(\gamma + i\omega)^2 + \omega_{01}^2} \tag{6}$$

There are identical spectra at positive and negative frequency in this cosine transform. If we assume the delta pulse probing time to be shifted by an amount τ, we get a phase shift $\phi = \omega_{01}\tau$ and the real and imaginary parts of the spectrum become mixed illustrating how important is the choice of time zero in experiments. It is also important to avoid timing fluctuations $\delta\tau$ in such experiments since they give rise to phase fluctuations $\omega_{01}\delta\tau$. A brief discussion of some elementary aspects of signal processing that need to be considered in IR experiments is presented in Section III.

The polarization induced in a molecule by n successive interactions with a field $E(t)$ is termed the nth order polarization. Each interaction involves the field coupling to a transition dipole μ. The 2D IR involves a calculation of the third order polarization, $Tr\{\rho^{(3)}(t)\mu\}$, which requires a quantum dynamics derivation of the third order term, $\rho^{(3)}(t)$, in the expansion of the density operator as a function of the field. The quantum dynamics is accomplished by solving the Liouville equation for the density matrix: $\dot{\rho}(t) = i/\hbar[\mu(t) \cdot E(t), \rho(t)]$, which is often done by some type of iterative procedure. In the experimental methods described herein the field $E(t)$ is composed of up to three light pulses that can be separated in time and direction by the experimenter. But always there will be three interactions: either all interactions from one pulse; two from one and one from the other; or one from each of three pulses. Mukamel's book [29] contains a full account of the theoretical methods of nonlinear spectroscopy which will not be dealt with further in this article.

III. TIME-DEPENDENT AND SPECTRAL PHASE

The subject of phase and phase retrieval with pulsed optical signals, although it is textbook material and involves well-known signal processing concepts [64, 65], has impacted on molecular spectroscopy only recently [66] through consideration of optical control experiments. As we shall see the phase is a consideration in heterodyne laser experiments because it influences the mixing of fields incident on a square-law detector. It is well known that a quadratic phase alters the spectrum, the time envelope and the time–frequency bandwidth of a pulse. Consider a pulse:

$$E(t) = e^{-at^2} e^{i(\omega_0 t + bt^2)} = \varepsilon(t) e^{i\phi(t)} \tag{7}$$

and its Fourier transform:

$$\mathbf{E}(\omega) = \int_{-\infty}^{\infty} E(t)e^{-i\omega t}dt = (\pi/(a - ib))^{1/2}e^{-(\omega_0-\omega)^2/4(a-ib)} = \varepsilon(\omega)e^{i\varphi(\omega)} \quad (8)$$

with $\phi(t)$ the *time-dependent phase,* a real spectrum amplitude $\varepsilon(\omega)$, and a *spectral phase* $\varphi(\omega)$ which includes a constant part. The power spectrum $\mathbf{E}(\omega)\mathbf{E}^*(\omega)$ of the field is $\varepsilon^2(\omega)$ whose time–frequency bandwidth is 0.44 $\sqrt{1 + (b/a)^2}$. All signal fields representing input or output fields of nonlinear optical experiments can be written in the equivalent forms in the last steps of Eqs. (7) and (8), and we can discuss them either in terms of the time-dependent or spectral phase. Although $\phi(t)$ and $\varphi(\omega)$ are often awkwardly related, there is an exact connecting relationship between them [67]:

$$\int t\phi'(t)\varepsilon^2(t)\,dt = \int \omega\varphi'(\omega)\varepsilon^2(\omega)\,d\omega \quad (9)$$

where $\varphi'(\omega) = d\varphi(\omega)/d\omega$ and $\phi'(t) = d\phi(t)/dt$ is the instantaneous frequency. The interpretation of $\phi(t)$ is straightforward: The phase gives the variations of frequency across the pulse. Changes in the spectrum enter through $\varphi(\omega)$, which may cause shifts in the mean frequency of the field. These definitions are easily illustrated for a Gaussian pulse having both quadratic and cubic phase, which would be approximately the situation if the phase were determined by passing the beam through standard optical materials [68] as occurs in our 2D-IR experiments:

$$E(t) = e^{-at^2/2}e^{i\varphi(t)} = e^{-at^2/2}e^{i(\omega_0 t + bt^2/2 + ct^3/3)} \quad (10)$$

for which $\phi'(t)$is $\omega_0 + bt + ct^2$, manifesting both *linear* and *quadratic* chirp. Its mean frequency of $(\omega_0 + c/2a)$ is calculated from the average over the envelope squared as $\int_{-\infty}^{\infty} \phi'(t)e^{-at^2}dt$, illustrating that the cubic phase shifts the mean frequency. The frequency bandwidth is computed from $\langle\phi'(t)^2\rangle - \langle\phi'(t)\rangle^2$, and only if there is no chirp do we get the expected variance of $a/2$. The complex spectrum of a linearly chirped pulse $(c = 0)$ is readily obtained analytically from Eq. (10) to illustrate some important aspects. Apart from constant phase and amplitude terms, it is

$$\mathbf{E}(\omega) = e^{-a(\omega-\omega_0)^2/2(a^2+b^2)}e^{-ib(\omega-\omega_0)^2/2(a^2+b^2)} \quad (11)$$

from which it is seen that the spectral phase is also Gaussian and it can cause the real part of the field to change its sign at certain frequencies, depending on the magnitudes of the factors a and b. The spectral phase at the $1/e$ points of the power spectrum of the pulse is $b/2a$. Although well known from conventional signal theory, these are important considerations for spectroscopies such as

2D-IR where the complex field is measured and where representations of the real and imaginary parts of a spectrum might be desired. A comparison of the result (11) to the time-dependent phase through expressions (10) and (9) is a useful exercise. In nonlinear spectroscopy the generated field may have a time-dependent frequency that manifests itself in much the same manner as these simple examples of chirp.

IV. THE EFFECT OF OPTICAL DENSITY

In many of the nonlinear IR experiments the samples might have to be optically dense. This presents challenges to the interpretation of multi dimensional spectroscopy as the following example describing the propagation of a Gaussian pulse through an absorbing medium shows. This question was treated sometime ago [69] for an input Gaussian pulse spectrum with spectral width σ:

$$\mathbf{E}(\omega, 0) = \left(\frac{1}{2\pi\sigma^2}\right)^{1/2} e^{-(\varpi-\omega)^2/2\sigma^2} \tag{12}$$

The output pulse after distance z is

$$\mathbf{E}(\omega, z) = \mathbf{E}(\omega, 0) e^{i\omega z n(\omega)/c} \tag{13}$$

where $n(\omega)$ is the complex refractive index through the resonance given by

$$n(\omega) = n_\infty - \frac{c\gamma\alpha(\omega_0)}{2\omega(\omega - \omega_0 + i\gamma)} \tag{14}$$

where ω_0 is the resonance frequency and γ is the resonance half-width (i.e., $1/T_2$, in angular frequency units). We assume that no other resonances need to be considered, which would be good approximation for an isolated vibrational transition. The field suffers loss with absorption coefficient $\alpha(\omega)/2$ as a result of the imaginary part of $n(\omega)$. The outgoing pulse in the time domain is then given by

$$\begin{aligned} E(z,t) &= \left(\frac{1}{2\pi\sigma^2}\right)^{1/2} \int_{-\infty}^{\infty} d\omega e^{i\omega t} \{ e^{-(\varpi-\omega)^2/2s^2} e^{-\alpha(\omega)z/2} \} e^{i\left\{\omega n_\infty z/c - \frac{\gamma(\omega-\omega_0)\alpha(\omega_0)z/2}{(\omega-\omega_0)^2+\gamma^2}\right\}} \\ &\equiv \int_{-\infty}^{\infty} d\omega e^{i\omega t} \varepsilon(\omega)\, e^{i\varphi(\omega)} = \int_{-\infty}^{\infty} d\omega e^{i\omega t} \mathbf{E}(\omega, z) \end{aligned} \tag{15}$$

where we have used the curly brackets to clarify our definition of the field amplitude $\varepsilon(\omega)$ and the spectral phase $\varphi(\omega)$, where both ε and φ are real and t is now a reduced time $(t - n_\infty z/c)$. In this case the spectral phase is a Lorentzian having a different sign on either side of the resonance. An important point about the integral in Eq. (15), according to Garrett and McCumber, is that a correct description is not obtained by expanding the Lorenzian phase and absorption factors about ω_0 up to quadratic or cubic terms *except when the spectral width of the light pulse is much less than the resonance width*. This limit is not useful when we use femtosecond pulses and vibrational resonances having dephasing times comparable or longer than the pulse widths—which are the only cases of much modern interest. Thus the integral must be evaluated numerically. When the optical density of the sample at the peak, given by $\alpha(\omega_0)z/2.303$, is large and the peak is relatively narrow compared with the bandwidth of the pulse, the integrand only has value on either side of the pulse. $E(z, t)$, as shown in Fig. 1(b). The Wigner spectrogram, $W(\omega, t)$, offers a useful representation of the time-dependent frequency of this signal. It is convenient to use the frequency

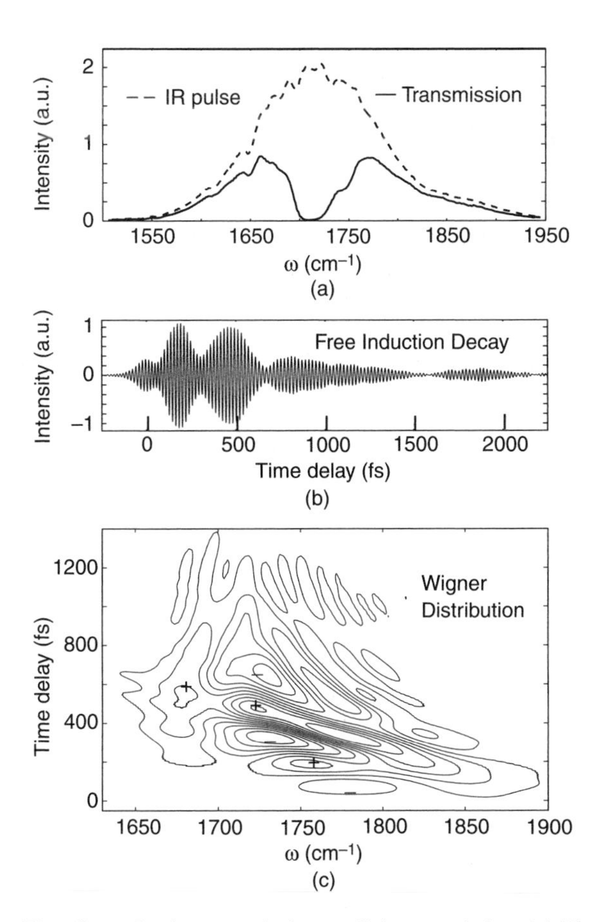

Figure 1. The effect of coherent excitation on light transmission. (*a*) The incident and transmitted pulses through a sample having an optical density of 1.0. (*b*) The free induction decay created by the coherent excitations by the pulse in (*a*). (*c*) The Wigner distribution (see text) of the FID shown in (*b*).

definition of the spectrogram:

$$W(\omega, t) = \int_{-\infty}^{\infty} d\theta \mathbf{E}^*\left(\omega + \frac{\theta}{2}, z\right)\mathbf{E}\left(\omega - \frac{\theta}{2}, z\right)e^{-i\theta t} \tag{16}$$

which is shown in Fig. 1(c) for a pulse passing through a sample with an optical density of 1.0. These are results for liquid acetone, illustrating the pulse reshaping that occurs because the sample has finite optical density. Of course this effect is related to continued reemissions of the type illustrated at first order in Eq. (5). The generated field measured by heterodyne detection is shown in Fig. 1(b). The effects of high optical density on heterodyned 2D spectroscopy have recently been discussed [70, 71].

V. HETERODYNE SPECTROSCOPY

In order to obtain multidimensional spectra, it is necessary to measure the amplitude and the phase of the signal generated by a sample in response to some incident fields. Directing the generated field to a square law detector such as a photomultiplier or photodiode measures only the amplitude squared of the generated field. However, if the generated field is combined collinearly with a reference field and both are incident on the detector, the current in the detector circuit has a component that depends on the product of the two fields and it determines the signal field if the reference is known. This procedure is termed *heterodyning* and the principle has been employed in nonlinear spectroscopy, particularly in Kerr effect measurements [21]. There are two principal methods of obtaining heterodyned spectra in the IR region: time domain and spectral interferometry.

VI. SPECTRAL INTERFEROMETRY

In spectral interferometry, two IR pulses separated by time τ are sent to a monochromator and the total spectrum is measured. By definition the two fields are the Fourier transforms:

$$\mathbf{E}_1(\omega) = \int_{-\infty}^{\infty} \varepsilon_1(t)e^{i\phi_1(t)-i\omega t}dt \quad \text{and}$$

$$\mathbf{E}_2(\omega) = e^{-i\omega\tau}\int_{-\infty}^{\infty} \varepsilon_2(t-\tau)e^{i\phi_2(t-\tau)-i\omega(t-\tau)}d(t-\tau) \tag{17}$$

so that $\mathbf{E}_1(\omega) = \varepsilon_1(\omega)e^{i\varphi(\omega)}$ and $\mathbf{E}_2(\omega) = \varepsilon_2(\omega)e^{i\varphi_2(\omega)-i\omega\tau}$. The latter form is a general way of expressing a field in the frequency domain having a particular time shift. The total field incident on a detector at setting ω of the monochromator is the sum of these two fields, and the current in the detector circuit $S(\omega)$ is proportional to the absolute square of that sum:

$$\begin{aligned} S(\omega) &= \left|\varepsilon_1(\omega)e^{i\varphi_1(\omega)} + \varepsilon_2(\omega)e^{i\varphi_2(\omega)-i\omega\tau}\right|^2 \\ &= \varepsilon_1(\omega)^2 + \varepsilon_2(\omega)^2 + 2\varepsilon_1(\omega)\varepsilon_2(\omega)\cos[\varphi_{21}(\omega) - \omega\tau] \end{aligned} \tag{18}$$

Often in an experiment it is possible to eliminate the contributions from the two power spectra leaving only the interference term. It is only this interference term that is dependent on phase and phase fluctuations. Note that for two identical pulses the signal is simply proportional to $2\cos^2[\omega\tau/2]$, which is a series of peaks in the frequency domain separated by $2/c\tau$ cm^{-1}. Thus a $\tau = 1$ ps delay yields a peak separation of 67 cm^{-1}. In general the peak separations in the frequency domain are not independent of frequency and instead depend on the spectral phase difference at each frequency. Therefore spectral interferometry presents a method by which to determine the phase differences of two pulses. When the pulses are the same, we can use spectral interferometry to determine their time separations. The inverse Fourier transforms of the first two contributions to the spectrogram in Eq. (18) peak at $t = 0$ whereas the cross term peaks at $t = \pm\tau$. Therefore Fourier transformation of $S(\omega)$ can permit a separation of the cross term from the power spectra of the signal and reference fields [72].

VII. TIME-DOMAIN INTERFEROMETRY

In time-domain interferometry the two pulses are sent collinearly to a square law detector which responds equally to all the frequencies in the pulses. The current in the slow detector circuit $S(\tau)$ is measured as a function of the delay, τ, between the two pulses. A common but not necessary situation in heterodyning is that one field, $E_1(t')$ is very weak so that its square can be neglected while the other, the local oscillator field, $E_2(t' - t)$ is much larger. The signal is time integrated by the slow detector:

$$S(t) = \int_{-\infty}^{\infty} dt' |E_1(t') + E_2(t' - t)|^2 \tag{19}$$

By intermittent chopping of the beams, the constant local oscillator background signal can be eliminated and a Fourier transform along t yields a spectrum that by the convolution theorem is the product of the spectra of the local oscillator and the signal:

$$S(\omega_t) = \varepsilon_2(\omega_t)\varepsilon_1(\omega_t)\cos\varphi_{21}(\omega_t) \tag{20}$$

which is the same as the result, Eq. (18), obtained by spectral interferometry at $\tau = 0$. Thus the two methods of spectral and time-domain interferometry are equally suitable for obtaining the spectra of pulsed fields.

VIII. THE PHOTON ECHO EXPERIMENT

Traditionally the two-pulse photon-echo of a two-level system is described in terms of dynamics where there is a separation of the frequency fluctuations into two widely separated time scales, one of which is much faster and the other much slower than the time that characterizes the inhomogeneous distribution of frequencies. This gives rise to a fixed distribution of homogeneously broadened transitions for each spectral transition of the solute. The echo electric field generated from two very short pulses interacting with a molecule but separated by an interval τ is, apart from constant factors, given by

$$e^{(i\omega_{10}-\gamma)\tau}e^{(-i\omega_{10}-\gamma)t} \tag{21}$$

where γ is the homogeneous width, t is the time between the excitation and detected fields, and $\hbar\omega_{10}$ is the energy of the molecular transition. The radiating polarization is induced by a single interaction with the field of the first pulse and two field interactions with the second, coherence transferring pulse. In relationship (21) the generated signal field is presented as a complex function. The real field generated in the laboratory is the real part of this function, apart from multiplicative factors. The conventional echo signal from an ensemble is detected on a square law detector and therefore involves the integral over the detection time t of the squared average over the distribution of frequencies, namely,

$$\int_0^\infty \left|\left\langle e^{i\omega_{10}(\tau-t)}e^{-\gamma(\tau+t)}\right\rangle\right|^2 dt \tag{22}$$

By assuming a Gaussian frequency distribution with fluctuations δ about a mean, along with standard deviation σ, the echo signal becomes

$$\begin{aligned}&\int_0^\infty dt\, e^{-2\gamma(\tau+t)}\left|1/\sigma\sqrt{2\pi}\int_{-\infty}^\infty d\delta e^{i\delta(\tau-t)}e^{-\delta^2/2\sigma^2}\right|^2\\ &\quad=\int_0^\infty dt\, e^{-2\gamma(\tau+t)-\sigma^2(t-\tau)^2}=\sqrt{\pi}/2\sigma\, e^{\gamma(\gamma/\sigma^2-4\tau)}\left(\operatorname{erf}\left(\frac{\gamma}{\sigma}-\sigma\tau\right)-1\right)\end{aligned} \tag{23}$$

As is well known, when the fixed inhomogeneous distribution is very large compared with the homogeneous width, this echo signal occurs around $t=\tau$ and decays with a time constant $\frac{1}{4}\gamma$. However, as σ approaches zero the time constant becomes $\frac{1}{2}\gamma$ and the signal then peaks at $t=0$. The limits are most readily seen from the second integral in (23) since $\exp[-\sigma^2(t-\tau)^2]$ only exists for $t\approx\tau$ in the limit of large σ, while for very small σ/γ the integral is an exponential decay with time constant $\frac{1}{2}\gamma$. For many vibrational systems the dynamics are more complex than assumed in this simple example as discussed later. The spectrum of the conventional echo is obtained by recording the absolute square of each frequency component in the frequency average of (21), obtained by Fourier transforming the emitted signal with a monochromator followed by square law detection of each frequency component.

IX. THE THREE-PULSE HETERODYNED ECHO

In the three-pulse heterodyned echo the generated fields are measured directly, so the signal is not given by expression (22). The amplitude and phase of the generated field depends on the amplitude and phase of the three pulses that induce the third-order polarization in the sample and on the local oscillator pulse. Thus the directions and timing of these pulses are important in the experiment. The physical interpretation of the experiment in Eq. (23) is also inappropriate for systems that have more than two levels. We will see that the rephasing process causing the echo to appear at $t \approx \tau$ in Eq. (23) may or may not act to rephase the polarization in a multilevel system, depending on the relationships between the distributions of frequencies of the various levels that fall within the bandwidth of the pulses.

In a three-pulse heterodyned echo measurement we obtain $S(\tau, t; T)$, where τ, T, and t are the time *intervals* between the first and second fields, between the second and third fields, and between the third field and the detected field. These experimentally controllable intervals are often referred to as the *coherence evolution time*, the *waiting time*, and the *detection time*, respectively. For convenience we assume in the following that the driving fields are much shorter than all the dynamical processes. For the infrared spectra of nearly degenerate groups of modes, such as the amide modes of peptides, the pulses would have to be around 100 fs. This spectral bandwidth brackets the complete distribution of amide-I modes found in the majority of secondary structures. The time-domain signal $S(\tau, t; T)$ can be obtained directly in a time-domain interferometry experiment in which, for a given waiting time, the time t becomes the interval between the third pulse and a short local oscillator pulse. A scan of the separation between the third and local oscillator pulses completes the data set $S(\tau, t; T)$. Alternatively, one can carry out spectral interferometry in which the generated field is sent to a monochromator along with a local oscillator with a flat spectrum which is advanced on the signal by a time interval d. The heterodyne signal is a real oscillatory signal that is related to $S(\tau, t; T)$ through

$$\mathrm{Re}\left[e^{-id\omega_t} \int_0^\infty dt e^{i\omega_t t} S(\tau, t; T)\right] \tag{24}$$

X. SELECTION OF PATHWAYS AND PHASE-MATCHING

The direction of the outgoing wave in a nonlinear experiment is determined by the wave vector k_p of the induced polarization, which depends on the directions and frequencies of the driving fields. In all experiments the outgoing signal field with wave vector k_s propagates along the vector sum of the ingoing wave vectors responsible for the nonlinear effect, and it is maximized in intensity when $|k_s - k_p| = 0$. The situation with pulses having different frequencies requires care because the phase mismatch $|k_s - k_p|$ depends on which frequencies are observed. This analysis implies that during interferometric experiments the generated field intensity from certain diagrams is phase-mismatched. These effects can be computed from standard nonlinear optical considerations. For example, the two pulse echo (22) is in direction $2k_2 - k_1$. A useful approach involves the double-sided Liouville path diagrams that track the evolution of both sides of the density operator through the successive interactions of the system with electric fields. Particular Liouville paths that are involved in the responses that determine $S(\tau, t; T)$ can be selected by the phase-matching conditions [29]. This procedure corresponds to a selection of the phases of the input pulses by selecting particular directions of the signal field. In our experiments we have mainly chosen a signal propagation direction of $-k_1 + k_2 + k_3$, where the indices specify the three incident pulses. Usually we examine signals at only four time orderings of these pulses: (1, 2, 3), (1, 3, 2), (2, 1, 3), and (2, 3, 1) all generated in the direction $-k_1 + k_2 + k_3$. Ippen and co-workers [73] have presented an insightful analysis of the directional properties of these signals. When these choices are combined with the rotating wave approximation that eliminates from consideration severely nonresonant processes, many of the theoretically possible pathways giving rise to the generated fields in the third order are eliminated. The (1, 2, 3) and (1, 3, 2) sequences generate the rephasing signal, while the others contribute the nonrephasing signal. When two frequencies are employed, other diagrams disappear because of phase mismatching. These considerations indicate that different diagrams may be accessed by using combinations of frequencies and phase-matching conditions. The pathways that need to be considered are shown in Fig. 2. In this figure the lowercase k's are used to index vibrational states in the one quantum regime and the uppercase K refers to two-quantum vibrational states. For $T > 0$, the diagrams R_4 to R_6 would be omitted. When summed over all the relevant complete sets of system vibrational states the total nonrephasing (diagrams R_7 to R_{11}) and rephasing (diagrams R_1 to R_6) contributions generate equal integrated signals in different quadrants of the frequency space.

Diagrams R_9 and R_{11} in the nonrephasing configuration are interesting because in a two-pulse experiment where k_2 an k_3 arrive first, the coherence during the interval prior to the third pulse oscillates at roughly twice the frequency of the vibrational mode. Studies of these diagrams permit direct measurements of the total dephasing dynamics of the two-quantum states.

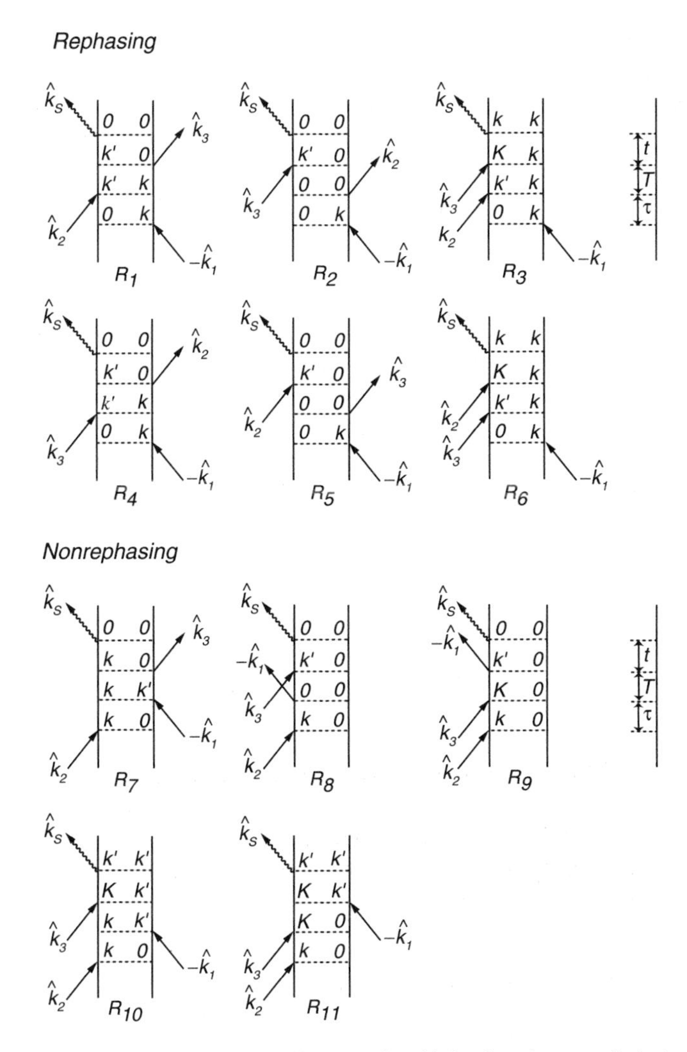

Figure 2. The Liouville pathway diagrams for third-order phase-matched signals (i.e., k_s = signal wavevector = polarization wavevector) into the direction $-k_1 + k_2 + k_3$. The lower- and uppercase k and K symbols represent one- and two-quantum states of a molecule, respectively.

XI. PUMP-PROBE 2D-IR SPECTRA

The pump-probe configuration provides the classic approach to 2D spectroscopy since it involves two independently tunable frequencies [1]. In NMR the analogous approach is double resonance Fourier spectroscopy. In fact the method is analogous to any of the many double resonance techniques used in spectroscopy: One example is dynamic hole-burning spectroscopy. The pump pulse requires to be centered at a particular frequency. Its spectral bandwidth is narrowed by means of a tunable Fabry–Perot filter. This narrowing is optimal in the sense that the time bandwidth is chosen to be shorter than the dynamical response of the system. The spectrally broad probe pulse is delayed by a time interval T and dispersed by a monochromator after the sample to generate the second frequency axis. A contour plot of the optical density as a function of the two frequencies provides the 2D-IR spectrum at each value of T. The attenuation or gain of the probe field is directly measured in such an experiment by self-heterodyning of the generated third-order probe field with the incident field, as is also the case in any linear absorption experiment. Therefore it is only the part of the generated field that is *in phase* with the incident field that appears in the signal. Thus the signal corresponds exclusively to the real part of the generated third-order field. In many situations involving molecular vibrations and their dynamics the real part of the generated field may be all that is required to completely characterize the response. Therefore this pump-probe approach is extremely powerful especially when the spectra being examined are diffuse. The pump-probe signal, while similar to the real part of the echo spectrum, actually incorporates an average of echo spectra over a range of τ values within the time bandwidth of the frequency-narrowed pump pulse. This will be discussed later.

XII. TWO-DIMENSIONAL INFRARED ECHO SPECTRA

The two-dimensional spectrum is defined as the complex 2D Fourier transform of the time-domain signal $S(\tau, t; T)$ or as the single Fourier transform of (24), the spectral interferogram $S'(\tau, \omega_t; T)$:

$$\mathbf{S}(\omega_\tau, \omega_t; T) = \int_0^\infty \int_0^\infty d\tau dt e^{i\omega_\tau \tau + i\omega_t t} S(\tau, t; T) = \int_0^\infty e^{i\omega_\tau \tau} S'(\tau, \omega_t; T) \tag{25}$$

For a given setting of the input pulses specified by the time intervals (τ, T) between the pulses generating the signal, the sample emits a field $\mathbf{E}_s(\tau, \omega; T) = \varepsilon_s(\tau, \omega; T)e^{i\varphi_s(\omega)}$ and the local oscillator pulse is $\mathbf{E}_L(\omega) = \varepsilon_L(\omega)e^{i\varphi_L(\omega) - i\omega t}$, where t is the delay between the LO and emitted field. In the time-domain interferometry we scan a complete range of t while detecting the signal at all frequencies in a square law detector. So the detector signal is equivalently written as

$$S_{\text{total}}(\tau, t; T) = \int_{-\infty}^{\infty} \left|(\varepsilon_s(\tau, \omega; T)e^{i\varphi_S(\omega)} + \varepsilon_L(\omega)e^{i\varphi_L(\omega) - i\omega t}\right|^2 d\omega \tag{26}$$

Usually the detection procedure is arranged so that the signal is free from the separate local oscillator, so that the remaining real signal is

$$S(\tau, t; T) = \int_{-\infty}^{\infty} \varepsilon_s(\tau, \omega; T)\varepsilon_L(\omega)\cos[\varphi_{LS}(\omega) - \omega t]\, d\omega \tag{27}$$

If we now assume that the local oscillator spectrum is very broad, corresponding to a very short pulse in the time domain, we can take it out of the integral and divide throughout by the square root of the power spectrum of the local oscillator giving the two definitions of $S(\tau, t, T)$:

$$S(\tau, t; T) = \int_{-\infty}^{\infty} \varepsilon_s(\tau, \omega; T)\cos[\Delta\varphi_{LS}(\omega) - \omega t]\, d\omega = \mathrm{Re}[E_s(\tau, t, T)e^{i\Delta\varphi_{LS}(t)}] \tag{28}$$

where the analysis is appropriate at each value of T. Both time-domain and spectral interferometry methods of obtaining 2D-IR spectra have been documented as indicated in the introduction. Figure 3 shows the 2D-IR spectra of dialanine obtained from time-domain and spectral-domain interferometry on the same apparatus.

A wide variety of signal manipulations analogous to those that are known from NMR [20], and field polarization conditions that are better known from multiple-pulse high-frequency spectroscopy [74–76] have been explored with IR pulse configurations. The basic concepts of 2D-IR spectroscopy have been frequently reviewed [11, 19, 43], and very recently there was another excellent survey of the methodology of 2D spectroscopy [77].

XIII. COMPARISONS WITH OTHER NONLINEAR EXPERIMENTS

We now we summarize some of the procedures that are used in analyzing multidimensional IR data. Constants factors are often omitted from the formulas as are the transition dipole factors which are easily incorporated [74] when the modes are a collection of coupled harmonic oscillators. More generally the variations of transition dipole with nuclear displacement should be incorporated. It is often useful to compare the 2D-IR results with the results of other nonlinear experiments because it turns out that various manipulations of these multidimensional signals provide all of the common nonlinear results such as echoes, gratings, degenerate four wave effects, and pump-probe spectroscopy.

The 3D-IR data set consisting of a cube of time, frequency or mixed time/frequency points encodes the information obtained from all other third-order nonlinear resonant experiments. When the time τ is chosen as zero, the variation of the signal with T is a heterodyned transient grating experiment. The detection on a slow detector of the generated field is a conventional transient grating in that case. When $T = 0$ the generated signal with sequence (1, 2, 3) is a

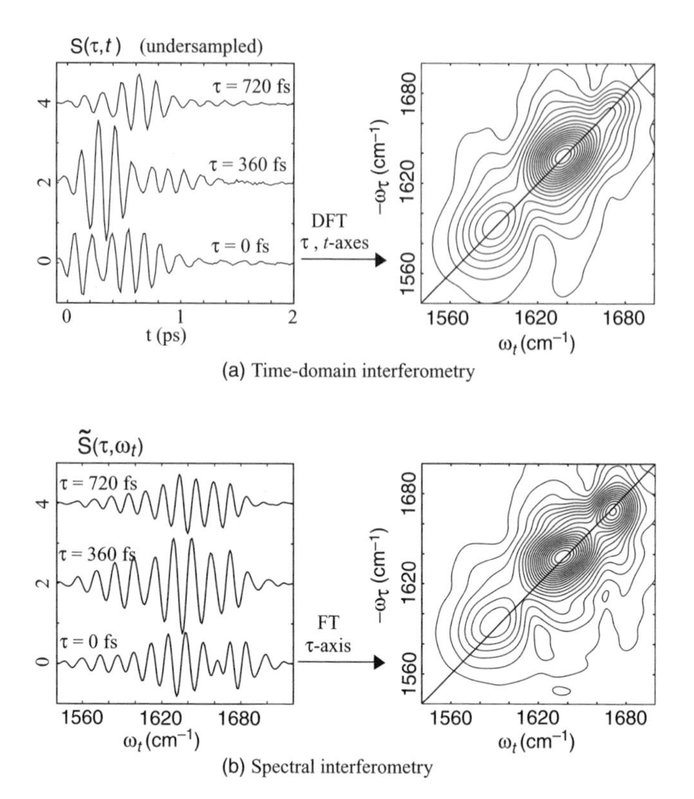

Figure 3. The 2D-IR spectra of a broad-band system (dialanine) recorded by means of (*a*) time-domain interferometry and (*b*) spectral interferometry. The spectra are the same with the signal to noise available, although the spectral resolution is slightly better for the spectral interferometry in our arrangement.

conventional two-pulse photon echo or reverse photon echo when the sequence (2, 1, 3) is used. We often use the terms rephasing or nonrephasing spectra for these two signals. The projection of the heterodyned signal $S(\omega_\tau, \omega_t; T)$ onto the ω_t axis [20], Eq. (30), is defined as

$$\int_{-\infty}^{\infty} d\omega_\tau S(\omega_\tau, \omega_t; T) = S'(\tau = 0, \omega_t; T) \tag{29}$$

the real part of which is the broad-band pump/broad-band probe spectrum in which two counterrotating pulses arrive at time zero and the free decay spectrum is measured after time T. In NMR the projections are defined generally for any skew axis, not just the ω_t axis [20], and this has been emulated in other multidimensional spectroscopies. The comparison of such projections with independent measurements of pump-probe spectra can be very useful. A narrow-band pump/broad-band probe experiment comparable with the earliest of the 2D-IR spectra to be reported is obtained from the data set as the real part of

$$\int_{-\infty}^{\infty} d\omega_\tau S(\omega_\tau, \omega_t; T)(G(\omega_0 + \omega_\tau) + G(\omega_0 - \omega_\tau)) \tag{30}$$

where $G(\omega)$ is the power spectrum of the hypothesized narrow-band pulse centered at ω_0. In this case the signal must be obtained in two of the quadrants (ω_τ, ω_t) and $(-\omega_\tau, \omega_t)$. This corresponds to incorporation of a range of τ values around zero, implying the use of two narrow-frequency bandwidth excitation pulses. This hole-burning experiment is very useful for measurements of the angular parts of the response and for identifying clearly the presence of spectral diffusion. If the spectral bandwidth of the pulses is kept less than the motionally narrowed part of the linewidth, there is no significant loss of information introduced by the averaging over the coherence evolution time. The vibrational frequency correlation function can be real or complex and may need to be described in terms of a number of parameters. Therefore the linear spectroscopy does not have sufficient information to determine this function. The set of 2D data contains much more information. For example, the spectra $\mathbf{S}(\tau, \omega_t; T)$ yield a set of signal versus τ curves, one for each value of T, that can be fitted to sets of parameters. Of course these signals also can be obtained directly from the time-domain set of data by obvious manipulations. The echo peak shift experiment in the IR was carried out previously by integrating the echo signal over the detection time t by detecting it on a slow-response square law detector:

$$S_{\text{peps}}(\tau, T) = \int_0^{\infty} dt\, S^2(\tau, T, t) \tag{31}$$

but the same information can be obtained directly from the multidimensional data set from, for example, $|S'(\tau, \omega_t; T)|$ at each detection frequency ω_t. This function provides a complete set of T-dependent data at each frequency and hence for each emitting oscillator. One main point of this measurement is to provide as many independent observables as possible at each T with which to determine the parameters needed to obtain an accurate representation of the frequency correlation function.

XIV. THE NONRESONANT RESPONSE

When the pulses used in the experiment have finite spectral width the induced polarization in a third-order experiment is the convolution of the system responses discussed above with the three inducing fields. Therefore, if the overall system response is completed much more rapidly than the time width of the pulse envelopes, then the polarization is no longer dependent on the response but is determined by the envelopes and phases of the input pulses. For example, in most of the 2D-IR experiments the solvent generates an "instantaneous" signal of this type even when it appears to be transparent. This response is nonresonant and is in reality essentially instantaneous because its time response is roughly the inverse of the gap between vibrational and electronic transitions. Fortunately, this signal is very small for water and D_2O in which media most of the biological applications of 2D-IR are carried out. Nevertheless, because these signals are instantaneous, they are useful in determining phase properties of the excitation pulses and for timing the pulses at the sample and relative to the local oscillator. An example of this is shown in Figure 4, which is the signal in the echo direction $-k_1 + k_2 + k_3$ for a transparent liquid CCl_4 driven by a three-pulse sequence, each with a center wavelength of 3 μm: in the case of solvents like CCl_4 the nonresonant signals in the infrared are not small, but they are very useful. Figure 4 also shows the spectrum and phase of the signal, relative to that of the local oscillator, determined directly from this experiment. The analysis of the

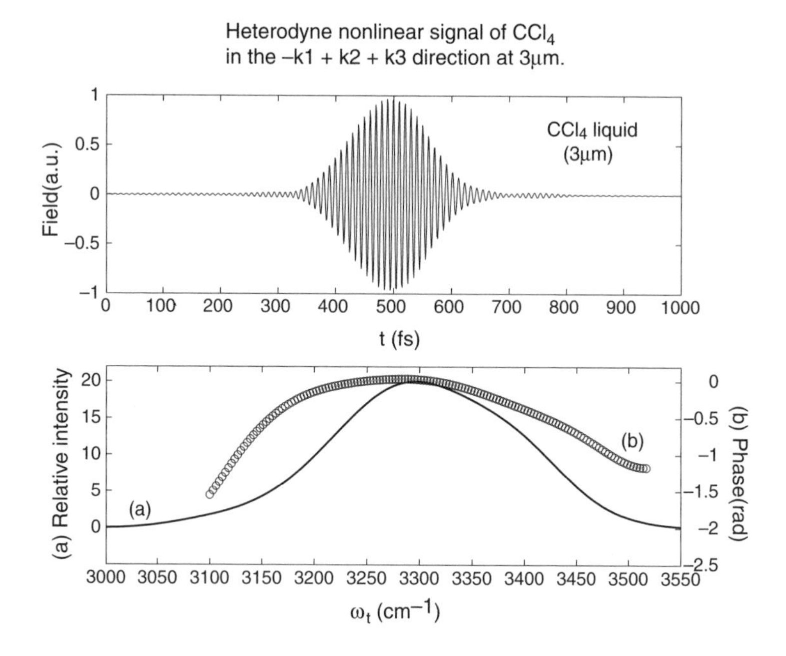

Figure 4. The nonlinearly generated signal from carbon tetrachloride in the echo direction using 3-μm pulses. (*a*) The heterodyned signal obtained by scanning the local oscillator. (*b*) The phase and power spectrum of the generated field.

pulses by this method proved to be very useful in our work on heterodyned photon echoes of pyrrole [78].

XV. SYSTEM RESPONSES IN THE VIBRATIONAL SPECTRUM

The standard approach to nonlinear spectroscopy is to write the polarization generating the signal field in terms of a sum of response functions. In a third-order nonlinear response a molecule interacts with three fields, each of which drives the system into a coherence or population state that depends on what was created by the previous interaction. The sequence of three molecular coherences or populations is usually called a *pathway*. The diagrams of Fig. 2 help in counting of the number of possible pathways and in deducing their formulas. Other diagrammatic methods based on conventional representations of transitions between states can also be useful [79].

There are a number of differences between the vibrational and the electronic spectrum that are important in considering the response functions. One is the relative magnitudes of pulse widths and linewidths. In the vibrational spectrum a typical linewidth corresponds to a dephasing time of 0.3 to 1 ps, whereas in the electronic spectrum the dephasing times are much shorter and only the shortest pulses known can bracket the complete electronic transition in the optical regime. In vibrational experiments where the spectra of individual modes are evident, a 100-fs pulse brackets the region of the response of one mode quite effectively. Exceptions to this are found in associated liquids, such as water, where vibrational spectra tend to be very diffuse representing the broad dynamical inhomogeneous distribution of structures that contribute to the spectra. Another important difference lies in the role of the population times which in the optical regime are usually much longer than the pure dephasing processes. This is not the case for vibrational transitions, which are frequently dominated by T_1 relaxation processes.

The signals referred to above can be calculated from theory in terms of the responses corresponding to the various Liouville pathways mapped out by the field interactions. A given signal can consist of many such pathways, but there are considerable simplifications introduced by adopting the rotating wave approximation and selecting particular terms by phase-matching as mentioned above [29]. The simplest types of two-level generated fields are those on the diagonal of the 2D IR spectrum having the form

$$\left\langle e^{\pm i\omega_{10}\tau-\gamma_1\tau} e^{-T/T_1} e^{-i\omega_{10}t-\gamma_1 t} \right\rangle \qquad (32)$$

where ω_{10} is the transition frequency, γ_1 is the motionally narrowed dephasing time, and T_1 is the population relaxation time. This field corresponds to either (a)

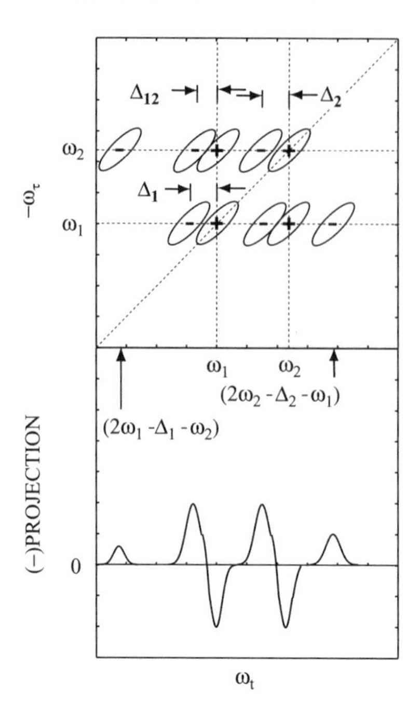

Figure 5. A typical 2D-IR real part of the rephasing spectrum for a pair of coupled oscillators. Shown at the bottom is the projection of this spectrum onto the ω_t axis obtained by summing contributions from all values of ω_τ at a particular ω_t.

R_1 of Figure 2 for a two-level system where the pathway is through a ground-state depletion or (b) R_2, which passes through a $v = 1$ population. Note that the location of the rephasing spectrum in the ω_τ, ω_t plane is evident from the signs of the frequency exponents in a complex response function such as Eq. (29). The nonrephasing spectrum is found in the other two quadrants. The real part of a typical 2D-IR spectrum of a pair of oscillators is depicted in Fig. 5. All the transitions are elongated along the diagonal because of inhomogeneous line broadening. This is a feature common to all the responses discussed below when the static inhomogeneous distribution of frequencies is included. Basically this elongation arises because each diagonal point represents the peak of a linear IR transition with its characteristic homogeneous width, has its own 2D-IR spectrum of four other similarly shaped transitions displaced along the ω_t axis. When there is a distribution of frequencies there are many diagonal points. The approximately elliptical shapes become nearly circular for purely homogeneously broadened transitions. The spectral line narrowing, intrinsic to the photon echo, is dramatized by the difference between the diagonal and antidiagonal widths of the transitions. The transitions along the $\omega_\tau = \omega_1$ line correspond to oscillation frequencies in the FID of $\omega_t = \omega_1 - \Delta$, $\omega_1, \omega_2 - \Delta_{12}, \omega_2$, and $2\omega_2 - \omega_1 - \Delta$, respectively. These signals correspond respectively to the Fig. 2 diagrams: R_3 and R_6 with $k = 1, K = 1 + 1; R_1, R_2, R_4$, and R_5 with $k = 1$ and $k' = 1; R_3$ and R_6 with $k = 1, K = 1 + 2; R_1, R_2, R_4$, and R_5 with $k = 1$ and $k' = 2$; and R_3 and R_6 with $k = 1, K = 2 + 2$. The signals at $2\omega_2 - \omega_1 - \Delta_2$ and $2\omega_1 - \omega_2 - \Delta_1$ are forbidden in the harmonic approximation since they involve $1 \rightarrow 2 + 2$ and $2 \rightarrow 1 + 1$ transitions, respectively, and they need not have the same dipole strength. Also sketched in Fig. 5 is what would be expected for the projection [20] of this real 2D-IR spectrum onto the ω_t axis,

$$\int_{-\infty}^{\infty} S(\omega_\tau, \omega_t)\, d\omega_\tau = \frac{S(\tau = 0, \omega_t)}{2} \tag{33}$$

which is by definition the pump-probe spectrum: in other words [20], it is "the Fourier transform along t of a single FID signal obtained with $\tau = 0$." This establishes a relationship between the pump-probe and the 2D-IR spectra that is discussed in a number of the reviews given earlier. In most of our experiments the phase is not precisely known because of timing inaccuracies, so that this projection is useful in finding which linear combination of real and imaginary parts of the observed spectrum, including its timing uncertainty, corresponds to the true real and imaginary parts of the phase selected 2D-IR spectrum. Responses of the type (32) becomes inaccurate when there is spectral diffusion or when relaxation can occur between different modes as discussed later. The averaging in (32) is over any distribution of motionally narrowed frequencies. Apart from the evolution during the waiting time T, this is the same response as presented earlier for the conventional photon-echo. The upper sign is the rephasing response whose spectrum is in the quadrant $(-\omega_\tau, \omega_t)$ and the lower sign gives the so-called nonrephasing response in the (ω_τ, ω_t) quadrant obtainable by interchanging the ordering of the first and second pulses used in the echo. The character of the 2D spectra of the homogeneous part of this echo response, which would correspond to a diagonal peak, is described in detail in books on NMR. The 2D-IR signal $S(\tau, t; T)$ defined by the time-domain interferogram is the real part of the echo field generated when the third pulse creates a coherence that is conjugate to the initial coherence that exists during the evolution time:

$$\begin{aligned} &\left\langle \mathrm{Re}\left[e^{+i\omega_{10}\tau - \gamma_1\tau} e^{-T/T_1} e^{-i\omega_{10}t - \gamma_1 t}\right]\right\rangle \\ &\quad = \langle(\cos\omega_{10}\tau\cos\omega_{10}t + \sin\omega_{10}\tau\sin\omega_{10}t)\rangle e^{-\gamma_1(\tau+t) - T/T_1} \end{aligned} \tag{34}$$

It is easy to discover by plots of (34) that the shape of this signal is not actually circular or elliptical, especially because of the presence of the sine term. The double Fourier transform of the signal (34) according the procedure (25) and the result in (6) is a 2D spectrum in the second and fourth quadrants of $\{\omega_\tau, \omega_t\}$. The real part of the sum of the responses (32), with different signs in ω_{10}, is the absorptive diagonal 2D-IR response, now free from the sine term:

$$\langle \cos \omega_{10}\tau \cos \omega_{10}t \rangle e^{-\gamma_1(\tau+t)-T/T_1} \tag{35}$$

A full discussion of the types of spectra obtained by adding the rephasing and nonrephasing parts was recently published [45]. For some applications it may be useful to display both rephasing and nonrephasing parts, especially when there are overlapping unresolved transitions, which will always be the situation for peptides and proteins [14, 43].

We now return to the cross peaks of the 2D-IR spectrum of a system having two coupled modes. The various Liouville paths that contribute to the signal in this case were discussed in regard to Fig. 5. One of the cross-peak responses is of the form

$$\left\langle e^{i\omega_{10}\tau-\gamma_1\tau}[e^{-T/T_1} + e^{i\omega_{21}T}]e^{-i\omega_{20}t-\gamma_2 t} \right\rangle \tag{36}$$

where ω_{20} represents the frequency of a second mode, γ_2 its dephasing time, and $\omega_{21} = \omega_{01} - \omega_{02}$ is the difference in frequency between the two modes. In that case the static average involves the joint distribution of the two frequencies [14]. The two terms involving T correspond to the two Liouville pathways that lead to the same coherence transfer. Using the result of Eq. (6) and (25), the 2D-IR spectrum corresponding to the field (36) can immediately expressed as follows:

$$\left\langle \frac{(e^{-T/T_1} + e^{-i\omega_{21}T})(\gamma_1 + i\omega_\tau)(\gamma_1 + i\omega_t)}{[(\gamma_1 + i\omega_\tau)^2 + \omega_{01}^2][(\gamma + i\omega_t)^2 + \omega_{01}^2]} \right\rangle \tag{37}$$

Since both frequencies are chosen to have the same sign in the absorptive spectrum, (37) becomes

$$\left\langle \frac{(e^{-T/T_1} + e^{-i\omega_{21}T})(\gamma_1 + i(\omega_{10} - \omega_\tau))(\gamma_1 + i(\omega_{20} - \omega_t))}{[\gamma_1^2 + (\omega_{10} - \omega_\tau)^2][\gamma_2^2 + (\omega_{20} - \omega_t)^2]} \right\rangle \tag{38}$$

This is the 2D-IR spectrum cross peak apart from transition dipole factors. The real part consists of a peak in the first quadrant at $\omega_\tau = \omega_{10}, \omega_t = \omega_{20}$. When δ, the mixed mode anharmonicity, is small, this cross term may be almost canceled by one involving the combination band of the two modes:

$$-\left\langle e^{i\omega_{10}\tau-\gamma_1\tau}[e^{-T/T_1} + e^{i\omega_{21}T}]e^{-i(\omega_{20}-\delta)t-\gamma_{1+2}t} \right\rangle \tag{39}$$

where γ_{1+2} is the dephasing time of the combination tone. The spectrum then becomes (38) with ω_{20} replaced by $\omega_{20} - \delta$, which is similar to (38) but with the peak shifted along ω_t by the off-diagonal anharmonicity. These two terms together constitute the dual frequency signal in the approximation where there is a distribution of motionally narrowed transitions. The first involves field-induced coherence transfer between the two modes, and the second involves transfer to a $(1+2, 1)$ coherence. Again the 2D-IR signal $S(\tau, t; T)$ is the sum of the real parts of these responses. As shown previously, the 2D-IR spectra of the cross peaks strongly depend on the correlations of the frequency distributions.

The absorptive cross-peak signal is

$$\left\langle \cos \omega_{10}\tau e^{-\gamma_1\tau}\left[e^{-T/T_1} + e^{i\omega_{21}T}\right]\left(e^{-i\omega_{20}t-\gamma_2 t} - e^{-i(\omega_{20}-\delta)t-\gamma_{1+2}t)}\right\rangle \tag{40}$$

which leads to the equation given in our article [17] on dual frequency 2D-IR when the average over the joint distribution of frequencies is carried out as described below.

XVI. SIMPLE MODEL OF THE 2D-IR SPECTRUM OF A PAIR OF RESONANCES

The previous discussions of the signal are nicely illustrated by an extremely simple model analysis using real fields and signals for two Lorenzian resonances at frequencies a and b. The sample is irradiated with two very short pulses whose spectra are flat. The real generated field from the sample is the real part of Eq. (21) or Eq. (33) with T set equal to zero for convenience since e^{-T/T_1} is in any case a multiplicative factor. In time-domain interferometry, this is measured directly along the indicated time axes as described above. In spectral interferometry the real generated field along with a real local oscillator field, delayed by time d, is dispersed (i.e., Fourier-transformed) by a monochromator, then squared by the detection to yield a spectrum on the array detector at each value of τ:

$$S(\tau, \omega_t) = I + 2\,\mathrm{sign}(R)e^{-\gamma_a\tau}\frac{\cos[\omega_a\tau - \omega_t d - \tan^{-1}(\Delta/\gamma)]}{(\Delta^2 + \gamma_b^2)^{1/2}} \tag{41}$$

where d is the delay of the local oscillator, $\Delta = \omega_b - \omega_t$, and I is the sum of the intensity spectra of the generated and local oscillator fields. The second term, up to a constant factor, is the interference of the flat local oscillator and the generated field, which shows oscillations along both ω_t and τ. The resonance pairs may have different signs, sign (R), which can be read off from the Liouville pathway diagram for the signal. The signs originate from the commutator in the Liouville equation and are negative when there is an odd number of interactions on one side, either bra or ket, of the density operator. Interference spectral components are obtained as per Eq. (38) for each pair of resonances of the sample, and they are additive since they correspond to the field amplitudes. The denominator of the cross peak in (40) is the square root of the power spectrum of the generated field. All resonances contribute a term of this type to add up to a two-dimensional spectrum once the I term is removed. Additional phase differences can easily be incorporated into the cosine. The Fourier transform of the cross term in (40) consists of two one-sided exponential decays, one extending from $t = d$ to infinity and the other from $t = -d$ to $-$infinity:

$$S(\tau, t) = 2\,\mathrm{sign}(R)[e^{-(\gamma+i\omega_a)\tau}e^{-i\omega_b(t-d)}\Theta(t-d)e^{-\gamma(t-d)} + e^{(-\gamma+i\omega_a)\tau}e^{-i\omega_b(d+t)}\Theta(-d-t)e^{-\gamma(d+t)}] \quad (42)$$

There is also a part at zero time from the terms represented by I and from the noise. Everything is discarded except the component at $t = d$, which when back Fourier-transformed yields

$$S'(\tau, \omega_t) = 2\,\mathrm{sign}\,(R)e^{-(\gamma_a+i\omega_a)\tau}e^{-i\omega_t d}/(\gamma_b + i(\omega_b - \omega_t)) \quad (43)$$

which is the complex spectrum of the generated fields at each value of τ. The absolute maximum of this spectrum decays along the τ axis with time constant $2\gamma_a$. No peak shift along the τ axis is predicted by this simple example because no inhomogeneous broadening was incorporated.

XVII. THE EFFECT OF CORRELATION OF VIBRATIONAL FREQUENCY DISTRIBUTIONS

Equation (32) can be rewritten as

$$e^{\pm i\omega_{10}^0\tau-\gamma_1\tau}e^{-T/T_1}e^{-i\omega_{10}^0 t-\gamma_1 t}\langle e^{\pm ix\tau-ixt}\rangle = e^{\pm i\omega_{10}^0\tau-\gamma_1\tau}e^{-T/T_1}e^{i\omega_{10}^0 t-\gamma_1 t}e^{-\sigma^2(t\mp\tau)^2/2} \quad (44)$$

where the mean frequencies ω_{10}^0 are taken out of the average which now involves the frequency deviations x from the mean of the equilibrium distribution with standard deviation σ. Simple forms are obtainable for the responses when the distribution of frequencies being averaged over is static [14]. A Gaussian distribution is assumed in the last step of Eq. (44). The results given below are obtained by inspection by assuming the frequency deviations (i.e., the x, y or their finite time integrals) are Gaussian variables and employing the well-known relationship between the average of the exponential and the exponential of the average obtained by integration over a multivariate Gaussian distribution:

$$\left\langle e^{\sum_i a_i x_i} \right\rangle = e^{1/2 \sum_{i,j} a_i a_j \langle x_i x_j \rangle} \tag{45}$$

We used this relationship to obtain the results of our first article on the spectra of the vibrational photon echo [80, 81] of the azide ion and hemoglobin.

If in Equation (44) we replace ixt by iyt to account for different resonant frequencies being involved in the τ and t evolutions, we can immediately carry out the Gaussian average:

$$\left\langle e^{\pm ix\tau - iyt} \right\rangle = e^{-\sigma_x^2 \tau^2/2 - \sigma_y^2 t^2/2 \mp f \sigma_x \sigma_y t\tau} \tag{46}$$

where f is the correlation coefficient of the two frequency distributions:

$$f = \frac{\langle xy \rangle}{\sqrt{\langle x^2 \rangle \langle y^2 \rangle}} \tag{47}$$

The result (46) for correlated Gaussian variables was used in our three-pulse-echo spectral study of hemoglobin [80, 81], where f was found to be close to unity.

When the distribution being averaged over is dynamic [29] but classical with Gaussian fluctuations $x(t)$, the gammas representing the homogeneous widths of different components should be omitted from the formulas and the average in Eq. (44) must be replaced by

$$\left\langle e^{\pm i \int_0^\tau x(t')dt' - i \int_{T+\tau}^{t+T+\tau} x(t')\, dt'} \right\rangle = e^{-g(\tau) \pm g(T) - g(t) \mp g(\tau+T) \mp g(t+T) \pm g(\tau+T+t)} \tag{48}$$

which reduces to Eq. (44) when x is time-independent. The dynamics is now described in terms of the correlation function of the frequency fluctuations, defined in

$$g_{xy}(t) = \int_0^t dt_1 \int_0^{t_1} dt_2 \langle x(t_1) y(t_2) \rangle, \qquad g_{xx}(t) \equiv g(t) \tag{49}$$

When the correlation function for Bloch dynamics, $\langle x(t)x(0) \rangle = \delta(t)\gamma + \sigma^2$, is used to obtain $g\,(t)$, Eq. (44) is recovered. This development is easily generalized to pairs of correlated variables in the τ and t domains as occur in Eq. (46) by replacing the $x(t)$ in the second integral of Eq. (48) by another fluctuation $y(t)$ occurring in the detection time domain. The cross correlation $\langle x(0)y(t) \rangle$, which gives the time dependence of the correlation between the variables, is then measured as part of the experiment:

$$\left\langle e^{\pm i\int_0^{\tau} x(t')\,dt' - i\int_{T+\tau}^{t+T+\tau} y(t')dt'} \right\rangle$$
$$= e^{-g_{xx}(\tau)\pm g_{xy}(T)-g_{xy}(t)-g_{yy}(\tau+T)+\binom{0}{2}g_{xy}(\tau+T)\mp g_{xy}(t+T)-g_{yy}(\tau+T+t)+\binom{2}{0}g_{xy}(\tau+T+t)} \quad (50)$$

which goes over to Eq. (48) when x and y correspond to the same variable and to (46) when both x and y define fixed distributions. Again the upper and lower signs and prefactors correspond to the rephasing and nonrephasing responses, respectively. In the two-pulse echo where $T = 0$, these results simplify considerably. Results for vibrational relaxation dynamics more general than the foregoing approaches, particularly for cases where g is complex and where noncommuting properties are important, have been given by Mukamel employing a cumulant expansion, which does not in general require the variables to be Gaussian, but implicitly does so when the expansion is terminated at second order. Results comparable to the above were also given in Ge et al. [14], where the correlation functions were written in terms of the anharmonicity fluctuations. An essential feature of these results is that the correlation function of the frequency fluctuations influences the evolution of the system during the waiting time T. The physical interpretation of this result is that the system maintains memory of the frequency imprinted on the original inhomogeneous distribution and that this can be rephased after the waiting period only if the spectral diffusion determined by the frequency correlation function is incomplete.

XVIII. DUAL-FREQUENCY 2D-IR

The use of two frequencies is a useful alternative to ultrashort pulse shaping and impulsive limit experiments. It can permit avoidance of pulse distortion that might arise in regions of strong solvent or solute absorption (see, for example, Fig. 1) when broad-band pulses were used. It also enables a broad selection of anharmonic couplings and an increased number of structural constraints to be obtained in a single measurement. Of course it may be convenient in some cases to use employ ultrashort pulses that cover the whole spectral range of all the modes of interest. In order to access both amide-I and N–H modes in proteins, a pulse having a bandwidth of greater than 1800 cm^{-1} would be needed. For a Gaussian pulse centered at 4 μm, this criterion implies a time width less than 10 fs. The 2D-IR spectrum is simplified considerably by the use of two frequencies as shown in our work on *N*-methylacetamide (NMA) [17]. In other work, joint nonlinear responses were stimulated from both the amide-A (N–H) and either the amide or ester carbonyl transitions or from amide-I and amide-II transitions [15]. When two center frequencies are employed, the signal may be interpreted by the cross-peak responses typified by Eq. (50). The x and y parameters then correspond to the correlated fluctuations in the frequencies the two resonances. From the foregoing analysis the absorptive cross-peak response in the Bloch approximation, which represents the complete dual frequency signal, is predicted to be [17]

$$\begin{aligned} S(\tau, t) = {} & [\cosh(f\sigma_{\mathrm{I}}\sigma_{\mathrm{II}}t\tau)\cos\omega_{\mathrm{II}}\tau(\cos\omega_{\mathrm{I}}t - e^{-t/T_1^{(\mathrm{II})}}\cos(\omega_{\mathrm{I}} - \Delta_{\mathrm{I,II}})t) \\ & + \sinh(f\sigma_{\mathrm{I}}\sigma_{\mathrm{II}}t\tau)\sin\omega_{\mathrm{II}}\tau(\sin\omega_{\mathrm{I}}t - e^{-t/T_1^{(\mathrm{II})}}\sin(\omega_{\mathrm{I}} - \Delta_{\mathrm{I,II}})t)]G_0(\tau, t)e^{-2D(t+\tau)} \end{aligned} \tag{51}$$

where the uncorrelated Bloch line shape function is $G_0(t, \tau)$ obtained from Eq. (46) with $f = 0$. The parameters ω_{I} and ω_{II} are the angular frequencies and σ_{I} and σ_{II} the inhomogeneous widths of the amide-I and amide-II transitions, the T_1's are the population relaxation times of the indicated states, D is the rotational diffusion coefficient, and the total homogeneous dephasing rates of the amide-I and amide-II fundamental transitions are included in G_0. The off-diagonal anharmonicity is $\Delta_{\mathrm{I,II}}$, chosen so that the frequency of the transition between II and I + II is $\omega_{\mathrm{I}} - \Delta_{\mathrm{I,II}}$. By inspection, the dual frequency 2D-IR spectrum, obtained from the double Fourier transform of $S(\tau, t)$, displays peaks $(\omega_{\mathrm{II}}, \omega_{\mathrm{I}})$ and $(\omega_{\mathrm{II}}, \omega_{\mathrm{I}} - \Delta_{\mathrm{I,II}})$. When $f = 0$, the case of uncorrelated distributions, the signal reduces to the difference between two components having opposite signs that are slightly displaced (by $\Delta_{\mathrm{I,II}}$) along the ω_t axis. The node separating the positive and negative parts of the 2D spectrum is then parallel to the ω_τ axis in the region of $\omega_\tau = \omega_{\mathrm{II}}$. The effect of finite f is to tilt this node one way or the other depending on the sign of f. Thereby, f is easily measured in this dynamic approximation.

The use of dual frequencies significantly enlarges the scope of 2D-IR vibrational spectroscopy. Interesting qualitative models of the anharmonic potential surfaces of peptides emerge from dual frequency results from which we find the amide-I (C=O mode) and amide II (C–N mode) frequency distributions having f-values significantly less than zero. We think of this in terms of solvent-induced mixing of valence bond structures where an increase in the zwitterionic form of NMA, as might occur for NMA molecules associated with particular solvent configurations that bind effectively to that form, causes a reduction in the C=O frequency and an increase of the C-N frequency.

XIX. POLARIZATION DEPENDENCE

The signals in 2D-IR experiments are fourth-rank tensor properties with indices corresponding to the four polar vector components of the incident and detected electric fields. In an isotropic medium there are only three independent fourth rank tensor components and they satisfy the relationship

$$\langle xxxx \rangle = \langle xxyy \rangle + \langle xyxy \rangle + \langle xyyx \rangle \tag{52}$$

where x and y are any two orthogonal directions in space. This result holds for all third-order nonlinear signals in isotropic media. The signal for a given set of fields chosen by the experimenter, say $\langle xyxy \rangle$, is proportional to the ensemble average of the product of four direction cosines taken at each of the times that the pulses interact with the system, namely $\langle \hat{\mu}_y(t_4)\hat{\mu}_x(t_3)\hat{\mu}_y(t_2)\hat{\mu}_x(t_1) \rangle$, where $\hat{\mu}_x(t_1)$ is the projection onto the x axis of the unit transition dipole interacting at time t_1. The time dependences of the signals are readily obtained by standard methods when the evolution of the system between the pulses obeys the diffusion equations for rigid bodies [74, 76]. Each of the diagrams in Fig. 2 is associated with a transition dipole factor of this type for any calculation of the 2D-IR signals. It is possible to find polarization conditions that enable the measurement of various linear combinations of the three independent tensors [73]. For example, when we choose the four polarizations as $\langle x+y, x-y, x, y \rangle$, the echo measures the linear combination $(\langle xyxy \rangle - \langle xyyx \rangle)$, which is free from the often dominant, intense diagonal contributions to the 2D-IR spectrum [75]. Figure 6 shows some examples of the use of different polarization conditions for the 2D-IR spectra of acylproline. The diagonal peaks in the bottom panel arise because the polarizers used in the experiment are imperfect. Some of the cross peaks are evident only in certain polarization conditions because of the angles between the transition dipoles of the coupled states and the distributions of these angles. For example, peaks A and C (Fig. 6a) correspond to the amide-I and amide-II transitions that have nearly perpendicular transition dipoles. The pump-probe method, in its traditional form, does not measure all the tensor components; rather it is confined to $\langle xxxx \rangle$ and $\langle xxyy \rangle$. The angular averages all decay with orientational relaxation dynamics and contain averages over angle distributions and the first and second Legendre polynomials of the cosines of the angles between the various transition dipole moments involved in the process. In order to increase the extent of the angular information available from 2D-IR, experiments would need to be carried out in anisotropic media, such as crystals and oriented films.

XX. PROJECTION ANALYSIS OF COMPLEX SPECTRA

When infrared spectra are diffuse and have overlapping bands, the 2D-IR method can be invaluable in characterizing the underlying structure and homogeneous dynamics. Equation (22) shows how the temporal decay of a photon-echo yields a homogeneous decay parameter in the presence of a broad frequency distribution. Therefore it is not surprising that the 2D-IR spectrum exposes the

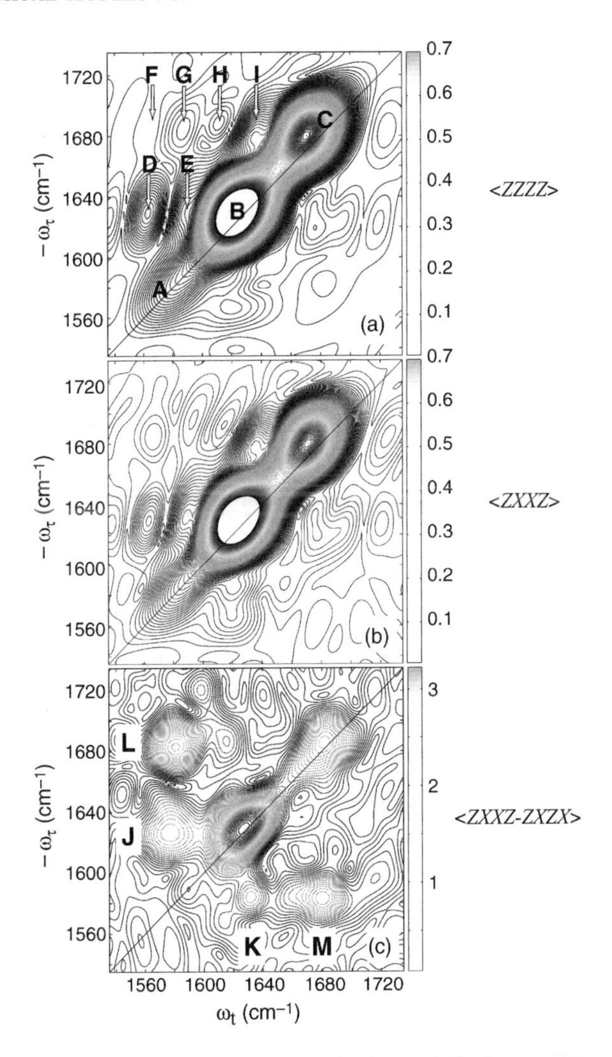

Figure 6. The 2D-IR spectrum of acylproline for three polarization conditions. These three measurements represent all possible orientational information on this isotropic system. This figure is adapted from published data (see References. 7 and 43). See color insert.

homogeneous width of the individual transitions, even when the spectrum is dominated by their inhomogeneous broadening. In a qualititative sense, this is shown already in Fig. 5, where the 2D-IR transitions are shown to be elongated along the diagonal. Correspondingly, these transitions are narrowed when traces are taken perpendicular to the diagonal; this is a manifestation of the line-narrowing capability of the photon-echo experiment. However, these antidiagonal widths do not give the homogeneous widths directly because of the contribution of the anharmonically shifted transitions. A formal way to generate traces of this type is by means of skew projections, which allow different aspects of the 2D-IR spectra to be emphasized.

We already discussed the projection onto the ω_t axis as being the pump-probe spectrum. The $\pm 45°$ projections can also be quite useful for analyzing the correlations in the inhomogeneous distributions. It is useful to note that Eq. (47) can be rewritten as

$$\left\langle e^{\pm ix\tau - iyt} \right\rangle = e^{-[(\sigma_x+\sigma_y)(\tau+t)+(\sigma_x-\sigma_y)(\tau-t)]^2/4+(1\mp f)\sigma_x\sigma_y t\tau} \tag{53}$$

Assuming for discussion that σ_x and σ_y are very similar, this average in the rephasing spectrum does not involve the inhomogeneous distribution if $t + \tau = 0$ and $f = 1$, but it does when $f = -1$. The opposite is the case in the nonrephasing spectrum. This implies that the Fourier transform of the projection of the time-domain data onto the line $t + \tau = 0$, which is equivalent to projecting the rephasing frequency-domain data onto the antidiagonal line $\omega_t + \omega_\tau = 0$, will show peaks that are very sensitive to the correlation. When the frequency dependences are incorporated, it is easy to see that the projected spectrum will display negative peaks at zero frequency and $\pm 2^{-1/2}\,\omega_{21}$ and will show positive peaks at $2^{-1/2}(\Delta_{12} \pm \omega_{21})$ and $-2^{-1/2}\Delta$. The spectrum obtained from the projection onto the line $t - \tau = 0$ or directly from the projection onto $\omega_t - \omega_\tau = 0$ has the full inhomogeneous width for $f = 1$ in the rephasing diagram. This is as expected because the signal at $t = \tau$ emphasizes the echo part of the free decay rather than the echo delay: If the inhomogeneous broadening is very large, a very short time spike is emitted at $t = \tau$.

XXI. WAITING TIME DEPENDENCE OF THE SIGNALS

After the second pulse of a 2D-IR experiment, either the system is put into a population state or an interstate coherence is developed, after which the coherence implanted during the first period is stored. During this storage period, the time T, the signal decays with the population or interstate coherence relaxation time. In the presence of an inhomogeneous distribution, this coherence can be rephased by the third pulse, which finally brings about the coherence transfer to a state that radiates near the vibrational transition frequencies. The variation of the signal with T therefore permits dynamical properties of the inhomogeneous distribution to be measured. The concepts behind this measurement have been well documented in the optical regime [82]. The forms of the responses given above simplify considerably when $T = 0$ as is the condition of a two-pulse echo. But for finite values of T, there are no

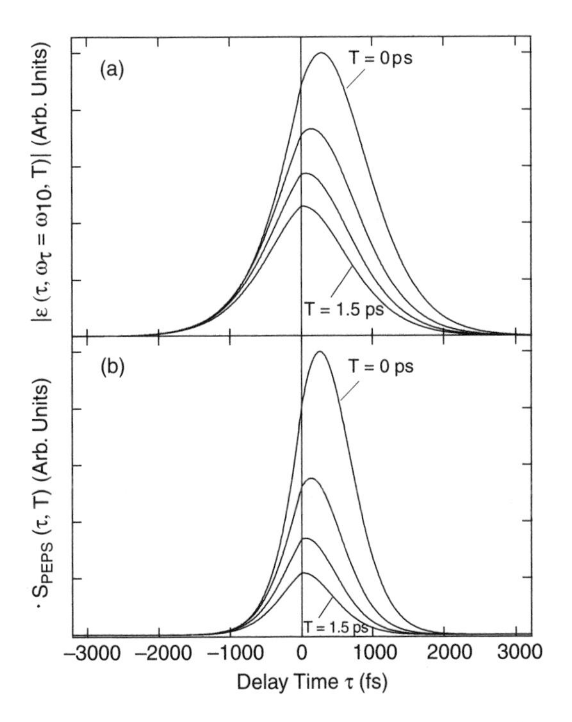

Figure 7. A comparison of a peak shift experiment by means of (*a*) the stimulated or integrated echo and (*b*) the heterodyned signal as discussed in the text. Note that the waiting time (T) dependence of the peak shift is the same in the two measurements, although the total signal decays faster in the integrated echo as a function of T.

simplifications in general and the evolution through the time period τ is different for every value of T and it depends on the choice of the vibrational frequency correlation function. A plot of $|S'(\tau, \omega_t; T)|$ for a single oscillator with an exponentially decaying correlation function is given in Fig. 7 as a function of τ for a few values of T. To produce that plot, some form for the correlation function has to be assumed. We have assumed simple phenomenological models for the correlation function:

$$C(t) = \sum_n \sigma_n^2 e^{-t/\tau_n} \tag{54}$$

Thus from (49) $g(t)$ is given by a linear combination of Kubo functions:

$$g(t) = \sum_n \sigma_n^2 \tau_n^2 (t/\tau_n + e^{-t/\tau_n} - 1) \tag{55}$$

Another possible two-parameter fitting function that has a fast decay followed by a slower decay [83] is

$$C(t) = \sigma^2 \cosh(bt)/\cosh(at), \qquad a > b \tag{56}$$

A common simplification of Eq. (54) is that one of the correlation times, say τ_1, is very short and gives a motionally narrowed contribution of $\sigma_1^2 \tau_1 t = \gamma t$ to $g(t)$, and one other exponential is included, leading to

$$g(t) = \gamma t + \sigma^2 \tau_c^2 (t/\tau_c + e^{-t/\tau_c} - 1) \tag{57}$$

This tends to a Bloch dynamics picture for large correlation times τ_c. A static inhomogeneous distribution of width σ_s can easily be added as $\sigma_s^2 t^2/2$ where $\sigma_s >> 1/t_{\exp}$, with $t_{\exp}$ the experimental time scale. The plots on Fig. 5 show that the peak in the $|S'(\tau, \omega_t; T)|$ versus τ signal does not occur at $\tau = 0$ for such correlation functions and that it gradually shifts to zero as T increases in analogy with what occurs in a 3PEPS [82] experiment. The 2D-IR experiment contains all this information about the equilibrium dynamics, and separate experiments such as the integrated three-pulse echo are an intrinsic part of it. In fact it is easy to show that for a separate resonance $|S'(\tau, \omega_t; T)|$ at the point $\omega_t = \omega_{10}$ is just the t-integrated absolute time-dependent signal, whereas the three pulse integrated echo, Eq. (31), is the t integrated absolute square of the signal. Therefore the two signals are simply scaled on the t axis. As shown in Fig. 5, the magnitude of the integrated echo signal decays more rapidly than the heterodyne signal, but the peak shift dynamics are comparable in the two cases. The first application of these types of correlation functions to vibrational spectra concerned the asymmetric stretching mode of the azide ion [40]. In that example the $g(t)$ had the form of Eq. (55) with an additional constant term. Recently there was a discussion of the applicability of separated time-scale models [84].

The response functions given above given here may be applicable to many vibrational problems. However, in all the foregoing examples the instantaneous frequencies along t and τ are simply the resonant frequencies. This is just another way of saying that the dynamical parameters are all real quantities, and that phase is only accumulated by the mean resonance frequencies. There are many circumstances when one would want to incorporate specific time-dependent frequencies. One is when there is strong coupling between low- and high-frequency modes [85]. In general the spectral density of the modes interacting with the driven mode may have detailed structure corresponding to a more discrete mode distribution, which will show up in the relaxation of the system.

Most vibrational excitations do not exhibit dynamic shifts having magnitudes that are remotely comparable with the large Stokes shifts of electronic spectra. However, if there is a change in the frequency with time as would occur if the system were undergoing relaxation into lower levels having altered anharmonicity, the signal could be processed as a time-dependent phase modeled from knowledge of the kinetics of the relaxation. This situation is discussed again later. Another example where there would be time-dependent frequencies is when the driven molecule is undergoing energy transfer into modes of other molecules having different frequencies. The small change of electric moment that may occur as a result of vibrational excitation will also generate time-dependent frequencies if the vibrational excited-state lifetime is sufficiently long. In these cases there is a need to incorporate the effect of the driven mode on the mode involved in the relaxation and hence generate a time-dependent oscillator frequency. The Brownian oscillator model of nonlinear optical spectroscopy [24] developed for electronic transitions has the correct form to account for both spectral diffusion and phase evolution. The line-shape parameter $g(t)$ then would incorporate an additional pure imaginary term:

$$g_b(t) = g(t) + iB \int_0^t dt' \langle x(0)x(t') \rangle \tag{58}$$

where the parameter B will be related to the anharmonic coupling between the high- and low-frequency modes. This modeling predicts an instantaneous frequency equal to the resonance frequency plus $B\langle \omega(0)\omega(t) \rangle$.

XXII. TRANSIENT GRATINGS

The four-wave mixing signals observed in 2D-IR echo experiments of the type outlined here can be described alternatively in the language of grating diffraction [29] using Bragg diffraction formulas instead of phase-matching conditions. This way of visualizing the signals could be quite useful in certain examples, and there is by now a vast literature on grating experiments on molecular systems [86–94]. Two pulses arriving at different times impress onto the sample a grating, which has a macroscopic spacing that depends on the geometry of the beams, and a well-defined frequency grating, dependent on the pulse delay; the third pulse can diffract from this grating into the signal direction to generate an echo that carries information regarding the phase. Complementary phase information is stored in the excited or ground vibrational states. This is important for vibrational echo experiments because the population relaxation times can be extremely fast, and gratings can be created during the standard evolution periods. The grating induced by the combined intensity of pulses $-k_1$ and k_2 could be impressed either on the solvent or on the probed molecule. For example, in water or D_2O the infrared pulses heat the solvent slightly, causing a grating to develop from the different absorption spectra (dichroism) of the solvent at different temperatures. There are also differences in the index of refraction of pumped and unpumped water. These gratings can be produced directly from the field interactions or indirectly by there first being light absorption by the molecule of interest followed by transfer of energy to the solvent. Also, the population of levels of the probed molecule that are not directly driven by the fields have spatial gratings impressed on them by the joint intensity of the first two pulses. These gratings, which are due to populations of low-frequency modes of the molecule, will diffract a pulse into the Bragg directions, but the spectral character of the diffracted light depends on the correlations of the nonradiatively coupled modes. When the grating is probed by resonant processes of the molecule, the resonance frequencies will be impressed on the generated field because these specific coherences are generated by the probing pulse. Characterizing these signals is an important experimental challenge in each of the examples we have studied. Fortunately, the thermal grating signal from D_2O is relatively small compared with the resonant signals from peptides so it does not present a barrier to applications of 2D-IR in aqueous biological systems. However, the thermal grating contributions grow in with increasing T and do persist for very long periods compared with the lifetimes of vibrationally excited states, and so they may ultimately dominate the signals when T becomes large enough.

The heterodyned transient grating analysis proceeds in the same fashion as just described. The pulses k_1 and k_2 create the grating from which k_3 is scattered into the direction $-k_1 + k_2 + k_3$. In a conventional transient grating, $\tau = 0$ so that $|S(0, \omega_t; T)|$ gives the transient grating at any detected frequency. The real part of the heterodyned signal can be examined directly; thus if the phases are known, there is no need to take absolute values, which introduce cross terms between the component responses oscillating at different resonance frequencies. As τ increases from zero, these gratings are still formed but they now store frequency information from the first interaction that is lost on the time scale of spectral diffusion.

XXIII. ULTRAFAST POPULATION DECAYS

In the case of vibrational responses the population relaxation times may be dominating the coherence decays. In addition, it can be essential to incorporate the multilevel nature of molecular vibrators into the response. The rate of repopulation of the ground state is seldom equal to the decay of the fundamental $v = 1$ state, so there can be bottlenecks in the ground state recovery. Following the coherence time, τ, the second pulse, k_2, may place the system into a ground $v = 0$ population state. A two-level system with $h\nu/k_{\mathrm{B}}T >> 1$ would evolve as $\exp[-T/T_1]$ during the waiting time since the ground state recovers at the rate of decay of the excited state. However, if there is a bottleneck in the relaxation, the evolution of the system during the waiting time will depend on the lifetime of the intermediate state: The evolution of the density matrix element ρ_{00} during the time T must be obtained from the master equations for the populations of the system. If the spontaneous exchange of populations and coherences is neglected, the T dependence for the population propagator for any state should be written as its survival probability that a system having population ρ_{00} at time 0 will retain this population after time T, $P(00;0|00;T)$, where

$$P(00;0|00;T) = e^{-\int_0^T k(t')\,dt'} = \frac{n_0(T)}{n_0(0)} \tag{59}$$

where $k(t)$, defined as $\dot{P}(00;0|00;T)/P(00;0|00;T)$, can be obtained by solving kinetic equations for the populations. For example, for a system of three levels $v = 0, v = 1$ and a third state, $|r\rangle$, with level spacings larger than k_BT the response function must contain the T-dependent factor:

$$P(00;0|00;T) = e^{\log\left[\frac{(k_{10}-k_{r0})e^{-k_1T}+k_{1r}e^{-k_{r0}T}}{k_1-k_{r0}}\right]} \tag{60}$$

where k_{ij} is the rate constant for relaxation from state $|i\rangle$ to state $|j\rangle$ and k_1 is the inverse lifetime of the $v = 1$ state. This example could be a common model in vibrational dynamics where relaxation of the ground-state depletion occurs via a state or states lying between the driven state and the ground state. The T dependence of the density matrix elements of the excited state ρ_{11} also contains terms that switch the population to other excited states; for the three-state model they involve $P(11;0|rr;T)$, the probability that a system in state 1 at time zero will be in state $|r\rangle$ at time t, where

$$P(11;0|rr;T) = (k_{1r}/k_1)(1 - e^{-k_1T}) \tag{61}$$

whereas $P(11;0|11;T)$ is simply $e^{-k_1T} = e^{-T/T_1}$ as was assumed in the response function given in Eq. (31) and would be the only term needed in a two-level system. If the relaxation switches the system to state $|r\rangle$, it is $|r\rangle$ that interacts with the third pulse k_3. But it may or may not be possible for the pulse to cause the transition $|r\rangle \rightarrow |0\rangle$, so diagrams of the type $R_1, R_2, R_4,$ and R_5 in Fig. 2 may be absent in this pathway. On the other hand, the responses of type R_3 and R_6 are always present because in all cases the mode r can form a combination band

$|r+1\rangle$ with the originally excited vibrational state of the molecule. In this case the frequency information from the coherence period becomes stored in the r state population. In the 2D-IR spectrum a peak appears at $\omega_t = \omega_{r+1,r}$ instead of ω_{10}. These considerations are not specific to the point $\tau = 0$. One can also think about the effects of these relaxation processes in terms of the frequency gratings induced by the first two fields as discussed earlier. The phase of the grating along the direction $k_2 - k_1$ induced by pulses k_1 and k_2 separated by τ is simply $\omega_{01}\tau$. The population relaxation into lower-frequency vibrational states or the ground state during the waiting time may preserve this phase. This means that the echo signals from the scattering of k_3 into the direction $-k_1 + k_2 + k_3$ may continue to appear long after the primary population state shown in the diagrams of Fig. 2 has gone. On the other hand the relaxation may cause a scrambling of the phase information when the coupled mode distributions are uncorrelated. Intermediate cases are readily dealt with using the dynamics approximations discussed earlier.

It will be very common in vibrational 2D-IR that vibrational relaxation will result in a frequency shift for the reasons given above. A very simple model for the phase that develops as a system evolves between two frequencies as a result of relaxation to modes with weaker anharmonicity is

$$\phi(t) = \int_0^t \omega(t')\,dt' = (\omega_1(1 - e^{\gamma t}) + \omega_2(\gamma t + e^{-\gamma t} - 1))/\gamma \tag{62}$$

in which the instantaneous frequency $d\phi(t)/dt$ moves from ω_1 to ω_2 on the scale of the overdamped relaxation time $1/\gamma$. It will not be clear *a priori* which of the two frequencies will be larger, but if the mixed mode anharmonicity is smaller than the diagonal anharmonicity, the FID frequency should increase with time as the two-quantum transition frequency approaches that of the fundamental. These relaxations give rise to peak shape changes in broad-band spectra because the spectrum moves along the t axis as the time T increases.

XXIV. SPONTANEOUS INTERCHANGES OF COHERENCE

Additional features arise from spontaneous transfers of coherence particularly, but not exclusively, when the population relaxations are very fast. The coherence equations, which are effectively the Redfield relations without interchanges of population with coherence, must be solved during each time interval and probabilities worked out for the appearance of coherences other than those that are driven by the excitation pulses. This procedure is particularly important for vibrational systems, where there are often a significant number of transitions having nearly the same frequency. For example, the coherence ρ_{v_i,v_i+1_j} oscillates at almost the same frequency, ω_{0j}, for all numbers of quanta, v_i, of all of the modes, i, when the anharmonic coupling is small. Furthermore, transitions between these coherences can be induced by spontaneous population relaxation. This implies that when a particular coherence is achieved by interaction of the system with a pulse, say the third pulse of the sequence that always generates a coherence, by the time the signal is detected it may have switched to a new coherence oscillating at a different frequency. The frequencies cannot differ significantly; otherwise the coherence transfer efficiency will vanish: they are then termed *nonsecular* in second-order perturbation theory. But in 2D-IR the frequency changes may be related to coupling or to anharmonicities, which are generally small. By small we mean that the period of the difference frequency is large or comparable with the time scale of the experiment. In this case each Liouville pathway divides spontaneously into a number of others so that the observed frequencies and/or the signal strengths become determined by the coherence transfer kinetics. In general the probabilities $P(ij;0|kl;t)$ should be found from Redfield or kinetic models and introduced formally into the system evolution during each period. In the example below we consider the effect of spontaneous coherence transfer in introducing cross peaks in to the 2D-IR spectrum of a dipeptide considered as a system with two oscillators with nearby frequencies coupled to a bath.

XXV. EXAMPLES OF SPONTANEOUS COHERENCE TRANSFER

We consider an oscillator having two frequencies ω_1, ω_2 which could represent the amide-I modes of a dipeptide. The difference frequency $\omega_1 - \omega_2$ is large compared with the bandwidth of each oscillator but small compared with the relaxation rates of the system. In other words the oscillation period $2\pi/|\omega_1 - \omega_2|$ is comparable or larger than the relaxation times of the system. The situation is depicted in Fig. 2. The diagram R_1 describes the usual diagonal peak of the 2D-IR spectrum. Its appearance in the spectrum depends on the probability that the created coherences persist throughout the τ and t periods. However, if the coherence $\rho_{10}(t)$ can spontaneously interchange with the coherence ρ_{20}, the cross peaks in the spectrum will appear. The coupling peaks in the spectrum then arise from the interactions of the oscillators with a bath having modes $p, q, \ldots$, because the coherences are exchanging in the equilibrium distribution according to

$$\begin{aligned}\dot{\rho}_{01} &= \gamma_{01}\rho_{01} + R_{0102}\rho_{02}e^{-i\omega_{12}t}\\ \dot{\rho}_{02} &= R_{0201}\rho_{01}e^{i\omega_{12}t} + \gamma_{02}\rho_{02}\end{aligned} \tag{63}$$

where

$$R_{10,20} = 2\pi/\hbar \sum_{p,q} P_p \langle q1|V|p2\rangle\langle p0|V|q0\rangle\delta(\varepsilon_p - \varepsilon_q + \varepsilon_2 - \varepsilon_1) \quad (64)$$

is a second-order (Redfield) coupling element. The lowest-order interaction potential V in Eq. (56) is bilinear in the molecular coordinates if the system is nearly harmonic. In that event, V destroys an excitation of one mode and creates the other one. Again we have assumed that coherences don't evolve into populations on the experimental time scale, although this latter assumption may not always be adequate for vibrational states covering a wide frequency range. It is also assumed that coherence transfers between conjugates vanish because they are highly nonsecular, meaning that they oscillate at about 2ω. The solutions to the Eq. (63) allow the calculation of the probabilities $P(ij; 0|kl; t)$.

One further example is found in experiments associated with the combination bands of two-mode systems. The coherences $\rho_{1,1+2}$ and $\rho_{2,2+2}$ (equivalently $\rho_{1,1+1}$ and $\rho_{2,2+1}$) are strongly mixed when the population relaxation between the two states is rapid. Their oscillation frequencies are both approximately ω_{20} adjusted by the slowly varying diagonal or off-diagonal anharmonicity, which for convenience are omitted in the following but are straightforward to incorporate. In the equilibrium distribution these coherences are dephasing and exchanging through the population relaxation $\gamma_{1\to 2}$, and they can be found by solving the equations

$$\begin{aligned} \dot{\rho}_{1,1+2} &= \gamma_{12}\rho_{1,1+2} + \gamma_{2\to 1}\rho_{02} \\ \dot{\rho}_{2,2+2} &= \gamma_{1\to 2}\rho_{1,1+2} - \gamma_{2+2,2}\rho_{2,2+2} \end{aligned} \quad (65)$$

where γ_{ij} is the total dephasing rate of the level pair ij and $\gamma_{1\to 2} = \gamma_{2\to 1}e^{-\hbar\omega_{21}/K_BT}$. A case of this type was evaluated by Wiersma and co-workers many years ago [95]. For peptides, these types of transfers can be very efficient because population relaxation times might dominate the relaxation in many cases. The relaxation time of amide modes are extremely fast and involve transfers to other internal modes of the molecules. There is also the strong likelihood that the 2D-IR active frequencies are time-dependent, being subject to coupling to lower-frequency modes having a range of anharmonic coupling constants. In terms of the response functions, such effects may be modeled in the frequency domain as spectral phase or in the time domain with time-dependent factors analogous to those used to describe dynamic Stokes shifts.

XXVI. 2D-IR OF MORE COMPLEX STRUCTURES

The 2D-IR of more complex structures such as helices consisting of many amide units require numerical simulations. Such systems are highly degenerate, and exciton models would seem to be appropriate to estimate their 2D-IR spectra. There are a number of simple direct approaches to this problem, one of which we shall discuss here. Mukamel and co-workers have introduced other approaches that save considerable computational time by solving the exciton equations of motion [44, 96–99].

The 2D-IR spectra $S(\mp\omega_\tau, \omega_t, T)$ again correspond to the rephasing (with $-\omega_\tau$) and nonrephasing (with $+\omega_\tau$) Liouville pathways shown in Fig. 2, where now the lowercase k indices correspond to the whole set of N one-exciton states while uppercase K indices correspond to the set of $N(N+1)/2$ two-exciton states. Therefore the overtone and combinations are labeled by K ($K = k + k'$). The use of the simulation results is made transparent if the frequency-domain responses are used. The overall profile of a 2D-IR spectrum is determined by contributions from the orientational prefactor of the vibrators and the signal strength. The simulation based on the diagrams of Fig. 2 and the choice $T = 0$ incorporates a dephasing and a fixed inhomogeneous distribution. General expressions for 2D-IR spectra in this Bloch limit [cf. Eq. (38)] can be written as

$$S(-\omega_\tau, \omega_t) = \left\langle 4\sum_{k,k'} \frac{\langle \vec{\mu}_{0k}\cdot\hat{a}\vec{\mu}_{0k'}\cdot\hat{b}\vec{\mu}_{k0}\cdot\hat{c}\vec{\mu}_{k'0}\cdot\hat{d}\rangle}{[-i(-\omega_{k0}-\omega_\tau)-\gamma_{k0}][i(\omega_{k'0}-\omega_t)+\gamma_{k'0}]} - 2\sum_{k,k',K} \frac{\langle \vec{\mu}_{0k}\cdot\hat{a}\vec{\mu}_{0k'}\cdot\hat{b}\vec{\mu}_{k'K}\cdot\hat{c}\vec{\mu}_{Kk}\cdot\hat{d}\rangle}{[-i(-\omega_{k0}-\omega_\tau)-\gamma_{k0}][i(\omega_{Kk}-\omega_t)+\gamma_{Kk}]} \right\rangle \tag{66}$$

$$\begin{aligned} S(+\omega_\tau, \omega_t) = \Bigg\langle &\sum_{k,k'} \frac{\langle \vec{\mu}_{0k'}\cdot\hat{a}\vec{\mu}_{0k}\cdot\hat{b}\vec{\mu}_{k'0}\cdot\hat{c}\vec{\mu}_{k0}\cdot\hat{d}\rangle}{[-i(\omega_{k0}-\omega_\tau)-\gamma_{k0}][i(\omega_{k0}-\omega_t)+\gamma_{k0}]} \\ &+\sum_{k,k'} \frac{\langle \vec{\mu}_{k0}\cdot\hat{a}\vec{\mu}_{0k}\cdot\hat{b}\vec{\mu}_{0k'}\cdot\hat{c}\vec{\mu}_{k'0}\cdot\hat{d}\rangle + \sum_K \langle \vec{\mu}_{Kk'}\cdot\hat{a}\vec{\mu}_{0k}\cdot\hat{b}\vec{\mu}_{kK}\cdot\hat{c}\vec{\mu}_{k'0}\cdot\hat{d}\rangle}{[-i(\omega_{k0}-\omega_\tau)-\gamma_{k0}][i(\omega_{k'0}-\omega_t)+\gamma_{k'0}]} \\ &-2\sum_{k,k',K} \frac{\langle \vec{\mu}_{0k'}\cdot\hat{a}\vec{\mu}_{0k}\cdot\hat{b}\vec{\mu}_{kK}\cdot\hat{c}\vec{\mu}_{Kk'}\cdot\hat{d}\rangle}{[-i(\omega_{k0}-\omega_\tau)-\gamma_{k0}][i(\omega_{Kk'}-\omega_t)+\gamma_{Kk'}]} \Bigg\rangle \end{aligned} \tag{67}$$

where the numerators are the orientation factors, written for a given sequence of laboratory-fixed pulse polarizations $\hat{a}$ *to* $\hat{d}$ and molecule frame transition dipole directions. Each of the terms in (66) and (67) can readily be seen to be the Fourier transform of one of the generated field functions given earlier in the time domain. The γ_{Kk} are the homogeneous widths of the individual transitions at ω_{Kk}. An overall ensemble average of Eqs. (66) and (67) over the resonance frequencies can be carried out to simulate the inhomogeneous broadening and accompanying localization as manifest in the Hamiltonians given below. The total 2D-IR rephasing signal has two parts: One includes only the $|0\rangle \rightarrow |k\rangle$ transitions of pathways R_1, R_2, R_4, and R_5, and the other includes both $|0\rangle \to |k\rangle$ and $|k\rangle \to |K\rangle$ transitions, diagrams R_3 and R_6. The first term contributes to the positive peaks on the diagonal (when $k' = k$) and off the diagonal (when $k' \neq k$), and the second term contributes to all the negative peaks. The total 2D-IR nonrephasing signal in Eq. (60) has three parts: The first includes only the $|0\rangle \to |k\rangle$ and $|0\rangle \to |k'\rangle$ transitions (R_7 and R_8), which contribute to the positive diagonal peaks; the second term contributes to the positive peaks on the diagonal (when $k' = k$) and off the diagonal (when $k' \neq k$); the third term, R_{10} and R_{11}, contributes to all the negative peaks. Finally, as in 2D-NMR, the projected 2D-IR spectra can be constructed by adding the rephasing and nonrephasing terms.

The one-exciton Hamiltonian for a particular polypeptide, n in a distribution of structures, was chosen as M coupled harmonic oscillators:

$$H_n^{(1)} = \sum_m^M \left(\varepsilon_m + \xi_m^{(n)} \right) |m\rangle\langle m| + \sum_{m \neq l}^M V_{ml}^{(n)} |m\rangle\langle l| \tag{68}$$

where ε_m is the vibrational frequency of the relevant transition of the mth amide unit; this frequency could be dependent on the location of the residue in the structure or whether it is hydrogen-bonded to solvent or to other residues. The set of site energy fluctuations $\{\xi_m^{(n)}\}$ for a given H_n chosen to represent the energy disorder can be randomly selected from distribution functions, such as a Gaussian or special distributions that account for correlated energy fluctuations. The inter-site interaction terms $V_{ml}^{(n)}$ involve through-bond and through-space interactions between the sites. Models for the fluctuations of the interaction terms can also be included. The eigenstates of H_n for the nth polypeptide are labeled by the index k with their corresponding eigenvalues $E_k^{(n)}$, and ensemble properties can be obtained by averaging over n.

The transitions between the $|0\rangle \to |k\rangle$ and $|k\rangle \to |K\rangle$ manifold of vibrational states can be obtained by diagonalization of the two-quantum Hamiltonian in the site basis:

$$\begin{aligned} H_n^{(2)} = &\sum_{l,m}^M \left(\varepsilon_m + \varepsilon_l + \xi_l^{(n)} + \xi_m^{(n)} - \delta_{lm}\Delta \right) |lm\rangle\langle lm| \\ &+ {\sum_{l,m}^M}' \sqrt{2} V_{lm}^{(n)} \left(|lm\rangle\langle mm| + |mm\rangle\langle lm| \right) + {\sum_{l,m,p}^M}' V_{mp}^{(n)} |lm\rangle\langle lp| \end{aligned} \tag{69}$$

where $\sum'$ omits terms with equal indices. The site anharmonicity, Δ, only appears in the site overtone states, signified here as $|mm\rangle$. The transition dipole moments $\vec{\mu}_{kK}$ between one-exciton state $|k\rangle$ and two-exciton states $|K\rangle$ can be calculated by invoking a harmonic approximation:

$$\vec{\mu}(|m\rangle \rightarrow |ml\rangle) = \kappa\vec{\mu}_{01}^{(l)} \tag{70}$$

where $\kappa = 1(m \neq l)$ or $\kappa = \sqrt{2}(m = l)$. In the exciton picture, essentially all the interband transitions become allowed as a result of the anharmonicity that induces a redistribution of the transition dipole strengths amongst the excitonic states.

XXVII. COUPLING BETWEEN AMIDE UNITS

The shapes of the 2D-IR spectra provide signatures of the through space and through bond interactions amongst the amide units. These same interactions cause the vibrational excitations to jump from site to site and give rise to the exciton-state distributions typical of helical and other secondary structures. The lowest order of the through-space potential that exchanges excitations is the bilinear part of the expansion in terms of the normal mode displacements of the sites. When the spatial extent of the site normal mode is small compared with the distance between sites undergoing coupling, the bilinear coupling is often represented by a dipole–dipole interaction. The chemical bond network of the helix represents an anisotropic polarizable dielectric medium that must influence the interactions between transitions charge distributions on different units. Immersing the coupled charge distributions in an anisotropic dielectric would have the effect of reducing the interaction between them [100]. However, the coupling of two charge distributions *in vacuo* but having an anisotropic dielectric medium in the region between them will also modify their interactions. These effects require further theoretical input; however, no exact approaches for interactions at intermediate distances have yet been introduced [100–106]. This is partly due to the lack of experimental measurements of specific, pairwise interactions as opposed to ensemble dielectric properties.

To compute transition charge interactions the transition charge density can be approximated by a distribution of charges and charge fluxes [2, 107–109]. In our recent work [110] we have used Mulliken charges and transition charge fluxes from *ab initio* density functional theory (DFT) calculations on model compounds. The charges and fluxes are sensitive to basis set [111] so their reliability is called into question, but the charges and charge fluxes we obtained corresponded to a transition dipole magnitude of $0.38D$ having an orientation of $\sim 20^\circ$ to the C=O bond axis toward the nitrogen atom in the amide plane. Various values of this angle between -19° and 20° have been obtained when using different force fields even for the same model compounds. The issue of charge distributions of vibrational modes that form highly degenerate sets of almost independent motions would seem to present some interesting theoretical questions with a significant practical impact.

The 2D-IR spectrum is very sensitive to the couplings between the amide groups of the helix. Obviously, through-space coupling is important, but so are the through-bond interactions: couplings frequently referred to as "mechanical," but better termed through-bond couplings, because they are not incorporated in a multipole expansion of the intersite electrostatic potential. The overall coupling energy for two adjacent amides, evaluated from *ab initio* DFT calculations of a dipeptide *Ac-Gly-NMe* with its (ϕ, φ) dihedral angles at α-helical values, yields two amide-I normal modes split by 15.0 cm^{-1}. Equally weighted linear combinations of these calculated modes produce two degenerate localized modes. The coupling energy from the transition charge interaction alone is $+10.2\,cm^{-1}$ for the same configuration, so the nearest-neighbor coupling is not simply a through-space effect. The computation of the transition charge interactions—or, equivalently, at larger distances, the dipole–dipole interactions—must give only approximate results because neither the dipoles or charge distributions are known exactly and the effects of the polarizability of the helix and solvent are unknown. However, the large transition dipoles of the amide-I modes appear to be transferable from structure to structure and the magnitudes and directions can be chosen empirically in order to match the results of experimental splittings and anisotropies. Coupling constants chosen by these methods enable the simulation of 2D-IR spectra using the procedures outlined above.

An essential point about the 2D-IR of complex systems is that its goal is to find 2D-IR signatures of anharmonicities and, hence, of conformations. The experiment measures anharmonic coupling constants. The very existence of these parameters implies that the modes are coupled, which means that excitation of one of them is dependent on whether the other is excited. The actual mode coupling may be obtained from the anharmonicity under certain conditions. For example, if the exciton model is valid for both the one- and two-quantum states the off-diagonal anharmonicities may be expressed in terms of intermode coupling constants [11, 107]. In many applications of 2D-IR to complex biological systems the precise value of the coupling constant may not be so important to obtain as is the demonstration of coupling which would imply proximity. This "top down" approach to peptide structure determination is already very useful and hopefully will make contact with and be given a much firmer basis by "bottom up" approaches involving detailed theories of anharmonicity and high-resolution experiments on peptides of ever-increasing size.

XXVIII. THE 2D-IR SPECTRA OF HELICES WITH INHOMOGENEOUS BROADENING

The simulated rephasing 2D-IR spectra are shown in Fig. 8 for the α- and 3_{10}-helices including inhomogeneous broadening. Obviously they are very different.

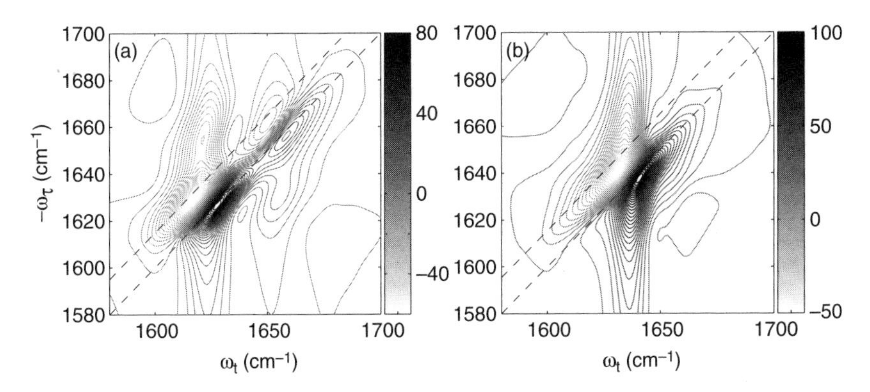

Figure 8. The calculated 2D-IR rephasing spectra of the amide-I mode for (*a*) a 20-unit 3_{10}-helix with dihedral angles of $\{\phi = -50°, \psi = -25°\}$ and (*b*) an α-helix $\{\phi = -58°, \psi = -47°\}$. The Gaussian random frequency fluctuation has $\sigma = 12\,\text{cm}^{-1}$ and $\gamma = 5\,\text{cm}^{-1}$. Some coupling constants are +1.0, −7.1, −0.4, −0.8, and −0.7 cm^{-1} for the 3_{10}-helix and +7.5, −4.7, −6.1, −0.5, and −0.7 cm^{-1} for the α-helix, corresponding to the coupling between the nearest neighbors, next nearest neighbors, and so on. The two major components on the diagonal for the 3_{10}-helix are the A-mode centered at ~1633 cm^{-1} and the E-mode at ~1655 cm^{-1}. The site anharmonicity is 15 cm^{-1} (see Ref. 110).

These spectra now show the well-known elongation of the diagonal peaks, including the positive diagonal peaks at $(\omega_\tau = \omega_{0k}, \omega_t = \omega_{k0})$ due to the $|0\rangle \rightarrow |k\rangle$ coherences on both axes as in diagrams R_1 and R_2 of Fig. 2 and the negative diagonal peaks at $(\omega_{0k}, \omega_{Kk})$ from pathways R_3. The negative peaks near the diagonal at $(\omega_{0k}, \omega_{Kk})$ are not so line-narrowed, because in the calculation the fluctuations in ω_{Kk} are uncorrelated with those in ω_{k0}. The remaining peaks at $(\omega_{0k}, \omega_{k'0})$ from R_1 and R_2 and $(\omega_{0k}, \omega_{Kk'})$ from pathway R_3 are also uncorrelated and are very much scattered in the (ω_τ, ω_t) space. We can see that the overall shape and intensity of the 2D-IR stick spectra is determined by the intrinsic properties specified in the helix Hamiltonian, including the coupling energies, anharmonicities, and correlations.

The 2D-IR spectra of simple molecules having a few resolvable modes usually display the anharmonicity very clearly along the ω_t axis. However, in an aggregate the number of two-particle states, which correspond approximately to combination bands of modes on different sites, is roughly *N* times the number of overtone states. All the two-quantum states are mixed by the diagonal anharmonicity, which results in the anharmonic shift of any given level being effectively diluted by roughly *N*. Nevertheless, the site anharmonicity is still manifested in the 2D-IR spectra. The separation between the positive diagonal peaks at $(\omega_{0k}, \omega_{k0})$ and the negative diagonal peaks at $(\mp\omega_{0k}, \omega_{Kk})$ increases toward the bottom of the helix exiton band. The overtone of the state at the bottom of the band resembles that of the local anharmonic state, whereas the states at the higher-energy side of the exciton band appear more harmonic. The two diagonal dashed lines in Fig. 8 represent the local anharmonic shift.

XXIX. ISOTOPICALLY LABELED HELICES

As an approach to using 2D-IR for the examination of the structures and dynamics of individual residues of extended structures, it is natural to turn to isotopic substitution. The use of isotopic replacements has been essential in the interpretation of vibrational spectra and their relationship to structure. Isotopomers have frequencies, force fields, and anharmonicities that are different from one another. Of particular interest in applications of 2D-IR is the use of isotopes to shift frequencies into regions where their couplings can be measured, free from interference by other modes of the system. For the amide-I mode, which is mainly a C=O stretching coordinate, the shifts by $^{13}C{=}^{16}O$ and $^{13}C{=}^{18}O$ substitution are large enough to displace the substituted amide group frequencies beyond the range of the natural distribution of frequencies found in most secondary structures.

Our strategy [112] is to insert both $^{13}C{=}^{16}O$ and $^{13}C{=}^{18}O$ labels into the helix. Since their isotope shifts are different, a pair of isotopic peaks will be created, separated by approximately 25 cm^{-1}, and 2D-IR spectroscopic methods can then be used to analyze the coupling between that specific pair of molecular transitions. If $^{13}C{=}^{16}O$ is substituted for one of the residues of a 25-residue helix, a $^{13}C{=}^{16}O$ diagonal peak will appear to the lower frequency of a much stronger and broad $^{12}C{=}^{16}O$ diagonal peak, corresponding to the set of helical exciton states of the remaining 24 residues. The couplings within the band and between the $^{13}C{=}^{16}O$ label and the band states will also show up in the off-diagonal regions of the 2D-IR spectrum. If we insert an additional $^{13}C{=}^{18}O$ label into the helix, another vibrator will be shifted out of the exciton band and appear at even lower frequency. The cross peaks between these two isotopic allies labeled amide-I modes assist in the measurement of the vibrational coupling between the two labeled vibrators. Because those labels can be inserted into various positions of the helix chain, the 3D structure of the molecule can be revealed, as reflected in the distances between the modes and relative orientations of the pairs of transition dipoles. In addition, the population relaxation times, the inhomogeneous distributions, and the correlation function of the fluctuations of the vibrational frequencies at the various sites can be measured as described above. However, the 2D-IR signals from different levels interfere with one another. So we need to develop efficient simulation methods to process and manipulate the data.

Figure 8 shows a cartoon of the zero order and delocalized energy levels involved in the 2D-IR experiments. The shaded regions represent the one- and two-exciton bands of the tagged helix, and the isotopomer levels are located below these bands. The band states and the trap states are mixed and shifted from their zero-order positions by the interactions between the residues. The two isotopomer levels below the one-exciton band are those seen in the linear-IR spectrum. A typical Liouville pathway, R_3, to the echo signal is depicted by the dashed and solid arrows using the diagrammatic method of Ref. 79. The first pulse creates a coherence with one of the isotopomer transitions and the second two pulses transfer this vibrational coherence via the other isotope to a two quantum coherence. In this example, k represents a state of one isotope and k' that of the other. The free induction decay initiates from the combination band of the isotopomers. The five upper levels are the two isotopomer overtones and their combination band whose energies are again determined by their coupling to each other and to the helix band states. In addition, the combination bands of the isotopomers with the remaining band states must be considered. The levels are identified by their zero-order energies, and the signal is formed by the sum of contributions from many pathways involving all the one- and two-quantum states arising from the isotopomers and their couplings to the band states as implicit in the spectra given as Eqs. (62) and (63). The experimental and calculated rephasing spectra are shown in Fig. 8.

The simulation of the isotopically substituted linear and 2D-IR spectra of helices is based on one- and two-exciton Hamiltonians, Eq. (61), which describe the frequencies and delocalization of amide-I modes of a helix with $N = 25$ coupled harmonic oscillators and two isotpomers. The zero-order isotope shifts were incorporated into the energy of the residues of the isotopomer modes ε_m and the naturally abundant $^{13}C{=}^{16}O$ modes also included by sampling techniques. A comparison between the calculated and observed 2D-IR spectra of these helices having $^{13}C{=}^{16}O$ and $^{13}C{=}^{18}O$ substitutions at the residues indicated, is shown in Fig. 9. The general features of the spectra are captured by the simulation, which provides a reasonable description of the results. For example, it is possible to show that the vibrational frequency of a residue is changed by placing a vibrational excitation on the other isotopomer. This off-diagonal anharmonicity depends mainly on the direct coupling of the two residues rather than on their coupling to the exciton band states of the helix. These beautiful 2D-IR spectra [113] in the first column of Fig. 9 demonstrate all the power of this method. The main helix bands are in the region of $(\omega_\tau = 1625, \omega_t = 1625)$. One can see the strong inhomogeneous broadening through the elongation of these peaks along the diagonals. The couplings between the isotopomer modes are evident as cross peaks in the 2D-IR. Furthermore, the coupling between the isotopomers and the exciton band states is evident in the spectra.

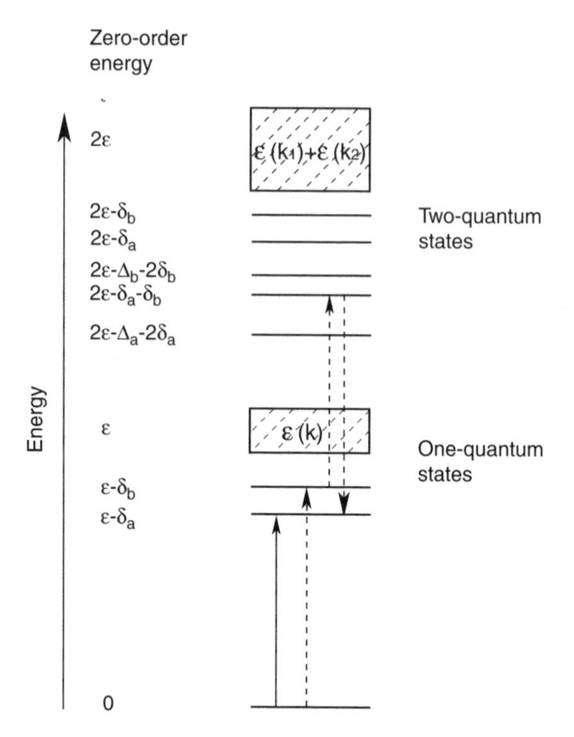

Figure 9. An overview of the energy levels of the helix containing two isotopically substituted residues. The zero-order isotope shifts are δ_a (for $^{13}C{=}^{18}O$) and δ_b (for $^{13}C{=}^{16}O$) while Δ_a and Δ_b are their unperturbed diagonal anharmonicities. The shaded areas represent the helix one- and two-exciton bands that become perturbed by the two isotopomer levels. (After Ref. 112.)

Such measurements as this are enabling the determination of the magnitudes and signs of the coupling between different amide units in helices. The 2D-IR spectra proved that the amide vibrations of the α-helix are delocalized. Cross peaks, originating from the isotopomer pairs, are in good agreement with a set of couplings that were derived from the transition charge-transition charge interactions. The magnitudes of the three largest coupling constants β_{12} (nearest neighbor), β_{13}, and β_{14} were found to be $|\beta_{12}| = 8.5 \pm 1.8$, $|\beta_{13}| = 5.4 \pm 1.0$, and $|\beta_{14}| = 6.6 \pm 0.8\ \text{cm}^{-1}$. The signs were independently indicated to be $\beta_{12} > 0$ while $\beta_{13} < 0$ and $\beta_{14} < 0$. The signs follow expectations from the dipole–dipole interaction between amide-I modes of a helix of this structure. One can read the signs of these coupling constants directly from the relative strengths of the two isotopomer transitions in the 2D-IR spectra shown in Fig. 10.

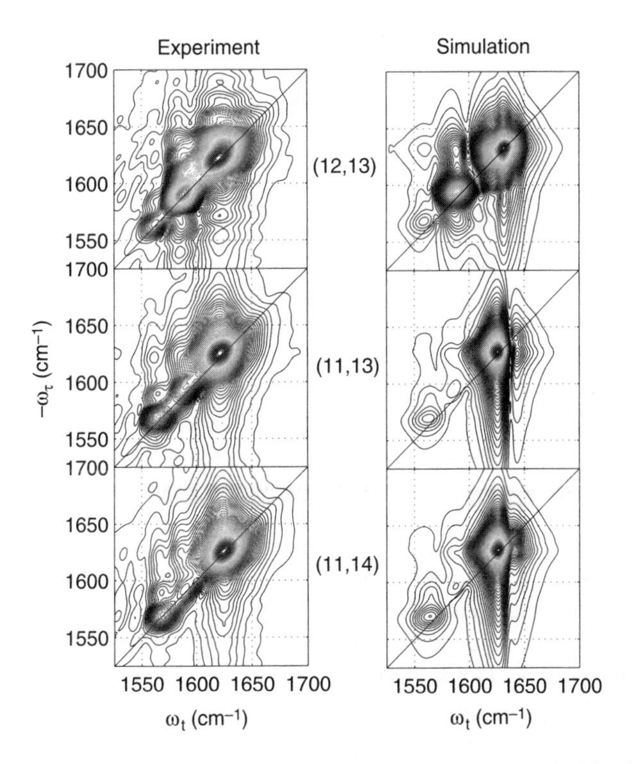

Figure 10. The absolute magnitudes of the experimental 2D-IR spectra of a 25-residue α-helix with carbonyl isotopic substitution at residue positions numbers [$^{13}C{=}^{16}O$, $^{13}C{=}^{18}O$]. Simulated spectra are shown to the right of each spectrum. (After Refs. 112 and 113.) See color insert.

XXX. CONCLUSIONS

There is a twofold purpose in the ongoing 3D-IR research on biological structures. First, there is the goal of simplifying the broadband IR spectra of complex systems by stretching them into multiple dimensions. This permits the determination of underlying structure distributions, angular and distance constraints, and the measurement of anharmonic contributions to the potential surfaces of peptides and proteins. Second, there is the intrinsic time resolution of the method, which can permit kinetic studies on essentially any time scale from femtoseconds to minutes. These same methods are ideal for exploring the effects of interaction between peptides and water. They enable determinations of the frequency correlation functions for particular chemical bonds and correlations in the solvent dynamics—for example, through hydrogen bonding, occurring at different bonds. Multifrequency 3D-IR experiments will be a powerful alternative to ultrashort pulse shaping and impulsive limit experiments. Through the development of these methods, measurements of the correlations of frequency distributions of different modes in different spatial regions of the molecule have already been made, and the main parameters of the equilibrium dynamics have been determined in a number of cases.

As a comment on the applications of 3D-IR, femtosecond pulse methods are complementary to more conventional FTIR methods if the spectral lines are sharp and resolvable in the sense that their separations are all large compared with their linewidths. Double resonance experiments with narrower spectral bandwidth infrared lasers could easily be set up that measure anharmonicities of such systems to as high accuracy as would be required. In many sharp line examples, where assignments in the two-quantum region are straightforward, narrow-band spectroscopy is entirely adequate to obtain anharmonicities to high accuracy. The grand challenge, as stated at the beginning of the summary, is to unravel underlying structure from broad spectra involving overlapping vibrational states in the one- and two-quantum regions and that are undergoing fast dynamics. This is precisely the situation presented by the spectra of vibrators in biological systems, in which useful features are obscured by the complexity of the eigenstates, by the dynamical processes, and by the exchange of excitations between modes. Nonlinear spectroscopy with designed ultrashort pulses is a promising approach to see more sharply these underlying structures.

This summary is an introductory approach to a form of nonlinear spectroscopy growing rapidly in its applications. The object was to describe the principles behind the methodology rather than the experimental results, which can be found in the references. However, it is worth mentioning again the results for the helix. For an α-helix having as many residues as a small protein, Figure 10 demonstrates the promise that applications combining linear and nonlinear infrared spectroscopy hold for discovery in structural biology. In that example the intermode potential was proven to be approximately a dipole–dipole type. If such a potential had been assumed, the 2D-IR spectra could have been used to measure distances to a fraction of an Angstrom across about one helix turn. The promise is that measurements of this type will enable useful measurements of distances and angles within a range of $\sim$8–10 Å between amide units that are either chemically connected or simply nearby to one another.

There were significant advances in the rapidly growing field of 2D IR since this chapter has been written. A comprehensive update has not been provided.

Acknowledgments

This research was supported by NIH (GM12592 and RR03148) and NSF. The text is based on a lecture I presented at the University of Chicago in June 2002 with addition of some review of recent advances in the field. I thank Jaehun Park for his help with all the figures and preparation of the final format; N.-H. Ge and Y. S. Kim for the experimental results in Figure 1; J. Park and J. Chen for the results in Figure 4; J. Chen for Figure 7; J. Wang for Figures 8 and 9; and C. Fang and J. Wang for Figure 10.

References

1. P. Hamm, M. Lim, and R. M. Hochstrasser, *J. Phys. Chem. B* **102**(31), 6123 (1998).
2. P. Hamm, M. Lim, W. F. DeGrado, and R. M. Hochstrasser, *Proc. Natl. Acad. Sci. USA* **96**, 2036 (1999).
3. M. C. Asplund, M. T. Zanni, and R. M. Hochstrasser, *Proc. Natl. Acad. Sci. USA* **97**(15), 8219 (2000).
4. P. Hamm, M. Lim, W. F. DeGrado, and R. M. Hochstrasser, *J. Chem. Phys.* **112**(4), 1907 (2000).
5. S. Woutersen and P. Hamm, *J. Phys. Chem. B* **104**(47), 11316 (2000).
6. M. T. Zanni, M. C. Asplund, and R. M. Hochstrasser, *J. Chem. Phys.* **114**, 4579 (2001).
7. M. T. Zanni, S. Gnanakaran, J. Stenger, and R. M. Hochstrasser, *J. Phys. Chem. B* **105**(28), 6520 (2001).
8. M. T. Zanni and R. M. Hochstrasser, *Curr. Opin. Struct. Biol.* **11**(5), 516 (2001).
9. M. Zanni, S. Gnanakaran, J. Stenger, and R. Hochstrasser, *J. Phys. Chem. B* **105**, 6520 (2001).
10. O. Golonzka, M. Khalil, N. Demirdoven, and A. Tokmakoff, *Phys. Rev. Lett.* **86**(10), 2154 (2001).
11. P. Hamm and R. M. Hochstrasser, in *Ultrafast Infrared and Raman Spectroscopy*, M. D. Fayer, (ed., Marcel Dekker, New York, 2001), p.273.
12. S. Woutersen and P. Hamm, *J. Chem. Phys.* **114**(6), 2727 (2001).
13. S. Woutersen and P. Hamm, *J. Chem. Phys.* **115**(16), 7737 (2001).
14. N.-H. Ge, M. T. Zanni, and R. M. Hochstrasser, *J. Phys. Chem. A* **106**(6), 962 (2002).
15. I. V. Rubtsov, J. Wang, and R. M. Hochstrasser, *J. Chem. Phys.* **118**(17), 7733 (2003).
16. I. V. Rubtsov and R. M. Hochstrasser, *Trends Opt. Photon.* **72** (Thirteenth International Conference on Ultrafast Phenomena, 2002), 386 (2002).
17. I. V. Rubtsov, J. Wang, and R. M. Hochstrasser, *Proc. Natl. Acad. Sci. USA* **100**(10), 5601 (2003).
18. I. V. Rubtsov, J. Wang, and R. M. Hochstrasser, *J. Phys. Chem. A* **107**(18), 3384 (2003).
19. M. Khalil, N. Demirdoven, and A. Tokmakoff, *J. Phys. Chem. A* **107**(27), 5258 (2003).
20. R. R. Ernst, G. Bodenhausen, and A. Wokaun, *Principles of Nuclear Magnetic Resonance in One and Two Dimensions*, Oxford University Press, Oxford, UK, 1987.
21. M. D. Levenson and S. S. Kano, *Introduction to Nonlinear Laser Spectroscopy*, Academic Press, Boston, MA, 1988.
22. A. Tokmakoff, M. J. Lang, D. S. Larsen, G. R. Fleming, V. Chernyak, and S. Mukamel, *Phys. Rev. Lett.* **79**(14), 2702 (1997).
23. N. A. Kurnit, I. D. Abella, and S. R. Hartmann, *Phys. Rev. Lett.* **13**(19), 567 (1964).
24. Y. J. Yan and S. Mukamel, *Phys. Rev. A* **41**(11), 6485 (1990).
25. J. P. Gordon, C. H. Wang, C. K. N. Patel, R. E. Slusher, and W. J. Tomlinson, *Phy. Rev.* **179**(2), 294 (1969).
26. R. G. Brewer and R. L. Shoemaker, *Phys. Rev. Lett.* **27**, 631 (1971).
27. T. J. Aartsma and D. A. Wiersma, *Phys. Rev. Lett.* **36**(23), 1360 (1976).
28. J. Y. Bigot, M. T. Portella, R. W. Schoenlein, C. J. Bardeen, A. Migus, and C. V. Shank, *Phys. Rev. Lett.* **66**(9), 1138 (1991).
29. S. Mukamel, *Principles of Nonlinear Optical Spectroscopy*, 1995.

30. J. D. Hybl, A. W. Albrecht, S. M. Gallagher Faeder, and D. M. Jonas, *Chem. Phys. Lett.* **297**(3, 4), 307 (1998).

31. T. Joo, Y. Jia, and G. R. Fleming, *J. Chem. Phys.* **102**, 4063 (1995).

32. Y. Tanimura and S. Mukamel, *J. Chem. Phys.* **99**(12), 9496 (1993).

33. P. A. Anfinrud, C. Han, and R. M. Hochstrasser, *Proc. Natl. Acad. Sci. USA* **86**, 8387 (1989).

34. M. Li, J. Owrutsky, M. Sarisky, J. P. Culver, A. Yodh, and R. M. Hochstrasser, *J. Chem. Phys.* **98**(7), 5499 (1993).

35. J. C. Owrutsky, M. Li, J. P. Culver, M. J. Sarisky, A. G. Yodh, and R. H. Hochstrasser, *Springer Proc. Phys.* **74** (*Time-Resolved Vibrational Spectroscopy* VI), 63 (1994).

36. S. Maiti, G. C. Walker, B. R. Cowen, R. Pippenger, C. C. Moser, P. L. Dutton, and R. M. Hochstrasser, *Proc. Natl. Acad. Sci. USA* **91**, 10360 (1994).

37. K. Wynne and R. M. Hochstrasser, *Chem. Phys.* **193**(3), 211 (1995).

38. K. Wynne, G. Haran, G. D. Reid, C. C. Moser, P. L. Dutton, and R. M. Hochstrasser, *J. Phys. Chem.* **100**(12), 5140 (1996).

39. D. Zimdars, A. Tokmakoff, S. Chen, S. R. Greenfield, M. D. Fayer, T. L. Smith, and H. A. Schwettman, *Phys. Rev. Lett.* **70**, 2718 (1993).

40. P. Hamm, M. Lim, and R. M. Hochstrasser, *Phys. Rev. Lett.* **81**(24), 5326 (1998).

41. P. Hamm, M. Lim, W. F. DeGrado, and R. M. Hochstrasser, *J. Phys. Chem. A* **103**, 10049 (1999).

42. M. Lim, P. Hamm, and R. M. Hochstrasser, *Proc. Natl. Acad. Sci. USA* **95**, 15315 (1998).

43. N.-H. Ge and R. M. Hochstrasser, *Phys. Chem. Comm.* **5**, 17 (2002).

44. W. M. Zhang, V. Chernyak, and S. Mukamel, *J. Chem. Phys.* **110**(11), 5011 (1999).

45. M. Khalil, N. Demirdoven, and A. Tokmakoff, *Phys. Rev. Lett.* **90**(4), 047401 (2003).

46. J. B. Asbury, T. Steinel, C. Stromberg, K. J. Gaffney, I. R. Piletic, A. Goun, and M. D. Fayer, *Chem. Phys. Lett.* **374**(3,4), 362 (2003).

47. N.-H. Ge and R. M. Hochstrasser, *Trends Opt. Photon.* **72** (Thirteenth International Conference on Ultrafast Phenomena, 2002), 255 (2002).

48. C. J. Fecko, J. D. Eaves, J. J. Loparo, A. Tokmakoff, and P. L. Geissler, *Science* **301**(5640), 1698 (2003).

49. K. Okumura, A. Tokmakoff, and Y. Tanimura, *Chem. Phys. Lett.* **314**(5,6), 488 (1999).

50. W. P. de Boeij, M. S. Pshenichnikov, and D. A. Wiersma, *Chem. Phys. Lett.* **253**(1, 2), 53 (1996).

51. N. Demirdoven, M. Khalil, and A. Tokmakoff, *Phys. Rev. Lett.* **89**(23), 237401/1 (2002).

52. A. Tokmakoff and M. D. Fayer, *J. Chem. Phys.* **103**, 2810 (1995).

53. J. Stenger, D. Madsen, P. Hamm, E. T. J. Nibbering, and T. Elsaesser, *J. Phys. Chem. A* **106**(10), 2341 (2002).

54. G. Gallot, S. Bratos, S. Pommeret, N. Lascoux, J. C. Leicknam, M. Kozinski, W. Amir, and G. M. Gale, *J. Chem. Phys.* **117**(24), 11301 (2002).

55. N. Bloembergen, *Rev. Mod. Phys.* **54**(3), 685 (1982).

56. R. M. Hochstrasser and H. P. Trommsdorff, *Acc. Chem. Res.* **16**(10), 376 (1983).

57. B. Dick, R. M. Hochstrasser, and H. P. Trommsdorff, *Resonant Molecular Optics*, Academic Press, Orlando, FL, 1987.

58. K. Wolfrum and A. Lauberau, *Springer Ser. Chem. Phys.* **60** (*Ultrafast Phenomena* IX), 293 (1994).

59. K. Tominaga, *Adv. Multi-Photon Processes Spectrosc.* **11**, 127 (1998).

60. W. P. De Boeij, M. S. Pshenichnikov, and D. A. Wiersma, *Annu. Rev. Phys. Chem.* **49**, 99 (1998).
61. S. Woutersen and P. Hamm, *J. Phys.: Condensed Matter* **14**(39), R1035 (2002).
62. J. C. Wright, *Int. Rev. Phys. Chem.* **21**(2), 185 (2002).
63. D. Keusters, H. S. Tan, and W. S. Warren, *J. Phys. Chem. A* **103**(49), 10369 (1999).
64. R. Trebino, K. W. DeLong, D. N. Fittinghoff, J. N. Sweetser, M. A. Krumbugel, B. A. Richman, and D. J. Kane, *Rev. Sci. Instr.* **68**(9), 3277 (1997).
65. R. Trebino, *Frequency Resolved Optical Gating*, Kluwer Academic Publishers, Boston, 2000.
66. N. F. Scherer, R. J. Carlson, A. Matro, M. Du, A. J. Ruggiero, V. Romero-Rochin, J. A. Cina, G. R. Fleming, and S. A. Rice, *J. Chem. Phys.* **95**(3), 1487 (1991).
67. L. Cohen, *Time-Frequency Analysis*, Prentice-Hall, Madison, WI, 1995.
68. N. Demirdoven, M. Khalil, O. Golonzka, and A. Tokmakoff, *Opt. Lett.* **27**(6), 433 (2002).
69. C. G. B. Garrett and D. E. McCumber, *Phys. Rev. A* **1**, 305 (1970).
70. D. Keusters and W. S. Warren, *Chem. Phys. Lett.* **383**(1,2), 21 (2004).
71. D. Keusters and W. S. Warren, *J. Chem. Phys.* **119**(8), 4478 (2003).
72. J. P. Likforman, M. Joffre, and V. ThierryMieg, *Opt. Lett.* **22**(14), 1104 (1997).
73. S. D. Silvestri, A. M. Weiner, J. G. Fujimoto, and E. P. Ippen, *Chem. Phys. Lett.* **112**, 195 (1984).
74. R. M. Hochstrasser, *Chem. Phys.* **266**, 273 (2001).
75. M. T. Zanni, N.-H. Ge, Y. S. Kim, and R. M. Hochstrasser, *Proc. Natl. Acad. Sci. USA* **98**, 11265 (2001).
76. O. Golonzka and A. Tokmakoff, *J. Chem. Phys.* **115**(1), 297 (2001).
77. D. M. Jonas, *Annu. Rev. Phys. Chem.* **54**, 425 (2003).
78. J. Chen, J. Park, and R. M. Hochstrasser, *J. Phys. Chem. A* **107**(49), 10660 (2003).
79. B. Dick and R. M. Hochstrasser, *J. Chem. Phys.* **78**(6, Pt. 2), 3398 (1983).
80. M. C. Asplund, M. Lim, and R. M. Hochstrasser, *Chem. Phys. Lett.* **323**, 269 (2000).
81. M. C. Asplund, M. Lim, and R. Hochstrasser, *Chem. Phys. Lett.* **340**, 611 (2001).
82. T. Joo, Y. Jia, J. Y. Yu, M. J. Lang, and G. R. Fleming, *J. Chem. Phys.* **104**, 6089 (1996).
83. A. Piryatinski and J. L. Skinner, *J. Phys. Chem. B* **106**(33), 8055 (2002).
84. J. R. Schmidt, N. Sundlass, and J. L. Skinner, *Chem. Phys. Lett.* **378**(5,6), 559 (2003).
85. J. Stenger, D. Madsen, J. Dreyer, P. Hamm, E. T. J. Nibbering, and T. Elsaesser, *Chem. Phys. Lett.* **354**(3,4), 256 (2002).
86. H. Eichler, G. Enterlein, J. Munschau, and H. Stahl, *Z. Angew. Phys.* **31**(1), 1 (1971).
87. K. A. Nelson and M. D. Fayer, *J. Chem. Phys.* **72**(9), 5202 (1980).
88. M. D. Fayer, *Annu. Rev. Phys. Chem.* **33**, 63 (1982).
89. K. A. Nelson, R. Casalegno, R. J. D. Miller, and M. D. Fayer, *J. Chem. Phys.* **77**(3), 1144 (1982).
90. D. W. Phillion, D. J. Kuizenga, and A. E. Siegman, *Appl. Phys. Lett.* **27**(2), 85 (1975).
91. H. J. Eichler, P. Guenter, and D. W. Pohl, *Springer Series in Optical Sciences, Vol. 50: Laser-Induced Dynamic Gratings*. (1986).
92. J. R. Andrews and R. M. Hochstrasser, *Chem. Phys. Lett.* **76**(2), 213 (1980).
93. J. R. Andrews and R. M. Hochstrasser, *Chem. Phys. Lett.* **76**(2), 207 (1980).
94. D. A. Wiersma and K. Duppen, *Science* (Washington, DC, United States) **237**(4819), 1147 (1987).

95. P. De Bree and D. A. Wiersma, *J. Chem. Phys.* **70**(2), 790 (1979).
96. A. M. Moran, S.-M. Park, J. Dreyer, and S. Mukamel, *J. Chem. Phys.* **118**(8), 3651 (2003).
97. A. Piryatinski, S. A. Asher, and S. Mukamel, *J. Phys. Chem. A* **106**(14), 3524 (2002).
98. C. Scheurer, A. Piryatinski, and S. Mukamel, *Springer Ser. Chem. Phys.* **66** (*Ultrafast Phenomena* XII), 507 (2001).
99. C. Scheurer, A. Piryatinski, and S. Mukamel, *J. Am. Chem. Soc.* **123**(13), 3114 (2001).
100. D. Van Belle, I. Couplet, M. Prevost, and S. J. Wodak, *J. Mol. Biol.* **198**(4), 721 (1987).
101. N. K. Rogers and M. J. E. Sternberg, *J. Mol. Biol.* **174**(3), 527 (1984).
102. A. Warshel and S. T. Russell, *Q. Rev. Biophys.* **17**, 283 (1984).
103. D. J. Lockhart and P. S. Kim, *Science* (Washington, DC, United States) **260**(5105), 198 (1993).
104. G. Loffler, H. Schreiber, and O. Steinhauser, *J. Mol. Biol.* **270**(3), 520 (1997).
105. P. L. Taylor, B. C. Xu, F. A. Oliveira, and T. P. Doerr, *Macromolecules* **25**(6), 1694 (1992).
106. K. Sharp, A. Jean-Charles, and B. Honig, *J. Phys. Chem.* **96**(9), 3822 (1992).
107. J. Dybal, T. C. Cheam, and S. Krimm, *J. Mol. Struct.* **159**, 183 (1987).
108. H. Torii and M. Tasumi, *J. Mol. Struct.* **300**, 171 (1993).
109. P. Hamm and S. Woutersen, *Bull. Chem. Soc.* (*Japan*) **75**, 985 (2002).
110. J. Wang and R. M. Hochstrasser, *Chem. Phys.* **297**, 195 (2004).
111. W. Qian and S. Krimm, *J. Phys. Chem.* **100**, 14602 (1996).
112. C. Fang, J. Wang, A. K. Charnley, W. Barber-Armstrong, A. B. Smith, III, S. M. Decatur, and R. M. Hochstrasser, *Chem. Phys. Lett.* **382**(5,6), 586 (2003).
113. C. Fang, J. Wang, Y. S. Kim, A. K. Charnley, W. Barber-Armstrong, A. B. Smith III, S. M. Decatur, and R. M. Hochstrasser, *J. Phys. Chem. B* **108**, 10415 (2004).

ENERGY TRANSFER AND PHOTOSYNTHETIC LIGHT HARVESTING

GREGORY D. SCHOLES

Lash Miller Chemical Laboratories, University of Toronto, Toronto, Canada M5S 3H6

GRAHAM R. FLEMING[†]

Department of Chemistry, University of California, Berkeley; and Physical Biosciences Division, Lawrence Berkeley National Laboratory, Berkeley, California 94720-1460, USA

CONTENTS

[†]We dedicate this chapter to Stuart Rice, who has provided inspiration and friendship that have enriched our lives and careers.

Adventures in Chemical Physics: A Special Volume in Advances in Chemical Physics, Volume 132, edited by R. Stephen Berry and Joshua Jortner. Series editor Stuart A. Rice

I. INTRODUCTION

The collection of solar energy by photosynthetic plants, algae, and bacteria and the subsequent transfer of that energy to reaction centers is known as *light harvesting*. The pigment–protein complexes responsible for light harvesting are often collectively referred to as antennae [1–4]. Despite the variety of structures and diversity of pigment cofactors used through the plant and bacterial kingdoms, light harvesting is universally almost 100% efficient at low light levels. A further ubiquitous feature is the implementation of protective mechanisms to guard against damage that would result from singlet oxygen sensitization. One obvious key to the efficacy of light-harvesting antennae, which have large spatial cross sections for light absorption, is to ensure that the elementary energy transfer processes that transport excitation to the reaction center (RC) are ultrafast. For example, typically there are about 200 (bacterio)chlorophyll pigments associated with each reaction center. Then, if energy simply hops randomly from pigment to pigment until reaching the RC trap, we can estimate that on average $(0.72 \times 200 \log 200 + 0.26 \times 200) = 363$ hops are required prior to trapping. Given the fluorescence lifetime of (bacterio)chlorophyll, this simple picture tells us that the average time for each hop must be $< \tau_{\text{flu}}/(9 \times 363) \approx 300$ fs in order to achieve a quantum yield of excitation trapping greater than 90%. Thus over the past years there has been a happy conjunction between femtosecond spectroscopy and high-resolution structural models [2, 5, 6] which has enabled some systems—in particular the peripheral light harvesting antenna (LH2) [7–11] and the RC of purple bacteria [12–17] and Photosystem I of cyanobacteria and green plants [6] to be modeled at a reasonable level of sophistication. Likewise, the availability of detailed structural and dynamical information has spurred the development of improved methods for calculating molecular interactions and energy transfer mechanisms.

To survey in detail the current state of knowledge of photosynthetic light harvesting would require an encyclopedic article. In this article we focus on the general principles that have been learned from studies of the purple bacterial and cyanobacterial systems. We discuss briefly the implications for green plant photosystems. We conclude with a discussion of questions that highlight areas that we feel are currently in need of investigation or resolution.

II. DYNAMICS OF ENERGY TRANSFER IN PHOTOSYNTHESIS

A. Structure and Dynamics

Light-harvesting pigment–protein complexes are employed by photosynthetic organisms to increase the spatial and spectral cross section per RC for collection of solar energy. In Fig. 1, we show structures of the peripheral light-harvesting complex LH2 of *Rps. acidophila* determined from X-ray crystallography [18]. In Fig. 2 we show the structure of the core light-harvesting complex LH1 of *Rps. rubrum* measured by low-resolution electron diffraction and a model based on the LH2 structure [19]. Recent studies have shown that LH1 may not always form a closed ring and may exist in a dimeric form. Figure 3 captures the layout of the antenna in purple bacteria, but should not be taken as a detailed model of the morphology of the entire photosynthetic unit. When grown under low light conditions the entire system contains about 250 bacteriochlorophyll (BChl) molecules RC. A stoichiometry of one LH1 per RC has been noted, the remaining BChl being contained in multiple LH2 complexes.

The structure of LH2 is known to 2.5 Å and 2.4 Å for *Rps. Acidophila* [18, 20] *and Rs. molischianum* [21, 22], respectively. The *Rps. acidophila* structure is based on subunits consisting of two trans-membrane α-helices (labeled α and β), which are arranged in a highly symmetric ring motif (C_9 symmetry in *Rps. acidophila* and C_8 symmetry in *Rps. molischianum*). This antenna complex contains a number of bound cofactors: two distinct rings of BChl *a* pigments, labeled B800 and B850, and at least one carotenoid per subunit, which makes a close approach to chromophores from each of these rings. In *Rps. acidophila* the B800 ring contains nine BChl *a* molecules while the B850 ring contains 18 BChl *a*. In *Rps. molischianum* the symmetry is eightfold, so the number of BChls is correspondingly reduced to 24 in total. LH1 is believed to be very similar in structure to LH2, but lacks an equivalent of the B800 ring, containing of single ring of 32 BChl *a* molecules known as B870 or B875 [23].

The overall timescale for trapping an excitation in the reaction center (and thereby initiating charge separation from the special pair) is 50–60 ps. The slowest step in this process is the final step from LH1 to the RC which takes

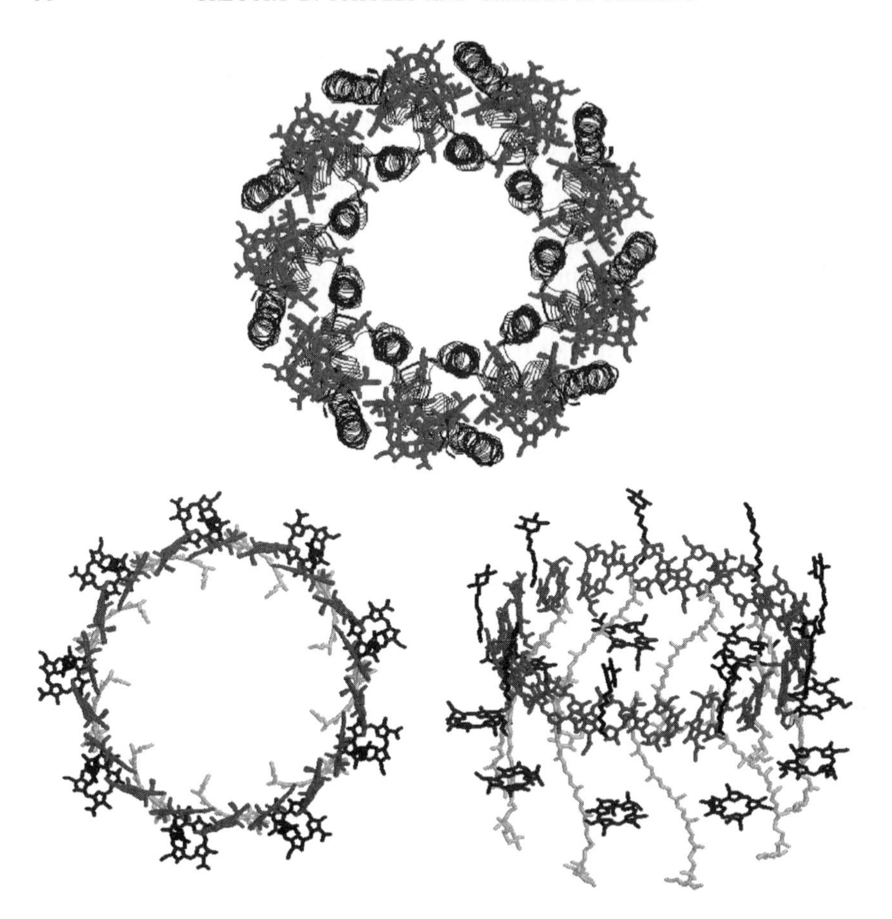

Figure 1. Illustration of the structure of the peripheral light-harvesting complex LH2 of the purple bacterium *Rps. acidiphila* strain 10050 [18]. The top view with α-helices represented as ribbons is shown at the top of the figure. The same view, but without the protein, leaving just the bacteriochlorophyll and carotenoid pigments, is shown at the lower left. On the lower right, this structure is shown tilted on its side, revealing the upper B850 ring of 18 Bchl pigments, the lower B800 ring of 9 Bchl pigments, and the carotenoids that weave their way between these rings. See color insert.

about 35 ps [24–26]. Transfer between LH2 complexes and from LH2 to LH1 takes 1.5–5 ps [27–29] and the transfers within each complex are much faster. For example, the transfer time between B800 and B850 in LH2 is about 700 fs at room temperature, [10, 30–35] while transfer between B800 molecules occurs on an average time scale of about 500 fs [35–38]. The dynamical time scale associated with the excited states in B850 and B875 is around 100 fs [39, 40], although the close proximity and strong electronic coupling (*vide infra*) of the

Figure 2. An illustration of the proposed structure of the LH1 ring of purple bacteria based on the LH2 structure [19, 22]. The protein has been removed from part of the ring to expose the B875 Bchl pigments and the carotenoids.

monomers comprising B850 and B875 make it far from clear that this process can be thought of as simple "hopping" of excitation between sites. Before attempting to provide a detailed picture of the energy transfer with B850 and B875, we need to understand the complex interplay between electronic coupling, electron–phonon coupling, and disorder.

The carotenoid molecules play dual roles as both light-harvesting and photoprotective pigments. We will briefly address the photoprotective role in Section VII.B. The overall efficiency of light harvesting from carotenoids (Cars) varies substantially from species to species [39–42]. In *Rb. sphaeroides* more than 95% of the photons absorbed by the Cars are transferred as excitation energy to the RC, while in strain 7050 of *Rps. acidophila* the overall efficiency is about 70%. Two electronic states of the Cars are involved in the energy transfer to the BChls. Energy transfer from the S_2 state of the Car is extremely rapid (50–100 fs) in all species studied so far, whereas the Car S_1 to BChl energy transfer time scale varies from $\sim$3 ps in *Rb. sphaeroides* to $>$25 ps and occurring with negligible quantum yield in *Rps. acidophila*.

Energy transfer processes have also been observed within the RC of purple bacteria [43–54]. The RC has 10 cofactors bound in a twofold symmetric arrangement: two closely spaced BChl *a* molecules (P_L and P_M) that comprise

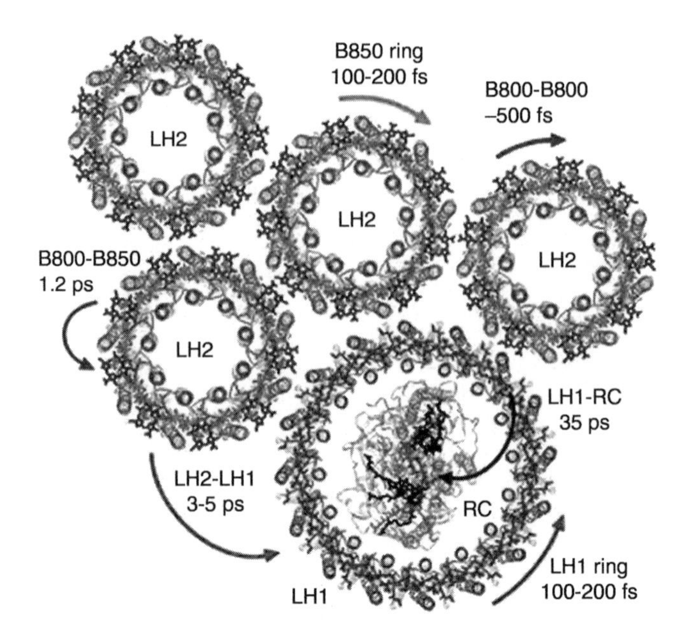

Figure 3. A schematic picture of the light-harvesting funnel in purple bacteria (*left*) and an illustration of how this corresponds to the layout of pigment–protein complexes (*right*). The approximate time scales of the various energy transfer processes are indicated.

the special pair or primary electron donor, two monomeric "accessory" BChl *a* molecules (B_L and B_M), two bacteriopheophytins (H_L and H_M), two ubiquinones (Q_A and Q_B), a carotenoid, and a nonheme iron as shown in Fig. 4 [55–58]. As early as 1972, Slooten [59] proposed that electronic energy transfer occurs from H and B to P in the *Rb. sphaeroides* RC. In the mid-1980s, ultrafast spectroscopy demonstrated that B to P energy transfer occurred in about 100 fs at both 300 and 10 K [43, 44]. More recently, it was shown that the appearance of P following excitation of H was 50% slower, than when B was excited directly, suggesting that B is a real intermediate in the H to P transfer process.

Many of the time scales described above have proven difficult, or impossible to obtain, using the standard Förster model [3, 60–64] of resonance energy transfer (coupling between point dipoles in donor and acceptor, overlap of measured donor emission spectrum with acceptor absorption spectrum, separation distance and mutual orientations specifiable by a simple parameter). Examples of processes where conventional calculations do not agree well with experiment are Car S_1 to BChl Q_y (the calculated rate would be zero!), B800 to B850 in LH2, and B to P in the RC, where calculated rates are always significantly

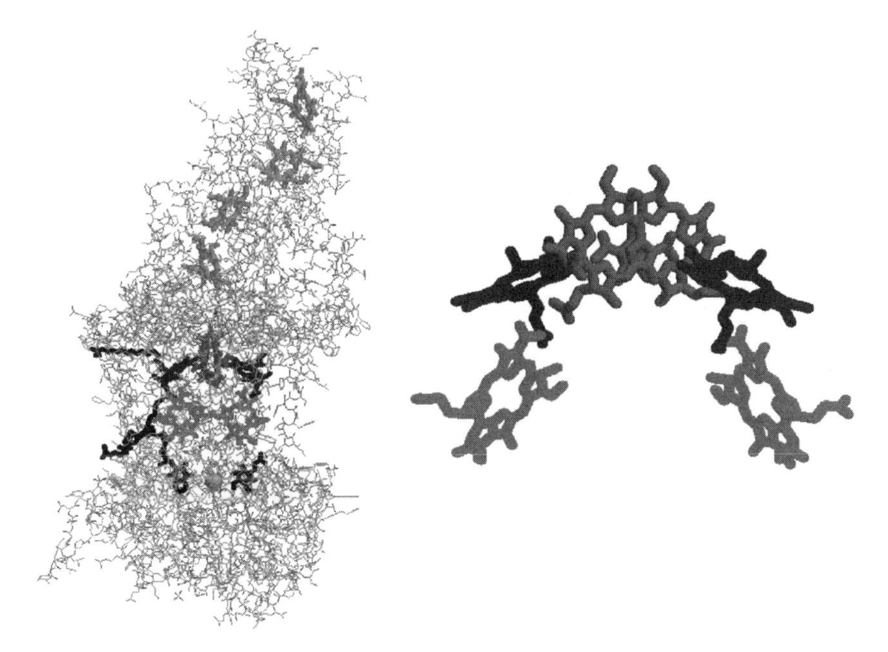

Figure 4. *Left*: Structural model of the photosynthetic reaction center of *Rps. viridis* from crystal structure data. *Right*: Arrangement of the special pair (dark gray), accessory bacteriochlorophyll (black), and the bacteriopheophytin (light gray) pigments.

slower than their measured values. In addition, the temperature dependence of these later two processes is strikingly weak and not predicted by simple calculations. Such discrepancies between theory and experiment have led to much speculation that new mechanisms are required to understand photosynthetic energy transfer. In this review, we will attempt to show that each of "troublesome" processes described above can be explained quantitatively by generalizing the conventional Förster description to include the effects of (a) multiple donors and acceptors, possibly with strong coupling between members of each group, (b) closely spaced donors and acceptors, and (c) energetic disorder among the donors and acceptors. We find that very weak or even normally forbidden transitions in a molecular aggregate may participate in efficient energy transfer via the Coulombic coupling mechanism, rather than by orbital overlap (e.g., the exchange or Dexter mechanism) [4] as it is often supposed. The remarkable efficiency of energy transfer in photosynthetic pigment–protein complexes of both plants and bacteria seems likely to be understandable in this context. A primary implication of our work is that optical spectroscopy is limited as a tool to determine electronic couplings in molecular aggregates. This means that, at the present time, general design principles for light-harvesting structures can only be revealed by a combination of experiment and theory.

More complex dynamics underlie the energy transfer processes that have been observed within the B850 and B875 rings of LH2 and LH1. In this case, an excitonic theory is required in order to relate linear and nonlinear spectroscopic observables to the underlying dynamics. Finally, in the case of Photosystem I from a cyanobacterium, both weak and strong coupling cases exist within the set of 96 nonequivalent Chls comprising the core antenna/RC complex.

B. Spectra of Purple Bacterial LH Complexes

Before turning to a description of electronic coupling in the LH complexes of purple bacteria, it is appropriate to review briefly the mechanisms used in the natural system to shift the absorption frequencies of the various components, such as B800, B850, and so on (Fig. 5).

In relative isolation the BChl-*a* molecules absorb at 772 nm (the Q_y band), 575 nm (the Q_x band), and 360 and 390 nm (the B bands). The carotenoids have a strong $S_0 \rightarrow S_2$ absorption in the region 450–550 nm, while their $S_0 \rightarrow S_1$ transition is dipole-forbidden and it is not found in the one-photon absorption spectrum.

However, each of these states plays a role in gathering light. The operation of the light-harvesting antenna of purple bacteria is based on an energy funnel to focus excitation energy to the reaction center. There are two obvious ways to construct such a funnel: (1) Select different chemical species that absorb at the

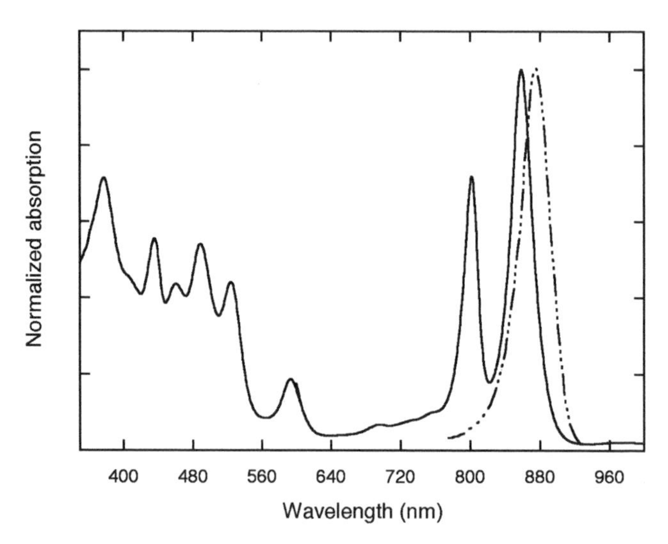

Figure 5. The absorption spectra of LH2 of *Rps. acidophila* (solid line) and LH1 (dash–dotted line).

required wavelength or (2) use exciton (and solvation) interactions to progressively red shift the spectra of the same chemical species. In purple bacteria, nature has adopted the second strategy for the BChl components: They use both interactions between the pigments ("excitonic coupling") and interactions between individual pigments, along with the protein environment to modify the spectroscopy of antenna components. In LH2, the BChl *a* absorption bands are located at 800 nm (B800) and at 850 nm (B850), while in LH1 the BChl *a* absorption has a maximum at 875 nm (B875). The B800 molecules interact weakly with each other and with the B850 molecules. In contrast, the B850 and B875 molecules are fairly strongly coupled amongst themselves to produce at least a significant portion of the red shift. However, the role of the protein is also evident here, as demonstrated by Fowler et al., who showed that site specific mutants of LH2 that removed specific hydrogen-bond interactions between the protein and the B850 chromophores, produced significant blue shifts of the absorption band [65–67]. One of the most intriguing questions in the study of light harvesting is whether there is an intrinsic advantage to the excitonic coupling strategy, which necessarily brings with it some degree of delocalization of the electronic states. In many other systems, chemical modification is also used to expand spectral coverage. For example, the binding of both chlorophyll *a* and chlorophyll *b* in LHC II [2].

The carotenoid constituents of the antenna systems also expand the spectral coverage of the antenna, although the efficiency of carotenoid to chlorophyll transfer varies significantly between species. The strongly allowed S_0–S_2 transition of carotenoid in the 450- to 550-nm spectral region significantly enhances absorption in this wavelength range and can transfer excitation to BChl or Chl molecules via the conventional Coulombic coupling mechanism. However, upper excited states are very short lived and rapid internal conversion to the S_1 state will occur in parallel with the energy transfer. The efficiency of S_1 to BChl or Chl energy transfer seems to vary significantly from complex to complex and species to species.

In addition to the systematic variations in transition frequencies of specific classes of pigments such as B800 or B850, there is significant disorder in the excitation energies from site to site and from complex to complex [10, 68–71]. This distribution of monomer energies can arise from (a) side-chain disorder in the protein, (b) deformation of the BChl macrocycle, (c) binding of ions, (d) ionizable side chains being near their pK_a values and thus existing in both neutral and ionized forms, (d) local or global distortions of the structure, and (e) the limited statistical sampling of full distribution of site energies possible in a complex of, for example, 9 or 18 monomer units. In addition to these types of disorder (generally referred to as *diagonal disorder*) in the excitonically coupled systems, variations in the electronic coupling between monomers (*off-diagonal disorder*) can also occur [72].

The fundamental question arising from the structure–function–dynamics relationships within a light-harvesting antenna is simply stated: What is the mechanism by which excitation moves in the antenna, and why is the overall process so wonderfully efficient? In this review we explore our knowledge of the ingredients required to formulate an answer to this question. We will describe our present understanding of the electronic interactions between the pigments, the line-broadening processes arising from electron–phonon coupling and disorder, and the implications of multiple, closely spaced chromophores for the dynamics. Finally, we will attempt to describe how the interplay between all these phenomena determines the dynamics of light-harvesting and funneling.

III. ELECTRONIC COUPLING AMONG THE CHROMOPHORES

A. Preface

When Förster [60] initially formulated his theory of energy transfer via the inductive resonance mechanism, he considered the interaction of single pairs of chromophores spaced by distances that are large compared to the size of the molecules. The situation in light-harvesting complexes is often rather different: Molecules are spaced by distances that are small compared to the overall molecular dimensions, making the definition of donor–acceptor separation and relative orientation ambiguous at best. In addition, there are often several or even many donor and acceptor molecules in close proximity, and these interactions may perturb the monomer spectral line-shape significantly. As we will describe in detail below, if any of the electronic couplings are strong enough to perturb significantly the spectral line shape or radiative rates, the standard Förster formulation of energy transfer becomes inadequate. An important example of this effect is the strong interaction between the two BChls of the special pair of the purple bacterial reaction center, which alter and shift the absorption spectrum dramatically compared to that of the monomer.

When the energy transfer involves one forbidden transition, it has been conventional to invoke mechanisms of electronic interaction other than Coulombic coupling, such as electron exchange via orbital overlap as originally formulated by Dexter [4]. Here an important general issue arises which relates to the length scale on which the molecular transition density is characterized by optical spectroscopy. In essence, the photon characterizes the molecular transition density in the far field, thus averaging over the entire molecular dimension. In the confined geometry of molecular aggregates, such as light-harvesting complexes, neighboring molecules may sense the shape of each other's transition density on a much finer scale [73]. It is clear that this effect will produce quantitative errors if the transition densities are approximated as point dipoles, but it is perhaps less obvious that qualitative mechanistic errors can arise if forbidden transitions are assumed to be incapable of Coulombic coupling to allowed transitions of molecules separated by distances smaller than the overall donor molecular dimensions. This type of symmetry breaking is important in, for example, the S_1 to Q_y transitions of carotenoids to BChls, and we will describe it in detail below. Both qualitative and quantitative aspects of this issue can be handled by explicitly calculating the Coulombic interaction between the transition densities of donor and acceptor, but now these must be obtained from electronic structure calculations, thus breaking the reliance on only experimentally determined quantities, which is the great strength of the Förster theory.

Our *ab initio* quantum chemical study of electronic interactions in LH2 [74] provides a starting point for the discussion. The calculated highest occupied molecular orbital (HOMO) for the intrapolypeptide BChl dimer in LH2 is shown in Fig. 6. It was anticipated that there would be significant orbital overlap between these two BChls, and this may assist delocalization of energy about the B850 ring. The calculated overlap density between the monomer BChl HOMOs is shown also in Fig. 6. The corresponding overlap integral was determined to be 1.72×10^{-3} (HF/3-21G*). Is overlap of this magnitude between two BChls significant for calculations of EET? According to a simple analysis based on the calculated overlap, it is not, unless the closest approach of the BChls is 3 Å or less. Note that there is only overlap between one of the four pyrrole rings of each macrocycle. If the BChl molecules were to be arranged in a more sandwich-like geometry, rather than being offset as in LH2, this overlap would increase owing to more orbital density being able to overlap. We can model this only by considering the distance-dependence of the overlap with respect to the spatial distribution of orbital density on each molecule. This is different from pushing the dimer together, which would increase the V^{short} coupling according to a simple exponential distance-dependence. Doubling the overlap would quadruple V^{short}, consequently having a significant effect of the EET rate at larger separations.

We have quantified the orbital overlap-dependent coupling for the BChl dimers in the B850 ring of LH2 [74]. Nonetheless, at closest approach separations of 4 Å or more, it seems reasonable to ignore the V^{short} contribution to the coupling. However, we have found that at typical interchromophore separations identified in light-harvesting complexes, the dipole approximation is unreliable for quantifying the Coulombic interactions. The dipole approximation completely ignores the shape of the interacting molecules—which turns out to be important in many cases. To overcome this barrier, we have developed the transition density cube method for calculating Coulombic interactions between electronic transitions, as we describe below. In addition, in multichromophoric systems, use of the dipole approximation can mask the way that energy transfer

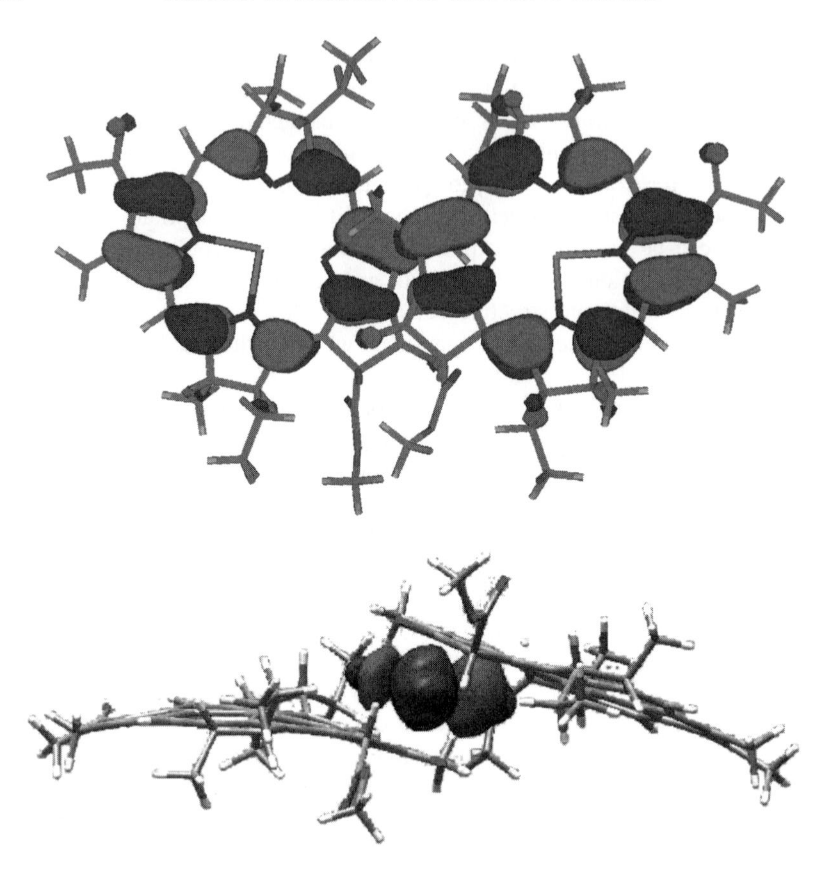

Figure 6. *Top*: A HOMO calculated for a B850 dimer of LH2 (HF/3-21G*). *Bottom*: The overlap density between the two Bchl chromophores is plotted.

dynamics are dictated by the arrangement of molecules in an aggregate. This latter point is rather subtle, and we will describe it further by way of the theory and examples in Section VI.

B. The Transition Density Cube Method

It is straightforward to show [75] that the Coulombic interaction that promotes excitation transfer between two two-level systems is given by the integral

$$\begin{aligned} V^{\mathrm{Coul}} &= 2\int \mathrm{d}\tau \; d'(1)a(2)r_{12}^{-1}d(1)a'(2) \\ &\equiv 2(d'd \mid aa') \end{aligned} \tag{1}$$

where d (d') is the HOMO (LUMO) of the donor, etc. d and a are doubly occupied in the ground state, while the excited state is represented as a single excitation from the ground state.

We have found it useful to express this Coulombic interaction in terms of transition densities (TDs) [73]. It has thereby been possible to calculate quite accurately, and with moderate computational effort, Coulombic interactions and energy transfer dynamics in rather complex light-harvesting assemblies. Furthermore, we have thus been able to gather several new physical insights into the mechanism of light harvesting. For example, we will describe here the physical, as well as practical, meaning of the dipole approximation with respect to energy transfer. We will show how and why the shape of molecules is just as important as their separation and orientation. Finally, we will show that the degree to which a transition is allowed or forbidden does not necessarily have direct implications for light-harvesting efficiency.

In general, the Coulombic interaction can be written in terms of the two-particle spinless transition density $\Pi_{KL,RS}$ that connects the states K and L on the donor (D) and connects the states R and S on the acceptor (A):

$$V^{\mathrm{Coul}} = \frac{e^2}{4\pi\varepsilon_0}\int \frac{\Pi_{KL,RS}(\mathbf{r}_1,\mathbf{r}_2)}{|\mathbf{r}_1 - \mathbf{r}_2|}\,\mathrm{d}\mathbf{r}_1\mathrm{d}\mathbf{r}_2 \tag{2}$$

Usually two-particle densities can only be written in terms of one-particle densities for single-configuration wavefunctions. However, because electron 1 and states K and L are localized on molecule D, whereas electron 2 and states R and S are localized on A, it is possible to factorize $\Pi_{KL,RS}(\mathbf{r}_1,\mathbf{r}_2)$ when nonorthogonality effects resulting from interpenetration of donor and acceptor electron densities are negligible. Thus we obtain:

$$V^{\mathrm{Coul}} = \frac{e^2}{4\pi\varepsilon_0}\int \frac{P^D_{KL}(\mathbf{r}_1)P^A_{RS}(\mathbf{r}_2)}{|\mathbf{r}_1 - \mathbf{r}_2|}\,\mathrm{d}\mathbf{r}_1\mathrm{d}\mathbf{r}_2 \tag{3}$$

The single-particle transition density matrix connecting states K and L of molecule D is defined as usual [76]:

$$P^M_{KL}(\mathbf{r}_1) = N\int \Psi_K(\mathbf{x}_1,\mathbf{x}_2,\ldots,\mathbf{x}_N)\Psi^*_L(\mathbf{x}'_1,\mathbf{x}'_2,\ldots,\mathbf{x}'_N)\mathrm{d}\mathbf{x}_2\ldots\mathrm{d}\mathbf{x}_N\mathrm{d}\mathbf{x}'_2\ldots\mathrm{d}\mathbf{x}'_N\mathrm{d}s_1 \tag{4}$$

where N is a normalization constant, $\mathbf{x}_i$ are the spatial and spin coordinates of electron i, and s_1 is the spin of electron 1.

Owing to the orthogonality between states K and L, $P_{KL}(\mathbf{r}_1)$ integrates to zero. Physically, this is because no net charge is gained or lost during an

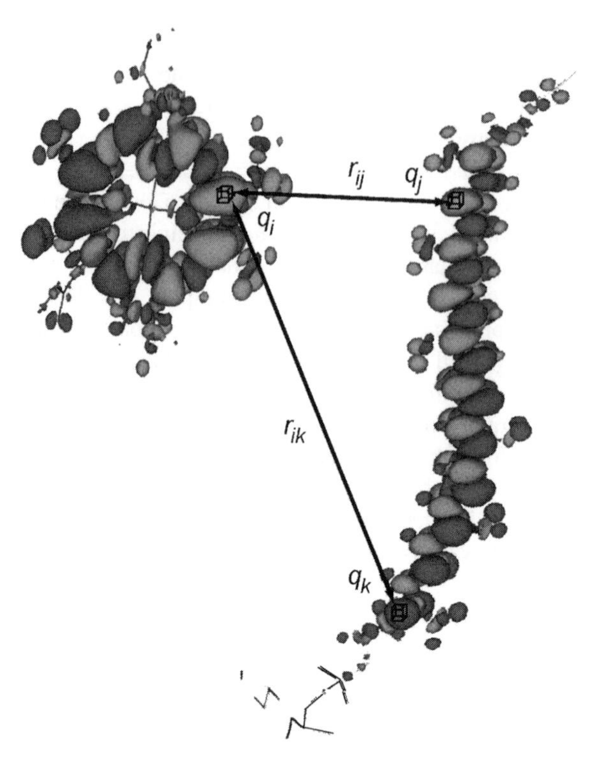

Figure 7. Transition densities calculated for a Bchl molecule and a carotenoid. Density elements, containing charge q_i, q_j, and so on, are depicted together with their corresponding separation r_{ij}. Summing the Coulombic interaction between all such elements gives the total Coulombic interaction, which, according to the TDC method, promotes energy transfer. See color insert.

electronic transition. A plot of the transition density, as shown in Fig. 7, reveals the manner in which the electron density of molecule D is polarized by interaction with light, in such a way as to induce a transition from L to K.

Since the wavelength of light is typically much larger than the physical size of a molecule, in optical spectroscopy it is usual to condense the information in the transition density of an allowed transition to its dipole moment,

$$\mu_{\alpha}^{LK} = \int (r_{\alpha})_1 P_{KL}^{M}(\mathbf{r}_1)\, d\mathbf{r}_1, \tag{5}$$

where the index α denotes the x, y, and z components of the vector. It is this quantity μ_{α}^{LK} that determines the strength of electric dipole-allowed electronic transitions between states L and K according to the dipole approximation.

Similarly, if we assume that the donor and acceptor electronic transitions are electric dipole-allowed, and the condition $|\mathbf{r}_1 - \mathbf{r}_2| \gg$ the spatial extent of D and A is satisfied, then Eq. (3) can be written in terms of transition multipoles (dipole, quadrupole, etc.) and a corresponding power series in $1/R$, where R is the center-to-center separation of the molecules [4, 75, 77]. Förster was the first to propose a connection between electronic spectra and the electronic coupling based on these arguments, so this is the approximation normally used in conjunction with Förster theory [60]. We note that this dipole approximation in the Coulombic coupling is a different kind of dipole approximation than that relating to the interaction between a molecule and light [77–80].

When the donor and acceptor molecules are nearby to each other, as is typically the arrangement in photosynthetic light-harvesting antenna complexes, the shape of the transition densities is very important in determining the electronic coupling. In that case, the correct physical picture is lost when the shape information in the transition density is averaged away by applying the dipole approximation. This idea was recognized by London in connection to van der Waals forces [81]:

> ... it is clear that even the dipole terms of this power series must turn out to be quite inappropriate if one has to consider oscillators of some length extended over a large region of a chain molecule. Another molecule would interact chiefly with one end of such a long virtual oscillator, and this situation would be completely distorted if one were to represent the oscillator by a decomposition into point-form multipoles, all located in the center of the molecule. It would obviously be much more appropriate in this case to represent each oscillator by several distinct poles, "monopoles," of different sign, suitably located in the molecule, thus directly taking account of the actual extension of the oscillator in question.

This idea is illustrated in Fig. 7, where we show the calculated (CI-singles/3-21G*) transition densities for the BChl-*a* Q_y transition and the Car (rhodopin glucoside) S_2 transition. The Coulombic interaction between these transition densities is the sum over all the interactions between charge "cells" on each transition density matrix, $q_i q_j / r_{ij}$. From inspection of this figure, it is evident that the topology of the transition densities cannot be ignored—for instance, r_{ij} is significantly different from r_{ik}. The only time that it is useful to calculate the interaction from mulipole moments of the transition densities and one average donor–acceptor separation is when the two molecules are sufficiently far apart that all the r_{ij} are similar. We suggest that a useful rule of thumb is to check whether or not the value of R is insensitive to the exact positions on the donor and acceptor molecules that are deemed to be the molecular centers. This will indicate that a multipolar expansion of the interaction potential provides a useful route to evaluation of Eq. (3).

We conclude that when the donor and acceptor molecules are closely located relative to molecular dimensions, the analogy between synergistic absorption and emission processes and the V^{Coul} interaction breaks down. We now need to think about V^{Coul} in terms of "local interactions" between the donor and acceptor transition densities because there is a distinct and important difference between (a) averaging over wavefunctions and then coupling them [Eq. (6a)] and (b) averaging over the coupling between wavefunctions [Eq. (6b)].

$$\frac{\left|\sum_i q_i \vec{\mathbf{r}}_i\right|\left|\sum_j q_j \vec{\mathbf{r}}_j\right|}{R_{DA}^3} \tag{6a}$$

$$\sum_{i,j} \frac{q_i q_j}{r_{ij}} \tag{6b}$$

Here we consider discrete charges q_i at position r_i on donor molecule D and charges q_j at position r_j on acceptor A. $r_{ij} = r_i - r_j$ and R_{DA} is the center-to-center separation between D and A, and κ_{DA} is the orientation factor between transition moments $\vec{\mu}_{\text{D}} = \sum_i q_i r_i$ and $\vec{\mu}_{\text{A}} = \sum_j q_j r_j$.

The key is that a single-center expansion of the transition density, implicit in a multipolar expansion of the Coulombic interaction potential, cannot capture the complicated spatial patterns of phased electron density that arise because molecules have shape. The reason is obvious if one considers that, according to the LCAO method, the basis set for calculating molecular wavefunctions is the set of atomic orbital basis functions localized at atomic centers; a set of basis functions localized at one point in a molecule is unsatisfactory.

To execute Eq. (3) numerically, we have used *ab initio* quantum chemical methods to calculate *transition density cubes* (TDCs) for the donor and acceptor from CI-singles or time-dependent density functional theory wavefunctions. A TDC is simply a discretized transition density,

$$\tilde{P}_{KL}^{M}(x,y,z) = V_\delta \int_z^{z+\delta_z} \int_y^{y+\delta_y} \int_x^{x+\delta_x} P_{KL}^{M}(\mathbf{r}_1) \tag{7}$$

where the δ_α denote the grid size of the transition density cube and $V_\delta = \delta_x \delta_y \delta_z$ is the element volume. In the TDC method the donor and acceptor transition densities are each represented in a 3D grid. Charge density in each cell of the donor q_i is coupled with that in each cell of the acceptor q_j via,

$$V^{\text{Coul}} \cong \sum_{i,j,k} \sum_{l,m,n} \frac{\tilde{P}_{KL}^{D}(i,j,k)\tilde{P}_{RS}^{A}(l,m,n)}{4\pi\varepsilon_0 \left|\mathbf{r}_{ijk} - \mathbf{r}_{lmn}\right|}. \tag{8}$$

Typically we use TDCs consisting of $\sim 10^6$ elements, each of volume ~ 0.23 bohr^3. The shape of a "cube" is chosen to contain best the shape of the molecule (it does not have to be a cube). Using Eq. (8), the donor–acceptor interaction topology is accounted for to a fine level of detail.

The accuracy of Eq. (8) is limited by the number of elements in the TDC, the size of each element, and the accuracy of the quantum chemical wavefunctions. A consequence of the first two factors is the problem of residual charge. That is, the sum of the charge over all cube elements is not zero, as it should be, but can be ~ 0.01 e. This residual charge can significantly affect the calculated coupling because, reverting to the language of the multipole expansion, it provides spurious charge–charge and charge–dipole interactions between the transition densities. To compensate for this residual charge q_R in a TDC with N elements, we subtract a quantity q_R/N from each element in the TDC, such that the residual charge is reduced to $\sim 10^{-14}$ e. Upon evaluating Eq. (8), one must also remove singularities that arise when cube elements of the donor TDC overlap with those of the acceptor TDC. We simply ignore these contributions to V^{Coul}, which is justified because the overlap density (i.e., the significance of overlapping transition density) must be small anyway when V^{Coul} dominates the electronic coupling.

A challenge for calculating the magnitude of electronic couplings accurately via the TDC method is to determine the ground- and excited-state wavefunctions as precisely as possible. However, this is generally an easier task than might be anticipated, for the reason that the most important result of the calculation is the shape of the TD. The shape of the TD is, of course, constrained by the shape of the molecule, and thus is easily obtained. Electronic couplings are overestimated by CI-singles TDCs, for the same reasons that transition dipole moments—for example, μ^{calc} from Eq. (5)—are overestimated. However, because the shape of the TD is well-calculated, it is possible to scale uniformly the TDC in such a way that Eq. (8) gives the experimental result for the transition dipole moment μ^{exp}. We do this by multiplying each element in the cube by $\mu^{\text{exp}}/\mu^{\text{calc}}$ (or equivalently post-processing the calculated coupling). This method does not work for forbidden transitions, of course. If TDs are calculated using time-dependent density functional theory or semiempirical methods like INDO [82], no scaling is necessary [83].

If wavefunctions are calculated using semiempirical methods that assume zero overlap between atomic orbitals (AOs) on different atomic centers, then a Mulliken population analysis [84, 85] can be applied to the calculated TD to yield transition monopoles distributed over each atomic center. Such an approach has proven to be effective [82] an advantage being that the interaction between distributed monopoles can be computed considerably faster then that between TDCs. At the same time, the basic topology of the donor–acceptor interaction is preserved. Here we describe this method in more detail for the more general case wherein differential overlap is preserved.

A general TD calculated in terms of a CI expansion of molecular orbitals (MOs) may be transformed to an AO basis P_{ij}^{KL} in terms of $\{\chi_i(\mathbf{r})\}$ such that [76],

$$P_{KL}(\mathbf{r}) = \sum_{i,j} P_{ij}^{KL} \chi_i(\mathbf{r}) \chi_j(\mathbf{r}). \tag{9}$$

The TD can then be analyzed in terms of the normalized orbital and overlap densities

$$d_i(\mathbf{r}) = (\chi_i(\mathbf{r}))^2, \quad d_{ij}(\mathbf{r}) = \frac{\chi_i(\mathbf{r})\chi_j(\mathbf{r})}{S_{ij}}, \tag{10}$$

where $S_{ij} = \langle \chi_j | \chi_i \rangle$ and the associated transition charges are

$$q_i = P_{ii}^{KL}, \quad q_{ij} = 2S_{ij} P_{ij}^{KL}, \tag{11}$$

leading to

$$P_{KL}(\mathbf{r}) = \sum_i q_i d_i(\mathbf{r}) + \sum_{i<j} q_{ij} d_{ij}(\mathbf{r}). \tag{12}$$

By summing over the AOs localized at each atomic center and integrating these over **r**, we can reduce Eq. (12) to a distribution of TD monopoles located at each atomic center, and a distribution of overlap-densities from the second term on the right-hand side of Eq. (12). When the overlap densities arise from overlap of AOs on different atomic centers, the resulting TD monopole can arbitrarily be placed halfway between the two atomic centers. More sophisticated reductions of Eq. (12) represent each TD monopole as a multipole expansion about the atomic center.

C. Coulombic Couplings in LH2

The availability of high-resolution structural data on various light-harvesting complexes has made it possible to relate spectroscopic observations of dynamics and their time scales to a detailed physical picture. An important link connecting the structural model to the dynamical information is the electronic Hamiltonian, containing the site transition energies for each chromophore and the electronic couplings between the chromophores. This information can be obtained from quantum chemical calculations, as has been described previously [9, 11, 13, 29, 74].

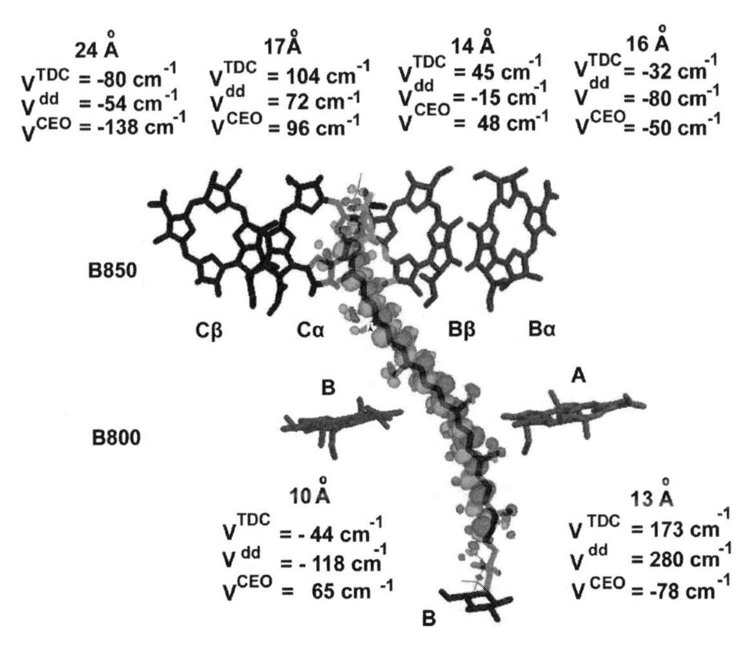

Figure 8. A summary of couplings calculated between the Bchl Q_y transition and the carotenoid S_2 transition for LH2. V^{TDC} are those calculated by the TDC method for *Rps. acidophila* [73], the V^{dd} provide a comparison with the dipole–dipole method, and V^{CEO} are those calculated for *Rs. molischianum* using the CEO method [11].

In Fig. 8 we depict part of the structure of LH2 from *Rps. acidophila* (strain 10050) showing four Bchls from the B850 ring and two from the B800 ring [18]. The associated rhodopin glucoside carotenoid threads its way past each of these rings. It is known that energy is transferred efficiently from the dipole-allowed S_2 state of the carotenoid to the Bchls—primarily via their Q_x states [21, 23, 25, 73]. Here we summarize electronic couplings we have calculated between the S_2 state of the rhodopin glucoside and the Q_x state of each Bchl for this complex using the TDC method, based on CIS/3-21G* wavefunctions, compared to electronic couplings estimated using the dipole approximation with respect to the same transition densities. It is immediately apparent that the dipole approximation will be problematic because it is not clear how best to define the interchomophore separations. We used the centers of each transition density, and we provide these distances in the figure. Quantitatively, we see that the results of the dipole approximation can, at best, be described as unpredictable compared to the TDC method and may not even predict correctly the sign of the coupling. This is because, for example, the Bchls in the B850 interact chiefly with just the top of the carotenoid transition density, but interact comparatively little with its other end. Apparently, the electronic coupling is determined by the shape and position of the carotenoid and the Bchl, which in turn dictates how their transition densities interact.

In Fig. 8 we also provide electronic couplings reported by Tretiak et al. [11] for the LH2 of *Rs. molischianum*. These couplings were calculated using the collective electronic oscillators (CEO) method [86, 87]. Note that the B800-carotenoid couplings differ between the two species, owing to the 90° difference in orientation of the B800 Bchls.

D. Carotenoid S_1 State and Electronic Coupling

The mechanism of electronic coupling that promotes Car $S_1 \rightarrow$ BChl EET has generally been rather mysterious, and is usually discussed in light of the relative merits of Förster versus Dexter energy transfer theories. The conventional wisdom is that, because Förster theory cannot be applied when the donor or acceptor transition is optically forbidden, Car S_1 to BChl coupling must be mediated by Dexter EET—and hence be dictated by the degree of orbital overlap between donor and acceptor states. Our recent *ab initio* calculations of B850 couplings in LH2 evince the possibility that V^{short} contributions to the Car S_2 to BChl coupling could reasonably have magnitudes of between 1 and 10 cm^{-1} [74]. However, according to our previous analysis of the microscopic mechanisms operative in Car S_1 to BChl coupling, V^{short} is likely to be much smaller than for Car S_2 to BChl coupling [77].

Considering that the shape and arrangement of transition densities of donor and acceptor were found to be so important for the carotenoid S_2 state interacting with the Bchl transitions in LH2 [73], the overall symmetry of the electronic transition may not be as restricting as might be supposed from the optical spectroscopy. In other words, there could likely be a significant Coulombic coupling between the carotenoid S_1 state and the Bchl transitions in LH2. By reasoning that the transition densities are mostly determined by the shape of the molecules, Walla et al. [83] estimated the approximate rhodopin glucoside to B850 Bchl couplings by scaling the S_2 to Bchl couplings uniformly such that the modified Förster theory for molecular aggregates (see Section VI. B) predicted their measured S_2-B850 EET rates. Very soon after this, Hsu et al. calculated these couplings using the TDC method (based on TD-DFT methods) and found remarkable agreement [1]. These results are collected in Table I, where they are also compared to Tretiak et al.'s [11] results for *Rs. molischianum*, determined from CEO calculations.

Hsu et al. [1] investigated the origin of the substantial Coulombic coupling between the carotenoid S_1 state and Bchls in LH2. They found that a significant contribution could be attributed to mixing of the $2A_g$ and the B_u carotenoid states, induced by distortion of the carotenoid structure. However, even for a completely planar carotenoid molecule, with a forbidden $S_0 \rightarrow S_1$ transition, the

TABLE I
Calculated Electronic Couplings (cm^{-1}) Between the Carotenoid S_1 State and Bchl Q_y State in LH2

	$B800_A$	$B800_B$	$\alpha B850_B$	$\beta B850_B$	$\alpha B850_C$	$\beta B850_C$
a	26	−7	−5	7	16	−12
b	31	−10	−5	9	32	−18
c	39	−0.4	3	−4	−4	−41

[a]Estimated by scaling the rhodopin glucoside S_1–Bchl Q_y electronic couplings in LH2 of *Rps. acidophila* calculated by the TDC method [83].
[b]Calculated (*Rps. acidophila*) by the TDDFT method [161].
[c]Calculated (*Rs. molischianum*) by the CEO method [11].

couplings were still found to be significant. The transition densities calculated for the rhodopin glucoside S_1 and S_2 transitions are plotted in Fig. 9. It can be conjectured that the Coulombic interaction with, for example a B850 Bchl, will be reasonable favorable for each of these states, given that the Bchl interacts principally with just the top end of the molecule. On the other had, it is evident that, overall, the symmetry of the transition densities differ: The S_1 state has a

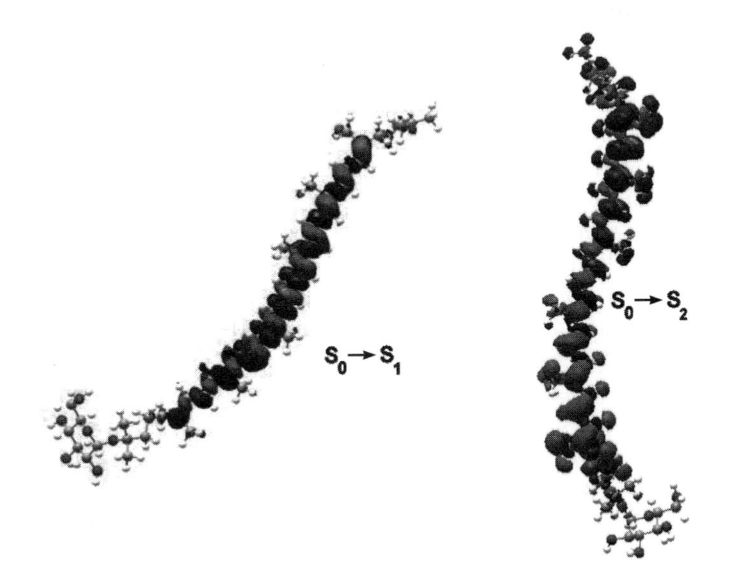

Figure 9. A comparison of transition densities for rhodopin glucoside calculated using TDDFT (6-31++g** basis set). On the right the $S_0 \rightarrow S_2$ transition is shown, with its large dipole transition moment being evidenced by the change in sign of this TD from one end of the molecule to the other. On the left the $S_0 \rightarrow S_1$ transition is shown. The symmetry of the TD causes the transition to be optically forbidden. See color insert.

symmetrically phased transition density relative to the middle of the backbone, whereas the S_2 state is antisymmetric and hence dipole-allowed.

E. Doublet States and Electronic Coupling

In isolated reaction centers, RCs, of photosynthetic purple bacteria, the primary electron donor P* can quench excitation by rapid EET from higher-energy RC pigments, either from the monomeric "accessory" bacteriochlorophyll-*a* molecules (B_L and B_M) or from the bacteriopheophytins (H_L and H_M). The photoexcited dimer P* then transfers an electron to H_L within three picoseconds with a quantum yield of nearly unity to form the radical dimer P^+ [88, 89]. If, however, the primary electron donor is oxidized (either chemically or under high light intensity) [90] to form P^+ before photoexcitation, electron transfer and thus photosynthesis are blocked. After excitation of B at 800 nm, the absorption of B recovers in ~130 fs in the neutral RC and in ~150 fs in the oxidized RC. In both cases, the ground-state recovery of B has been interpreted as energy transfer from B to P within the RC. Remarkably, EET to the oxidized primary electron donor (from the accessory bacteriochlorophyll or from the antenna) apparently still occurs [45, 46, 59, 91] even though the strong absorption band P has disappeared. Why the wild-type RC and oxidized RC primary electron donors are equally efficient quenchers of the excitation has been an unanswered question for the last 30 years. Recently, however, we have been able to explain this observation [17, 92].

We summarize below how we went about modeling EET in the neutral RC based on our model for EET in molecular aggregates. The most significant feature that differentiates the oxidized RC from the neutral reaction center, and any previously reported energy transfer systems we are aware of, is that the acceptor is a dimeric radical. Therefore, the focus of the problem was to determine the electronic energies and origins of the electronic transitions of the oxidized special pair acceptor and to quantify the electronic coupling between each of these relevant transitions and the donor transitions.

After recognizing that EET between a singlet state and a doublet state is spin-allowed, since no spin flips are necessary $\left({}^2({}^1B^*\ {}^2P^+) \rightarrow {}^2({}^1B^2P^{+*})\right)$, it was apparent that we could employ the TDC method to calculate the electronic coupling between B and P^+. Still, the acceptor states needed to be identified before calculating their transition densities. This was not a trivial problem, owing to (a) the complex internal spin structure in the P^+ electronic transitions and (b) the difficulty identifying the P^+ absorption bands in the experimental spectrum. It was possible to undertake these calculations using the method of Reimers and Hush [93]. A second challenge is that the first four excited states of P^+ borrow significant intensity from the fifth excited state, by vibronic coupling. Vibronic coupling mixes transition density from a more strongly allowed transition into that of the acceptor state with sufficient spectral overlap to acceptor excitation from B*, hence increasing the electronic coupling. In general, the Coulombic interaction between state i of molecule M and state k of molecule N, where i is vibronically mixed with state j according to the vibronic coupling parameter υ, is written as

$$V' = V_{ik}^{\text{Coul}} + \upsilon V_{jk}^{\text{Coul}} \tag{13}$$

where the normalization factor can be ignored when υ is small.

IV. THE PROTEIN ENVIRONMENT

A. Dielectric Screening

For a molecular aggregate, the dielectric screening must be incorporated at the level of the individual inter-site couplings. Each coupling V is multiplied by the screening factor D. Thus, for the modified Förster theory described in Section VI, dielectric screening cannot be simply incorporated in the final rate expression as can be done for a donor–acceptor system. Usually dielectric screening is assumed to have the form $D = n^{-2}$, where $n = \varepsilon_{\text{R}}^{1/2}$ is the refractive index of the medium at optical frequencies [94]. This limiting expression for D is justified when the disturbances induced in the medium are of much greater wavelength than the donor–acceptor separation. It is appropriate only when V is a dipole–dipole coupling and the two chromophores are separated by a distance large compared to their sizes in a nondispersive, isotropic host medium, and local field corrections are negligible [95]. If these conditions hold, then it is likely that the system cannot be a confined molecular aggregate.

In general we suggest that the corrections introduced by the dielectric medium will be fairly small, though certain specific interactions—for example, in a protein host—may be significant. A model for medium effects on closely spaced molecules has been developed recently in our laboratory. It is suggested that for molecules that are distant from one another we can enclose each in a cavity such that the two cavities are separated by the dielectric medium. Solution of this problem leads essentially to the result $D = n^{-2}$. However, when the molecules are closely spaced relative to their sizes, we need to reconsider such a treatment. Hsu et al. [96] enclosed the pair of molecules in a cavity. They then found that the electronic coupling could be either decreased or increased, depending upon the orientation of the molecules and their positions within the cavity. In any case, because the dielectric medium is now confined to the outside of the cavity containing the dimer, the screening is smaller than for the case of well-separated molecules.

A model for large complexes in which particular pairs of chromophores may be separated by transmembrane helices has been developed by Damjanovic and

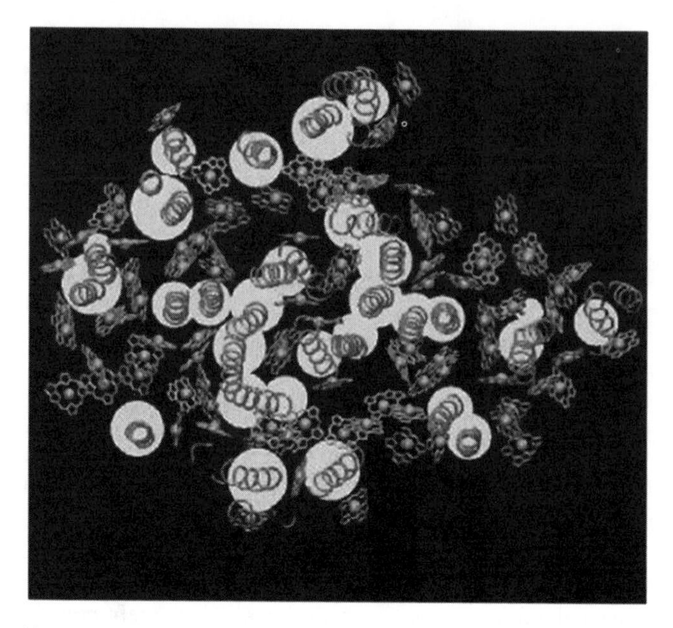

Figure 10. Dielectric model of the protein. Within this model, the protein medium (i.e., the medium with the refractive index of $n = 1.2$) is represented with a set of cylinders. The cross section of these cylinders is shown with white circles. The real location of the transmembrane part of α-helices in PSI are indicated by coiled structures. Chlorophylls are presented as Mg-chlorin rings, lacking the phytyl tail. Chlorophyll Mg atoms are shown in van der Waals representation. See color insert.

co-workers [97]. The model is best illustrated by reference to Fig. 10, which shows the distribution of transmembrance helices in Photosystem I, represented as cylinders. The heterogeneity of the protein environment is accounted for crudely by defining the site-dependent refractive index, n_{nm}, relevant for the Coulombic coupling between chlorophyll molecules n and m as follows. If the line connecting the Mg atoms of Chls n and m intercepts one of the cylinders, then $n_{\mathrm{nm}} = 1.2$; otherwise $n_{\mathrm{nm}} = 1$ in line with the arguments presented above. This model neglects possible screening by the Chl phytyl chains, and it clearly treats the protein in a highly simplified way. However, for a complex system such as PSI, it seems preferable to the standard approach of assigning a single value (usually ranging between 1.2 and 1.6) to n_{nm}.

B. Specific Interactions

Studies of the spectra and of energy transfer of site-directed mutant strains of the LH2 of *Rb. sphaeroides* have suggested that the influence of the H-bonding residues αTyr44, αTyr45 (αTyr44, αTrp45 in *Rps. acidophila*) from the adjacent α protein to the C3-acetyl group of B850 Bchl *a* contributes significantly to the spectral shift (compared to B800 or to 777 nm in organic solvent). It was found that single (αTyr44, αTyr45 $\rightarrow$ PheTyr) and double (αTyr44, αTyr45 $\rightarrow$ PheLeu) site-specific mutations produced blue shifts of 11 and 24 nm, respectively (at 77 K) of the B850 absorption band. It has also been reported that changing the charged residue βLys23 $\rightarrow$ Gln produces an 18-nm blue shift in the B850 absorption maximum. A similar situation has been found for the B800 absorption band. Furthermore, it is the hydrogen bond from the αTyr44 to the acetyl group of βB850 which is associated with the significant distortion of this BChl *a*. The consequence of the resultant saddle conformation is a further red shift of the absorption spectrum. It is also well known that the central Mg of BChl (or Chl) should be described by a coordination number of greater than 4; that is, the Mg is typically coordinated to a Lewis base. In the case of the B850 BChls *a* of LH2, the central Mg coordinates to a His ligand. Hence it is clear that specific interactions between the BChls and certain residues play an important role in tuning the absorption spectra and, therefore, for example, in the rate of B800 to B850 energy transfer via the resultant effect on the spectral overlap integral.

In our recent *ab initio* MO studies of LH2, we found that the calculated excitation energies of B800 and the αB850 BChls (i.e., those with planar structures) are approximately the same, whereas that of the βB850 BChl were noticeably lower, presumably owing to its distorted structure. It is interesting to note that this His residue red-shifts the spectra of each monomer significantly, whereas the H-bonding ligands (αTrp and αTyr) have a lesser effect.

V. ROBUSTNESS WITH RESPECT TO DISORDER

A. Disorder in Photosynthetic Proteins

It is well known that there is significant disorder in the excitation energies from site to site and from complex to complex. This distribution of monomer energies can arise from (a) side-chain disorder in the protein, (b) deformation of the BChl macrocycle, (c) binding of ions, (d) ionizable side chains being near their p*K* values and thus existing in both neutral and ionized forms, (e) local or global distortions of the structure, and (f) the limited statistical sampling of full distribution of site energies possible in a complex of, for example, 9 or 18 monomer units.

Figure 11 shows the mean energy of 9-mer aggregates (bars) selected from a Gaussian distribution of otherwise identical monomer energies with width Δ. Clearly the distribution of 9-mer energies can be characterized by a width Σ and the relation $\Delta^2 = \Sigma^2 + \sigma^2$ holds, where σ is the width of the energy distribution within a complex. Thus for pigment problem complexes with relatively small numbers of chromophores, the static site energy distribution within a complex does not cover the entire distribution, but rather samples a subportion of the

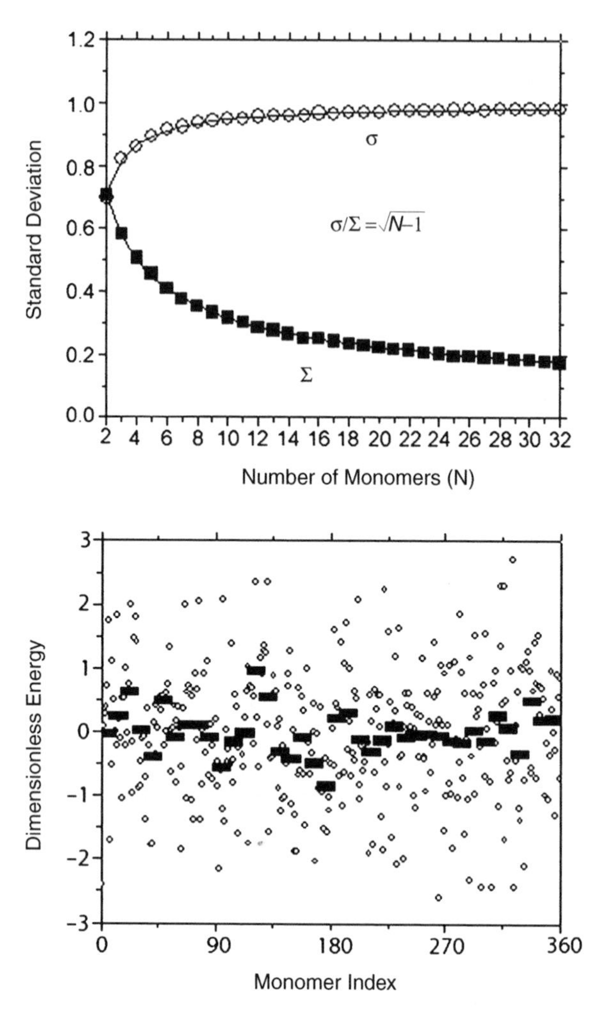

Figure 11. *Top*: Comparison of σ and the distribution width of the means, Σ, within an N-mer as a function of N. See text. *Bottom*: The mean energy of 9-mer aggregates (bars) compared to the momomer energies (circles) of distribution width σ.

total disorder. As the upper panel of Fig. 11 shows, the value of σ approaches zero for large values of the total number of pigments, N. In fact $\sigma/\Sigma = \sqrt{(N-1)}$ for $N > 1$. An absorption spectrum is only sensitive to the total value of the disorder, Δ and not the way it is partitioned between σ and Σ. Energy transfer or exciton relaxation processes within individual complexes depends only on σ, whereas Σ influences the energy transfer rate between

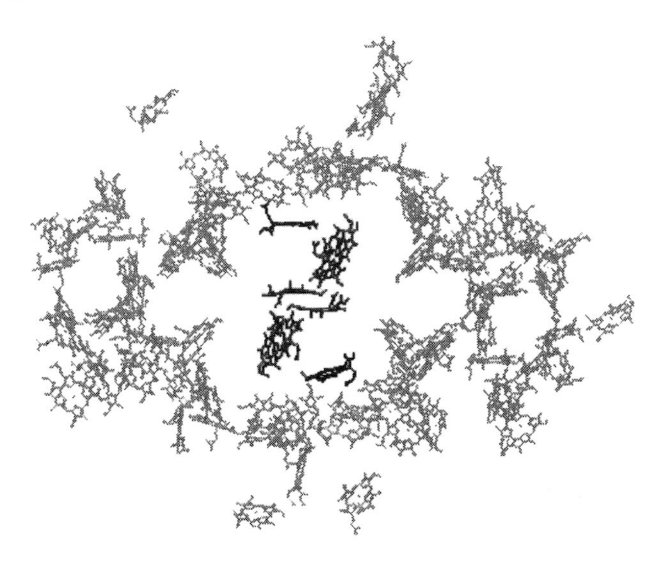

Figure 12. The organization of pigments in PS-I from the crystal structure data of Ref. 162. The reaction center is centrally located (the special pair is seen side-on). The darkest 8 Chls represent the six reaction center Chls and the two linker Chls. The linkers are the uppermost and lowermost dark Chls, respectively.

complexes. If these processes occur on different timescales, photon-echo measurements of the decay of the memory of the transition frequency can be exploited to separately determine intra- and inter-complex energy transfer dynamics as was demonstrated by Agarwal et al. [98]. The discussion above is most pertinent to complexes constructed from repeat units of one or a small number of chromophore structure "types" and environments such as LH2 based on an eight- or ninefold repeat of a structural element containing one B800 BChl and two B850 BChls. Photosystem I (PSI) of green plants and cyanobacteria provides a striking contrast to such a symmetric structure. In PSI, there are 96 nonequivalent Chla molecules, each in a different protein environment and with no obvious symmetry elements in the structure (Fig. 12). In addition, in contrast to LH2, the absorption spectrum of the complex is much broader than that of a dilute solution of Chla. In other words, PSI is both spatially and spectrally disordered. An additional difference between PSI and LH2 is that PSI is effectively a three-dimensional energy transfer system whereas LH2 is quasi-one-dimensional and one might expect that the disruptive effect of energy disorder on energy transfer may be different in the two cases.

This is indeed the case as the two histograms in Fig. 13 show [97]. The histograms are plots of overall excitation trapping times for 1000 different realizations of the transition energies of the Chla molecules. The transition

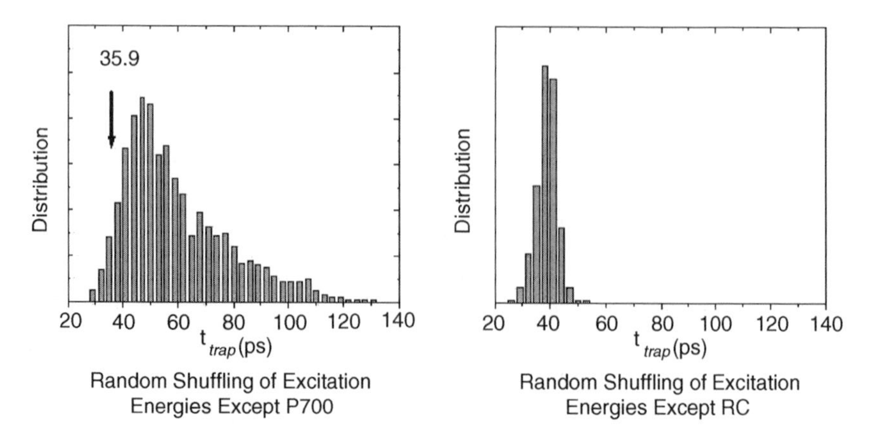

Figure 13. Distributions of trapping times (1000 samples) calculated by (*left*) a model that includes random shuffling of excitation energies in all the pigments except those of P700 and (*right*) random shuffling of excitation energies except those of the six RC pigments and the two linker Chls.

energies are taken as those obtained by Damjanovic et al. [99] via quantum chemical calculations for each individual Chla molecule. The arrow marks the calculated trapping time when the energies are assigned according to the calculation (which reproduces the absorption spectrum at both low and ambient temperatures). In the left panel, the transition energies of all Chls except P700 (the primary electron donor) are randomly shuffled. The mean value of the trapping time is 59.4 ps with a standard deviation of 18.9 ps. The distribution is asymmetric, and the energy configuration obtained by Damjanovic et al. is near the lowest bound of the distribution, suggesting that the PSI energy landscape is highly optimized for the given distribution of static energies.

A second distribution of 1000 replicas was obtained by random shuffling of the energies of the antenna Chls while the energies of the reaction center (RC) and linker Chls (see Fig. 12 and the caption) are fixed at the values calculated by Damjanovic et al. [99]. The resulting histogram of trapping times is shown in the right panel of Fig. 13. The distribution is much more symmetric and significantly narrower than the left panel. The mean and standard deviations are, respectively, 38.8 ps and 3.7 ps. From these results, we conclude that the calculated energy configuration of the reaction center and linker Chls is highly optimized and the energy configuration of the antenna Chls is highly tolerant of energy disorder because of the high connectivity of the structure. The optimality associated with the RC and the linker Chls seems to result from a quasi-energy funnel structure around the RC [97].

Of course, there must be a limit to the distribution of site energies that any given structure can tolerate at a given temperature, which relates to the ratio of the homogeneous width of the individual transitions and the energy disorder. In fact, disorder in the antenna Chls of PSI does influence the energy transfer kinetics. A trapping time of 25.4 ps is obtained when the energies of the antenna Chls are set identical at their mean value, and the quasi-funnel structure around the RC remains as calculated. However, against the increased trapping time in the actual system (35.9 ps) the system gains total absorption cross section by extending the absorption region. The reason that the homogeneous system is so effective lies in the very small reorganization energy of Chl/protein systems, and in the next section we briefly turn to a discussion of the Stokes shift and reorganization energy of Chl molecules in protein systems.

B. Stokes Shift and Reorganization Energy

Measurement of absorption and emission spectra of the pigments in photosynthetic antenna complexes has shown that Stokes shifts are typically small [2]. Presumably this is a consequence of the solvent environment of the protein that surrounds the pigments. However, it is somewhat surprising that coupling between the pigment electronic transitions and fluctuations of the protein are so small given the key role specific pigment–protein interactions can have, such as dictating the redox potentials of pigment sin the reaction center. There are two important consequences of the small reorganization energy associated with the Stokes shift. First, the spectral overlap between like pigments is large, meaning that energy migration among the pigments in an antenna complex is very efficient. Second, excitation can be more effectively delocalized among strongly coupled pigments—for example, in the B850 ring of LH2—since excitation is localized by spectral line broadening mechanisms.

Charge transfer interactions arise in closely spaced Chl pairs. Such pairs usually exhibit strong Coulombic coupling and in addition to heterogeneity in the Coulombic coupling, the magnitudes of the electron–phonon couplings (reorganization energies) have been shown to be heterogeneous. For example, Small and co-workers [100] have shown that there are at least two types of Chla molecules in the PSI complex which are characterized by larger electron–phonon couplings than the bulk Chla because of charge transfer character in their electronic excited states. Such a variation in electron–phonon coupling strength was taken into account in the calculations of energy transfer in PSI described below [97].

Most of the information about the electron–phonon coupling comes from low-temperature spectroscopy such as hole-burning. Extrapolating these low-temperature spectral densities to room temperature can be difficult, but no determination of a room temperature spectral density for Chla has been made as far as we aware. An alternative approach is to calculate the spectral density via combined quantum mechanical/molecular mechanics calculations [99, 101], although the quantitative reliability of such an approach is not yet adequate for detailed energy transfer calculations. At present, the best approach appears to be to use the low-temperature spectral density and make appropriate modification to fit the room temperature spectrum as described by Zucchelli et al. [102] for Chla.

C. Diagonal Disorder and Energy Transfer

To apply the Förster equation, the emission and absorption line shapes must be identical for all donors and acceptors, respectively. However, in many types of condensed-phase media (e.g., glasses, crystals, proteins, surfaces), each of the donors and acceptors lie in a different local environment, which leads to a distribution of static offsets of the excitation energies relative to the average, which persists longer than the time scale for EET. When such "inhomogeneous" contributions to the line broadening become significant, Förster theory cannot be used in an unmodified form [16, 63].

If there is just a single donor–acceptor pair, then we must ensemble average the nuclear spectral overlap—for example, using a Pauli master equation. One needs to think in terms of the inhomogeneous line broadening present in the donor emission spectrum and that present in the acceptor absorption spectrum leading to individually ensemble-averaged quantities. The spectral overlap is also an ensemble average quantity, and it is not related in a simple way to the overlap of the ensemble-averaged emission and absorption spectra.

In a chromophore aggregate, where there are couplings among the donor and/or acceptor chromophores, the site energy disorder affects both the electronic and the nuclear factors simultaneously. As we have described in Ref. 63, if there are m molecules that together make up the donor and n molecules that comprise the acceptor, then the EET dynamics must be determined by $m \times n$ electronic couplings. To introduce disorder properly into the EET rate calculation, each of the $m \times n$ electronic couplings $V_{\delta\alpha}$ must be associated a corresponding spectral overlap factor $J_{\delta\alpha}(\varepsilon)$. This provides us with the *dimensionless* coupling-weighted spectral overlap for each interaction, $u_{\delta\alpha}(\varepsilon) = |V_{\delta\alpha}|^2 J_{\delta\alpha}(\varepsilon)$. This quantity governs the mechanism by which EET is promoted in complex aggregates. For example, we can ascertain which electronic states most significantly mediate the EET by comparing the values of each of the $\int d\varepsilon u_{\delta\alpha}(\varepsilon)$, which are directly proportional to the rate for each pathway.

We incorporate disorder into the calculation by ensemble-averaging the set of coupling-weighted spectral overlaps for many aggregates using a Monte Carlo method [38]. In this way the effect of disorder on both electronic couplings and spectral overlap is properly accounted for by ensemble averaging $\Sigma_{\delta,\alpha} u_{\delta\alpha}(\varepsilon)$.

Energy migration among a number of chromophores with inhomogeneously broadened spectra can be modeled using a Pauli master equation approach [10, 27, 70, 71, 103–107] as long as the excitation is localized as it hops from molecule to molecule. In such a model the probability of finding the excitation on site i in the aggregate $P_i(t)$ is determined by solving the coupled differential equations,

$$\frac{dP_i(t)}{dt} = \sum_j \left[k_{ij}P_j(t) - \left(k_{ji} + \tau_i^{-1}\right)P_i(t)\right] \tag{14}$$

where the excited state lifetime is τ_i and uphill RET rates are calculated via detailed balance, $k_{ji} = k_{ij}\exp(-\Delta E_{ij}/kT)$, with ΔE_{ij} equal to the energy difference between donor and acceptor absorption maxima. The site–site rates are calculated according to a spectral overlap involving homogeneous line shapes. A Monte Carlo sampling procedure is used to account for disorder, typically with ~2000 iterations. At each iteration the site energy offsets for each molecule in the aggregate δ_i are chosen randomly from a Gaussian distribution of standard deviation σ, $w(\delta_i) = \exp\left(-\delta_i^2/2\sigma^2\right)/\left(\sigma\sqrt{2\pi}\right)$. A Gaussian distribution is in accord with the Central Limit Theorem. It is useful to note that the FWHM of the distribution $\Delta = \sigma(8 \ln 2)^{1/2}$. If the electronic coupling varies from aggregate to aggregate, because for example the molecules are oriented differently, then this can also be included.

The Pauli Master equation approach to calculating RET rates is particularly useful for simulating time-resolved anisotropy decay that results from RET within aggregates of molecules. In that case the orientation of the aggregate in the laboratory frame is also randomly selected at each Monte Carlo iteration in order to account for the rotational averaging properly.

D. Off-Diagonal (Coupling) Disorder

Disorder that affects electronic couplings is also present in chromophore aggregates. Such disorder arises from distributions of orientations and separations of the chromophores. One expects off-diagonal disorder to be most significant among closely coupled chromophores, such as those comprising B850, since orientation and distance dependencies of the coupling are most pronounced at close interchromophore separations. Once again, for EET within a chromophores aggregate, both the electronic couplings and the spectral overlaps will be by off-diagonal disorder. It is therefore rather difficult to differentiate the manifestation of diagonal from off-diagonal disorder.

Jang et al. [72] have systematically studied the effects of diagonal versus off-diagonal disorder in the B850 ring of LH2. They conclude that the diagonal disorder could be similar in magnitude to that in the B800 ring. In that case, the total disorder, observed spectroscopically to be significantly larger than that in the B800 ring, could be achieved through the addition of off-diagonal disorder.

VI. CALCULATIONS OF ENERGY TRANSFER RATES

A. Preface

Typically, light-harvesting complexes contain many chromophores in close proximity, among which energy is funneled. To relate the structures of photosynthetic antennae to a functional model requires a theoretical framework that is able to capture the essential physics. Förster theory for EET is very appealing because it is compact and simple and has few adjustable parameters. Förster theory has proven to be enormously successful for calculating EET rates between donor–acceptor pairs, but it has been observed numerous times that Förster theory cannot rationalize EET dynamics observed in chromophore aggregates. In recent work we have shown that the theory must actually be modified in order to model EET in chromophore assemblies. Photosynthetic light-harvesting proteins provided the inspiration for this work and have been a valuable testing ground to prove the quantitative utility and robustness of the theory.

B. Rate Expression for Singlet–Singlet Energy Transfer in an Aggregate

We have generalized Förster theory so that it is possible to calculate rates of energy transfer in molecular aggregates. The Generalized Förster Theory (GFT) was inspired by the ideas that (i) often weak coupling interactions promote energy transfer—because intramolecular reorganization tends to trap and localize excitation on a donor, simultaneously destroying memory effects—and (ii) in a molecular aggregate it is not necessarily clear what entities really are the energy donors and acceptors. We reasoned that there can be a mixture of weak and strong electronic couplings in a molecular aggregate. The strongly coupled molecules will exhibit collective spectroscopic properties, and the eigenstates of these coupled molecules thus collectively constitute energy donor or acceptor states. We refer to these collective states as the effective donor and acceptor states, δ and α respectively. By partitioning the Hamiltonian of the aggregate in this way, we find that the electronic couplings connecting the effective donor and acceptor states are indeed weak. Thus energy transfer from δ to α may be estimated by a Fermi Golden Rule expression, in the spirit of Förster theory. The GFT reveals that the donor emission and acceptor absorption spectra cannot be used to directly to quantify the rate of energy transfer in molecular aggregates, which has helped to explain much of the confusion in the literature regarding the explanation of observed energy transfer rates in photosynthetic proteins, as we describe in the following sections. Instead, we must turn to a "electronic coupling-weighted spectral overlap" between effective donors and acceptors, as we describe below. A fundamental feature of the GFT is that the organization of the molecules in the aggregate is explicitly accounted for in the

Hamiltonian. The final ingredient in the GFT is to include a correct ensemble averaging procedure to account for static disorder in the donor and acceptor transition frequencies; we implement this in the site representation.

We showed that for an aggregate consisting of m donor molecules and n acceptor molecules, we can divide the problem into interactions between effective donor and acceptor eigenstates, so that at least $m \times n$ energy transfer pathways must be considered [16, 63]. In the limit that the interactions between each pair of molecules is very weak, such that the donor and acceptor absorption spectra are unperturbed from that of the monomers, then the energy transfer rate is a sum of Förster rates. Otherwise we consider electronic couplings and spectral overlaps for each pair of eigenstates. We still think in the Förster picture, but we explicitly account for each donor emission and acceptor absorption process. For example, in B800 → B850 energy transfer, the donor is a single B800 bacteriochlorophyll, and the acceptor is the ground state of the B850 ring of 18 bacteriochlorophylls. Thus we must consider the de-excitation of B800 and excitation into each of the 18 B850 eigenstates. We note that this energy transfer involves transfer into the B850 eigenstates, so discussion of delocalization length with respect to this process is redundant. However, as described by Kühn and Sundström [108], this transfer process does involve most of the B850 ring. Dynamic relaxation processes follow the energy transfer event. An expression for the rate of energy transfer from donor states δ to acceptor states α that incorporates all these concepts given by,

$$k = \frac{2\pi}{\mathrm{h}} \left\langle \int_0^\infty \mathrm{d}\varepsilon \sum_{\delta,\alpha} P_\delta |V_{\delta\alpha}(\varepsilon_\mathrm{d}, \varepsilon_\mathrm{a})|^2 J_{\delta\alpha}(\varepsilon, \varepsilon_\mathrm{d}, \varepsilon_\mathrm{a}) \right\rangle_{\varepsilon_\mathrm{d},\varepsilon_\mathrm{a}} \tag{15}$$

where $V_{\delta\alpha}$ are the electronic couplings between the effective donors and acceptors, as described in Refs. 17, 63, and 64, and ε_d and ε_a represent static offsets from the mean of the donor and acceptor excitation energies as described in the previous section. Thus it is emphasized that both the couplings and the spectral overlaps depend upon disorder. It is assumed that each $V_{\delta\alpha}(\varepsilon_\mathrm{d}, \varepsilon_\mathrm{a})$ does not vary across the energy spectrum of its corresponding $J_{\delta\alpha}(\varepsilon, \varepsilon_\mathrm{d}, \varepsilon_\mathrm{a})$. P_δ is a normalized Boltzmann weighting factor for the contribution of δ to the thermalized donor state,

$$P_\delta = \exp[(\varepsilon_{\delta=1} - \varepsilon_\delta)/kT] / \sum_\delta \exp[(\varepsilon_{\delta=1} - \varepsilon_\delta)/kT].$$

The angle brackets denote that an ensemble average is taken over many aggregate units (e.g., RC complexes) so as to account for static disorder in the monomer site energies. The spectral overlap between bands δ and α is defined in terms of donor and acceptor densities of states as in Eq. (6):

$$J_{\delta\alpha}(\varepsilon, \varepsilon_d, \varepsilon_a) = N_\alpha a_\alpha^{\text{hom}}(\varepsilon, \varepsilon_a) N_\delta f_\delta^{\text{hom}}(\varepsilon, \varepsilon_d) \tag{16}$$

Note that each $J_{\delta\alpha}(\varepsilon, \varepsilon_a)$ is associated with an electronic coupling factor $V_{\delta\alpha}(\varepsilon_d, \varepsilon_a)$ within the ensemble average. The $f_\delta^{\text{hom}}(\varepsilon, \varepsilon_a)$ and $a_\alpha^{\text{hom}}(\varepsilon, \varepsilon_d)$ specify the donor and acceptor densities of states (D.O.S.), as described in Ref. 63. The dependence upon disorder is assumed to introduce a static offset of the origin, as is usually assumed. These D.O.S. represent the emission (absorption) line shape of the donor (acceptor), calculated without disorder (hence the superscript "hom") and without dipole strength. N_δ and N_α are area normalization constants such that $1/N_\delta = \int_0^\infty d\varepsilon f_\delta^{\text{hom}}(\varepsilon)$ and $1/N_\alpha = \int_0^\infty d\varepsilon a_\alpha^{\text{hom}}(\varepsilon)$.

Our procedure requires as input a site representation of the electronic Hamiltonian that we can modify by adding disorder to the site energies. Using this "disordered" Hamiltonian, we find the set of effective donor states δ, effective acceptor states α, and the couplings between them $V_{\delta\alpha}(\varepsilon_d, \varepsilon_a)$. We can think of the $\{\delta\}$ as collectively comprising the donor emission spectrum, and we can regard the $\{\alpha\}$ as collectively comprising the acceptor absorption spectrum. For each δ and α we wish to calculate $|V_{\delta\alpha}(\varepsilon_d, \varepsilon_a)|^2 J_{\delta\alpha}(\varepsilon, \varepsilon_d, \varepsilon_a)$, the dimensionless quantity that defines the rate of $\delta \rightarrow \alpha$ EET. For this strategy to work, the $V_{\delta\alpha}$ must be classified as "weak." To determine $J_{\delta\alpha}(\varepsilon, \varepsilon_d, \varepsilon_a)$, we need electron–phonon coupling information together with intramolecular vibrational information in terms of a line-shape function or spectral density that relates to the eigenstate representation. We can input this information using explicit equations, as we do in Refs. 16, 17, and 63, but since the line-shape information is contained in experimental emission and absorption spectra (in the absence of significant inhomogeneous line broadening), experimental spectra may also be used in some cases (e.g., see Refs. 17 and 83).

C. Energy Transfer in a Complex with Heterogeneous Coulombic Coupling

Calculations of energy transfer rates and mechanisms are generally based on perturbation theory. In a dimer system with weak coupling, Forster theory can be successfully applied. However, when the Coulombic coupling is stronger than the electron–phonon coupling strength, Redfield theory [109–120] is more appropriate. In this case, the energy transfer (or exciton relaxation) is induced by the electron–phonon coupling. Yang and Fleming have shown how the Redfield and Forster theories can be combined to reasonably describe energy transfer dynamics over a wide range of parameters [121]. Based on these ideas, we describe below a strategy for calculating energy transfer dynamics in systems with a wide range of Coulombic couplings.

We begin with an overall molecular Hamiltonian $H = H^{\mathrm{el}} + H^{\mathrm{Coul}} + H^{\mathrm{el-ph}} + H^{\mathrm{ph}}$ where H^{el} and $H^{\mathrm{el-ph}}$ describe the static electronic excitations and the electron–phonon coupling respectively. H^{Coul} is the Coulombic coupling and H^{ph} is the phonon Hamiltonian.

$$H^{\mathrm{el}} = \sum_{n=1}^{N} |e_n\rangle \varepsilon_n \langle e_n|, \quad H^{\mathrm{el}-ph} = \sum_{n=1}^{N} |e_n\rangle u_n \langle e_n| \tag{17}$$

where $|e_n\rangle$ represents the excited electronic states of the nth monomer. Within the monomer n, ε_n is its excited state energy and u_n is the electron–phonon coupling.

$$H^{\mathrm{Coul}} = \sum_{n=1}^{N} \sum_{m>n}^{N} J_{nm}(|e_n\rangle\langle e_m| + |e_m\rangle\langle e_n|) \tag{18}$$

where J_{nm} is the Coulombic coupling between $|e_n\rangle$ and $|e_m\rangle$. Our strategy is to split the Coulombic Hamiltonian into two groups of pairwise interactions: the strong Coulombic Hamiltonian, $H^{\mathrm{Coul},S}$, and the weak Coulombic Hamiltonian, $H^{\mathrm{Coul},W}$,

$$H^{\mathrm{coul}} = H^{\mathrm{coul},S} + H^{\mathrm{coul},W}$$

where if $J_{\mathrm{nm}} \geq J_{cutoff}$, we have $H^{\mathrm{Coul},S}_{\mathrm{nm}} = J_{\mathrm{nm}}$, and $H^{\mathrm{coul},W}_{\mathrm{nm}} = 0$, and if $J_{\mathrm{nm}} < J_{cutoff}$, we have $H^{\mathrm{Coul},S}_{\mathrm{nm}} = 0$ and $H^{\mathrm{Coul},W}_{\mathrm{nm}} = J_{\mathrm{nm}}$. Thus the total Hamiltonian is rewritten as

$$H = H^{\mathrm{el}} + H^{\mathrm{Coul},S} + H^{\mathrm{ph}} + H^{\mathrm{el-ph}} + H^{\mathrm{Coul},W} \tag{19}$$

Next this Hamiltonian is expressed in the basis set of exciton states obtained by numerical diagonalization of $H^{\mathrm{el}} + H^{\mathrm{Coul},S}$.

$$|\mu\rangle = \sum_{n=1}^{N} \phi_{\mu n} |n\rangle \qquad \text{for } \mu = 1, \ldots\ldots, N \tag{20}$$

where $|n\rangle = |e_n\rangle \Pi_{M=1,\, M\neq n}^{N} |g_m\rangle$ represents a state where only the nth molecule is excited and all others are in their ground ($|g\rangle$) states. $\phi_{\mu n}$ is the amplitude of the nth Chl molecule's contribution to the μth exciton state. For example, if an exciton state μ is completely localized on one molecule, say M, then $\phi_{\mu m} = \delta_{\mathrm{nm}}$. This is the case for all Chls with $H^{\mathrm{Coul},S} = 0$. If the state is equally delocalized over two Chls, k and m, $\phi_{\mu n} = (\delta_{\mathrm{nm}} + \delta_{nk})/\sqrt{2}$. In the new representation we have diagonal $\left(H^0 = \sum_{\mu=1}^{N} |\mu\rangle H_\mu^0 \langle\mu|\right.$ and off-diagonal $\left(H' = \sum_{\mu,\mu'\mu\neq\mu'} |\mu\rangle\right.$ $H'_{\mu\mu'}\langle\mu'|$ Hamiltonians:

$$H_\mu^0 = E_\mu + H^{\mathrm{ph}} + \langle\mu|\left(H^{\mathrm{el-ph}} + H^{\mathrm{coul},W}\right)|\mu\rangle \tag{21}$$

$$H'_{\mu\mu'} = \langle\mu|\left(H^{\mathrm{el-ph}} + H^{\mathrm{coul,W}}\right)|\mu'\rangle \tag{22}$$

In the diagonal term $\langle\mu|H^{\mathrm{el-ph}}|\mu\rangle$ and $\langle\mu|H^{\mathrm{coul,W}}|\mu\rangle$ are responsible, respectively, for the energy fluctuation and energy shift of the state μ. These diagonal parts are treated nonpertubatively. The off-diagonal term is responsible for energy transfer between exciton states. The magnitude of the off-diagonal Hamiltonian involves the strengths of the electron–phonon (μ_n) and Coulombic (J_{nm}) couplings and also the overlap of the exciton wavefunctions $\phi_{\mu n}\phi_{u'n}$. A perturbative approach to the energy transfer calculation is justified even when $\mu_n$and J_{nm} are large, provided that the overlap of the two exciton wavefunctions is small. Thus, the energy transfer rate from a state μ' to a state μ, $k_{\mu\leftarrow\mu'}$, can be calculated via the Golden Rule.

$$k_{\mu\leftarrow\mu'} = 2\,\mathrm{Re}\int_0^\infty d\tau Trq\left(e^{iH_{\mu'}^0\tau} H'_{\mu'\mu} e^{-iH_\mu^0\tau} H'_{\mu\mu'}, \rho_{\mu'}^{eq}\right) \tag{23}$$

where Tr_q denotes a trace over the nuclear degrees of freedom and $\rho_{\mu'}^{eq} = e^{-\beta H_{\mu'}^0}/Tr_q\left(e^{-\beta H_{\mu'}^0}\right)$ with β the Boltzmann factor. Yang derived an expression for $k_{\mu\leftarrow\mu'}$ in terms of standard line-broadening functions of the Chl molecules [121]. The resulting expression satisfies detailed balance. The full expression is given in Ref. 121.

When the exciton states are localized on individual Chls, the expression reduces to the well-known Förster formula. If $J_{cutoff} \rightarrow \infty$ this holds for all transfer rates. On the other hand, when $J_{cutoff} \rightarrow 0$, the expression reduces to that derived from modified Redfield theory. It can be shown analytically that the forward and backward rates satisfy the detailed balance condition

$$\frac{k_{\mu'\leftarrow\mu}}{k_{\mu\leftarrow\mu'}} = e^{-\beta\left(E_{\mu'}^0 - E_\mu^0\right)} \tag{24}$$

where $E_\mu^0 = E_\mu + H_{\mu\mu}^{\mathrm{Coul},W} - \lambda_{\mu\mu,\mu\mu}$ corresponds to the $0-0$ transition energy of the state μ. Thus once downhill rates are calculated via numerical integration, the corresponding uphill rate can be calculated from the detailed balance expression.

The precise value of the cutoff interaction energy might be considered problematical. However, as Fig. 14 shows, for a model dimer system, the modified Redfield theory and Forster theory rates are very similar over a wide range of energy gaps (degrees of delocalization); and provided that J_{cutoff}

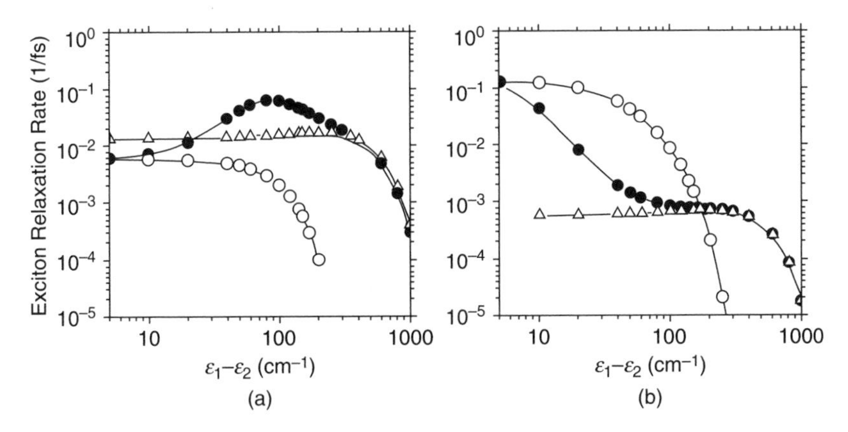

Figure 14. Downhill exciton relaxation rates as a function of energy gap between two monomers, predicted by the Förster model (open triangles), the traditional Redfield model (open circles), and the modified Redfield model (filled circles). In (*a*) the electronic coupling, $|J_{12}| = 100\,\text{cm}^{-1}$ and in (*b*) $|J_{12}| = 20\,\text{cm}^{-1}$. The reorganization energy, $\lambda = 100\,\text{cm}^{-1}$ and the spectral density is represented by a Gaussian correlation function with $\tau_g = 100\,\text{fs}$ [163].

corresponds to reasonably weak coupling, the precise value of J_{cutoff} is not critical.

This theory connects with the modified Förster theory for molecular aggregates as follows. The J_{cutoff} procedure partitions the system into the set of effective donors δ and the effective acceptors α. Whether a state is designated δ or α is determined by the excitation conditions. Energy transfer from δ to α occurs as described by Eq. (15). However, the Redfield theory can account for more complex dynamics that arise owing to competition between relaxation in the δ manifold and δ-to-α energy hopping. In other words, the multistep evolution of the excited-state population subsequent to excitation can be followed in an arbitrarily large molecular aggregate.

D. Energy Transfer to a Dimeric Acceptor: Bacterial Reaction Centers

The photosynthetic reaction center (RC) of purple bacteria is a pigment–protein complex present in the thylakoid membrane that efficiently accepts excitation energy from antenna complexes to initiate light-induced charge separation from the primary electron donor (P); this is the first step in photosynthesis. Excitation of the primary electron donor, a bacteriochlorophyll dimer, to form the lowest excited singlet state (P*) usually occurs by energy transfer from the antenna. In isolated RCs, P* can quench excitation by rapid EET from higher-energy RC pigments, either from the monomeric "accessory" bacteriochlorophyll-*a* molecules (B_L and B_M) or from the bacteriopheophytins (H_L and H_M). The

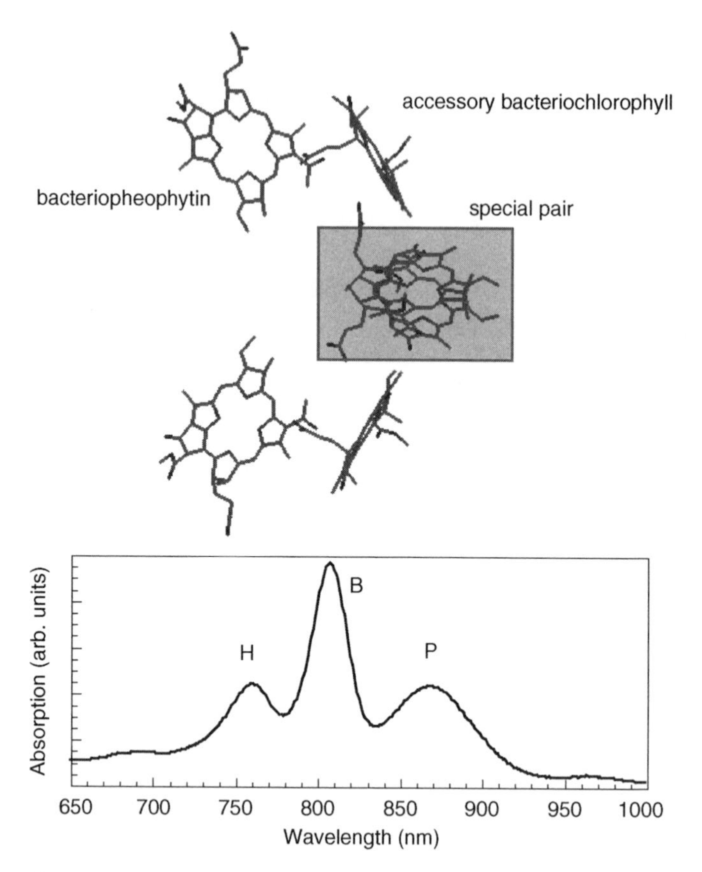

Figure 15. *Top*: Arrangement of pigments in the reaction center *Rb. Sphaeroides*. *Bottom*: Plot of the absorption spectrum of this RC, with absorption features attributed to the pigments H, B, and P indicated.

arrangement of these pigments and the absorption spectrum of the RC are shown in Fig. 15.

As early as 1972, Slooten, using absorption measurements, proposed that electronic energy transfer (EET) from H and B to P occurs in the *Rb. sphaeroides* RC [59]. In 1986 such energy transfer was shown to occur from B to P for the RC of *Rps. viridis* in less than 100 fs at 298 K. Two years later, Breton *et al.* [44] demonstrated a similar result at 10 K for the same species. Within the time resolution of these experiments, the energy transfer time was insensitive to temperature.

A consensus exists that the usual application of Förster theory is deficient by as much as an order of magnitude in accounting for the rate of EET between the

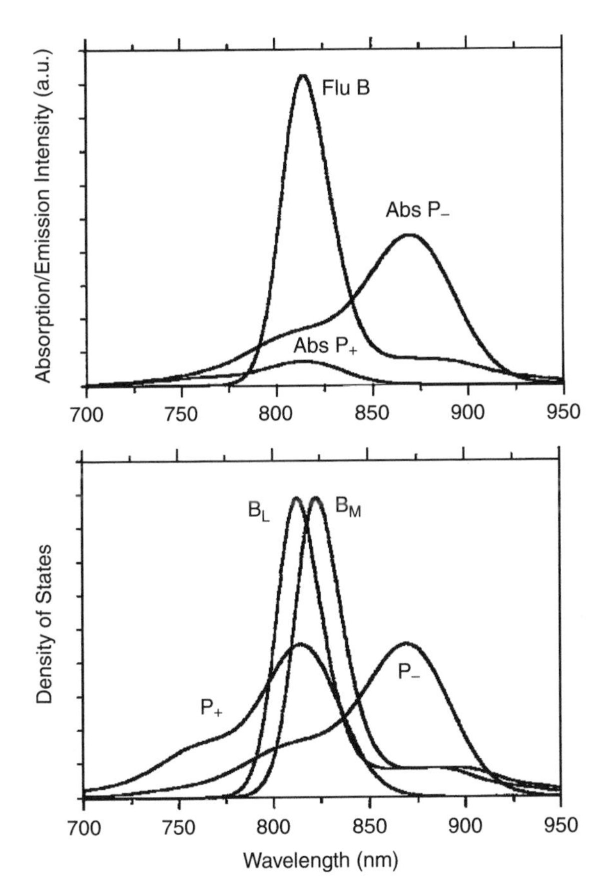

Figure 16. *Top*: The absorption spectra of the special pair acceptor states P_+ and P_- plotted over the fluorescence of the donor B. Although P_+ has superior spectral overlap with B, it has a small intensity because it is dipole-forbidden. *Bottom*: The density of states calculated for the donors and acceptors in B-to-P energy transfer. See Ref. 17.

RC cofactors. This comes about because the lower exciton state of P, P_-, carries 88% of the dipole strength and is therefore strongly coupled to B according to the dipole approximation, but has only a small overlap with the B emission, as shown in Fig. 16. The net effect is that P_- is not predicted to be an effective acceptor for B. On the other hand, the upper exciton state P_+ overlaps significantly with B emission (see Fig. 16), but since it carries only 12% of the total dipole strength, this state is predicted to be too weakly coupled to B to be an effective acceptor. The answer to the conundrum lies in the idea that the absorption spectra of the P acceptor states do not contain the relevant information for predicting the electronic coupling between B and P_+ or P_-.

Recently the theory for EET in molecular aggregates was applied to wild-type and mutant photosynthetic reaction centers (RCs) from *Rb. sphaeroides*, as well as to the wild-type RC from *Rps. viridis*. Calculations of EET in two mutants, (M)L214H or the beta mutant and (M)H202L or the heterodimer, were also reported. Experimental information from the X-ray crystallographic structure, resonance Raman excitation profiles, and hole-burning measurements were integrated with calculated electronic couplings to model the EET dynamics within the RC complex. To check the model, which contains no adjustable parameters, optical absorption and circular dichroism spectra were calculated at various temperatures between 10 K and room temperature and compared well with the experimentally observed spectra. The rise time of the lower exciton state of P, P_-, population, subsequent to the excitation of the accessory bacteriochlorophyll, B, in *Rb. sphaeroides* (*Rps. viridis*) wild-type at 298 K was calculated to be 193 fs (239 fs), which is in satisfactory agreement with experimental results. The calculations suggest that the upper exciton state of P, P_+, plays a central role in trapping excitation from B. Our ability to predict the experimental rates was partly attributed to a proper calculation of the spectral overlap $J_{\delta\alpha}(\varepsilon)$ using the vibronic progressions.

That work provided the following answers: (1) The EET dynamics in the RC are promoted via a weak-coupling mechanism. Most importantly, we had to adapt Förster theory so that it could be applied to molecular aggregates like the RC. Our model employed only Coulombic couplings (aside from the coupling between P_M and P_L), and we conclude that short-range interactions, depending explicitly on orbital overlap between the pigments, are relatively unimportant for promoting EET. Crucially, we had to calculate correctly the *effective* donor–acceptor couplings and their associated spectral overlaps. Simple application of Förster theory blurs the details of the aggregate and leads to physically incorrect results. (2) Energy is transferred according to the following scheme: $H \rightarrow B \rightarrow P_+(\rightarrow P_-)$. (3) Our calculations suggest that P_+ is the principal acceptor state involved in energy transfer from B to P in the wild-type RC. (4) The temperature independence of EET can be understood now that we have correctly calculated the spectral overlap between B and P_+. This overlap governs the rate, and we have found it to be insensitive to temperature. The overlap between the B emission and P_-, where P_- is peaked at 865 nm at 298 K and at 890 nm at 10 K, is significantly affected by temperature, but is relatively unimportant in the overall dynamic process. (5) The large displacements of the vibrational modes of P make an important contribution to the EET by increasing the spectral overlap between B and P_+, which, in turn, increases the rate and plays a role in the temperature independence. (6) The same weak-coupling mechanism (i.e, the generalized Förster theory presented here) provides an

TABLE II
Calculated Electronic Couplings (cm^{-1}) Between the Accessory Bchls and the Exciton States of the Special Pair in the RC of *Rb. sphaeroides* (see Text and Ref. 17)

	Effective Couplings	Monomer B to P	Dipole–Dipole
B_M–P_-	−67	−55	198
B_L–P_-	69	76	−182
B_M–P_+	67	78	−6
B_L–P_+	83	72	−29

adequate description of EET in both the beta and heterodimer mutant, although in the case of the heterodimer, it is dependent on diminished electronic coupling between D_M and the rest of the RC pigments. In the beta mutant, where the weakly coupled H_M chromophore is replaced with a $BChl_a$, energy transfer is both quantitatively and mechanistically similar to the wild-type. However, in the heterodimer mutant, where the mutated pigment is part of a strongly coupled special pair, the energy transfer proceeds at a quite different rate along each branch. This leads to a biexponential rise of population of the P state.

The main advance that was made, however, was to calculate the electronic couplings $V_{\delta\alpha}$ in terms of the molecular composition of donor and/or acceptor aggregates, rather than treating the acceptors P_+ and P_- as point dipoles associated with each spectroscopic band. It can be seen in Table II that the effective electronic couplings $(V_{\delta\alpha})$ calculated for B to P_+ and B to P_- are approximately equal in magnitude. These couplings were determined from the full Hamiltonian of the RC, but compare closely with the "monomer B to P" couplings, which are effective electronic couplings calculated for the system consisting only of one B and the special pair. These latter electronic couplings may be compared directly with analogous dipole–dipole couplings calculated for B to P_+ and B to P_-, showing that the dipole approximation fails completely, thus explaining why Förster theory cannot predict the rate of B to P EET for the RC. But, why precisely does the dipole approximation fail in this case?

In Fig. 17 we compare transition densities calculated for the special pair upper exciton state P_+ (lower panel) and lower exciton state P_- (upper panel). The P_+ transition density has many alternating positive and negative phase regions that are averaged away by the dipole operator to give a small transition dipole moment. The P_- transition density has one region of negative phase and another region of positive phase, indicative of a dipole-allowed transition, that are averaged by the dipole operator to give a large transition dipole moment. Such an averaging over the topology of the transition density is carried out by light, which has a wavelength large compared to molecular dimensions and separations, and is therefore manifest in the absorption and emission spectra. However, if one imagines the B chromophore located near the right-hand side of

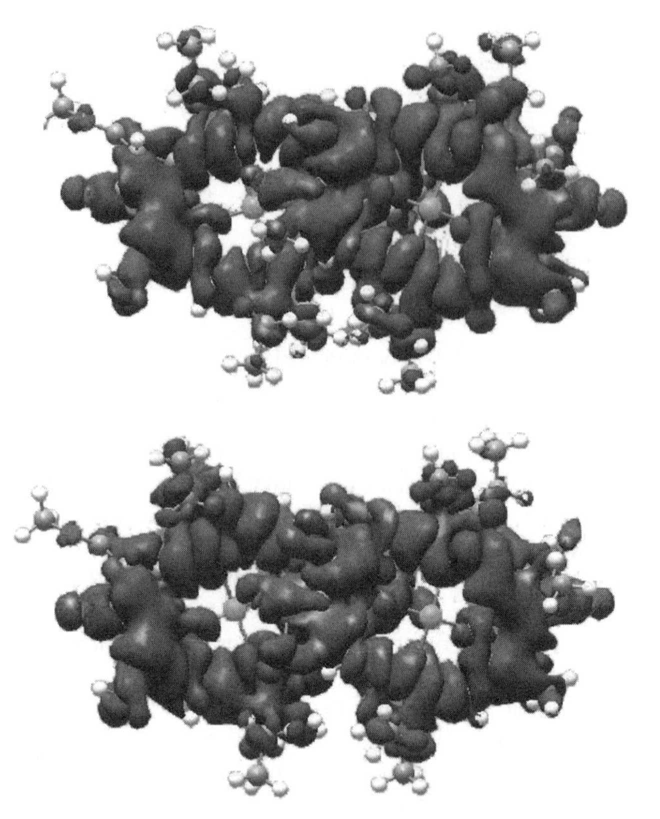

Figure 17. Transition densities calculated for the special pair. *Top*: Transition densities for P_-, *Bottom*: Transition density for P_+. See color insert.

P in Fig. 17, then it can be seen that the local P_+ and P_- transition densities that interact most significantly with the B transition are almost identical. Thus, from the viewpoint of the B donor, the electronic coupling to either P_+ or P_- should be similar, as indeed the calculations reveal.

We can summarize by stating that the averaging imposed on electronic couplings in a molecular aggregate by the dipole approximation is implemented on two levels. First, it is implemented with respect to the coupling between sites. This is the difference between panels a and b of Fig. 18, which depicts the special pair and one accessory bacteriochlorophyll of the photosynthetic RC of a purple bacterium. Panel a of Fig. 18 depicts an essentially exact calculation of the Coulombic coupling between a monomeric bacteriochlorophyll molecule k and the upper exciton state of a dimer formed by molecules m and n. This is accomplished by performing separate quantum chemical calculations of the

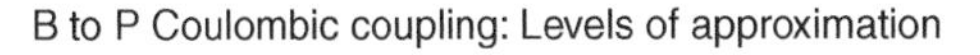

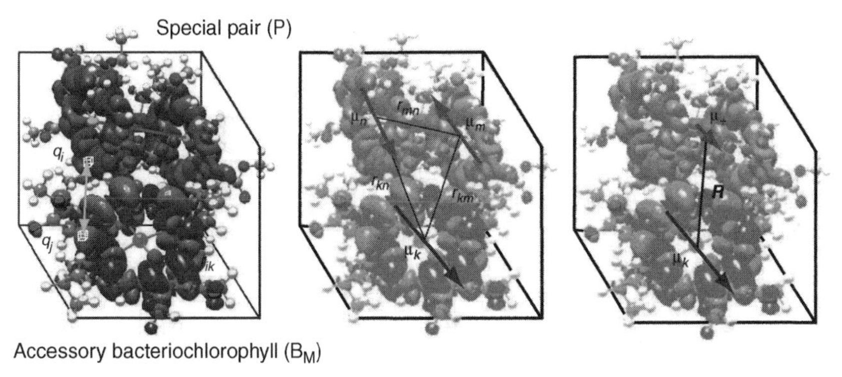

(a) Transition density cube (TDC) method. (b) Distributed transition dipoles. (c) Dipole approximation.

Figure 18. An illustration of the levels of approximation used in estimating the B to P_+ electronic coupling. (*a*) An essentially "exact" calculation can be made using the TDC method. (*b*) Distributed dipoles used in the GFT method (see Section VI.B) represent the minimal acceptable approximation. (*c*) The harsh dipole approximation, in which the correct physical picture of the system is completely washed away. See color insert.

ground and relevant excited states of k and the m-n dimer in order to obtain the corresponding transition densities, $P^{k}_{0\delta}(\mathbf{r}_1)$ and $P^{m-n}_{\alpha 0}(\mathbf{r}_2)$ respectively, which are plotted in the figure. These transition densities interact via the Coulomb potential to give the Coulombic interaction as described earlier in this review. Figure 18*b* depicts a simplification of this method, which we see as the minimal representation of this aggregate. Here the transition densities have been reduced to transition dipoles on each molecular center, according to Eq. (5). For the dimer, we need to ascertain the coefficients describing the admixture of monomer wavefunctions that comprise the dimer wavefunction, λ_m and λ_n. Then we can write $\mu^{\alpha 0}_{\varsigma} = \lambda_m \mu^{m0}_{\varsigma} + \lambda_n \mu^{n0}_{\varsigma}$.

Second, an averaging can be implemented with respect to the coupling within the donor or acceptor supermolecules (panel c of Fig. 18). In this case we would couple $\mu^{0\delta}_{\varsigma}$ and $\mu^{\alpha 0}_{\varsigma}$ directly. Such an averaging is invoked in analyses of RC energy transfer when, for example, either the P_- or P_+ special pair states are taken to be the energy acceptor in the Förster model, where donors and acceptors are treated as point dipoles associated with each spectroscopic band (i.e., P_+ and P_-). This approach fails to account for the true interactions within a multichromophoric assembly, as we have already described.

The Förster spectral overlap is an incredibly useful quantity for understanding EET in donor-acceptor pairs, but unfortunately it turns out to be useless for describing molecular aggregates and disordered systems. However, in the spirit of the Förster spectral overlap, we have introduced the electronic coupling-weighted spectral overlap between effective donor and acceptor states:

$$u_{\delta\alpha}(\varepsilon) = \left\langle |V_{\delta\alpha}(\varepsilon_\delta, \varepsilon_\alpha)|^2 J_{\delta\alpha}(\varepsilon_\delta, \varepsilon_\alpha) \right\rangle_{\varepsilon_\delta, \varepsilon_\alpha}. \tag{25}$$

This quantity allows us (1) to quantify the rate of EET, according to the summation of the area of each $u_{\delta\alpha}(\varepsilon)$, as in Eq. (15), (2) identify the dominant states that mediate energy transfer in a complex system, and (3) work in terms of correctly ensemble-averaged quantities. The spectrum of the quantity $u_{\delta\alpha}(\varepsilon)$ derives from overlap of the donor emission density of states with that of the acceptor, $J_{\delta\alpha}(\varepsilon)$. This can be nonintuitive in a disordered molecular aggregate owing to the interdependence of electronic couplings and site energies, as we describe in the following section. The intensity of each spectrum is adjusted by the donor–acceptor electronic couplings. This occurs *within the ensemble average over static disorder* in the transition frequencies, ε_δ and ε_α. Now the

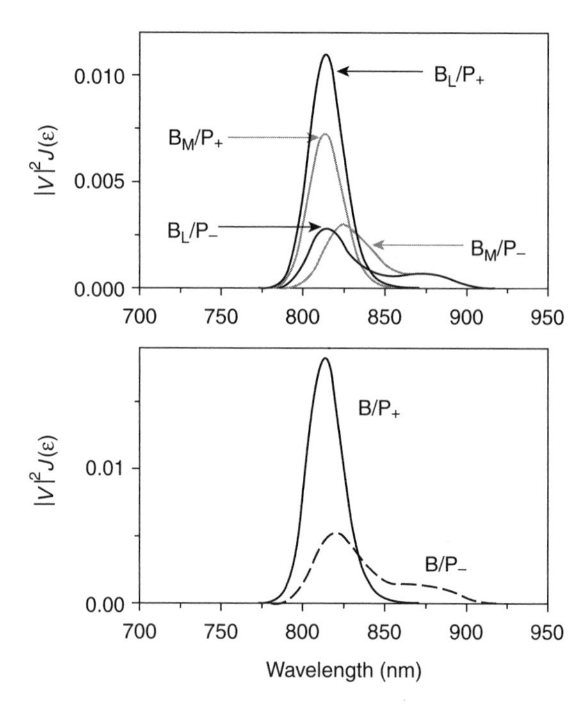

Figure 19. *Top*: Coupling-weighted spectral overlaps calculated using the GFT for all four interactions between B and P (see Ref. 17). *Bottom*: The average B to P_+/P_- picture of these coupling-weighted spectral overlaps. It is now evident that B to P_+ energy transfer dominates deactivation of initially excited B pigments.

overall area of each $u_{\delta\alpha}(\varepsilon)$ is proportional to the rate of EET via the corresponding δ–α pathway. The ensemble average coupling-weighted spectral overlaps for each of the four B to P EET pathways are plotted in Fig. 19. Inspection of the relative intensities of the $u_{\delta\alpha}(\varepsilon)$ for each pathway leads us to conclude that B to P_+ is the dominant EET channel, most likely as a consequence of better spectral overlap $J_{\delta\alpha}(\varepsilon)$.

E. Energy Transfer in LH2

1. B800 to B850 Energy Transfer

In the introduction to this review we have described in detail the structure and function of the peripheral light-harvesting antenna LH2 of purple bacteria. Light absorbed by the B800 ring is transferred rapidly to the B850 ring on a time scale of 800 fs in *Rps. acidophila* and 650 fs in *Rb. sphaeroides* at room temperature, increasing to just 1.2 ps at 77 K for both *Rps. acidophila* and *Rb. sphaeroides*. Förster theory, however, provides an unsatisfactory estimate of this time scale and, in particular, fails to elucidate the reasons for the remarkable insensitivity to temperature.

The donor molecule of the B800 ring is approximately monomer-like and is located ~18 Å away from the acceptor. The acceptor consists of the 18 bacteriochlorophylls of the B850 ring, part of which is shown in Fig. 20. The

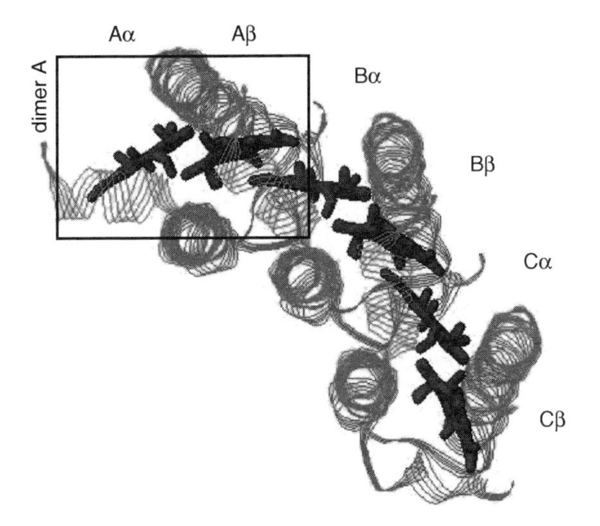

Figure 20. A depiction of part of the B850 ring from LH2 showing the α-helices as ribbons, labeled as α and β according to their position on the inside and the outside of the ring, respectively. The pigments are labeled A, B, C according to the α,β-subunit they belong to, and they are labeled individually as α or β according to whether they are coordinated to the α-helix labeled α and β. The close interaction (3–4 Å) between these Bchl chromophores is evident.

Bchl molecules that comprise the B850 ring are relatively strongly coupled to each other ($\sim$300 cm^{-1}), which has a significant impact on the nature and operation of the acceptor states. This also means that there are not just one or two acceptor states: There are 18 acceptor states that must be individually considered. Moreover, each LH2 complex is slightly different spectroscopically, owing to significant inhomogeneity in the site energies of each Bchl. The effect of this static disorder on the absorption is shown by the absorption spectra calculated for individual LH2 complexes (Fig. 21). These spectra are similar to the striking experimental observations reported by van Oijen et al. [16]. Any model for predicting the EET dynamics in this complex system must capture these essential features.

In order to model realistically the effective acceptor states of the B850 ring, it was first necessary to calculate the electronic couplings on the basis of the crystal structure data. Owing to the close approach of the Bchls in B850 (cf. Figs. 6 and 20), we decided to ascertain the significance of contributions to the electronic coupling that depend on orbital overlap, V^{short}. V^{short} mostly derives from interactions indicative of mixing of donor–acceptor wavefunctions

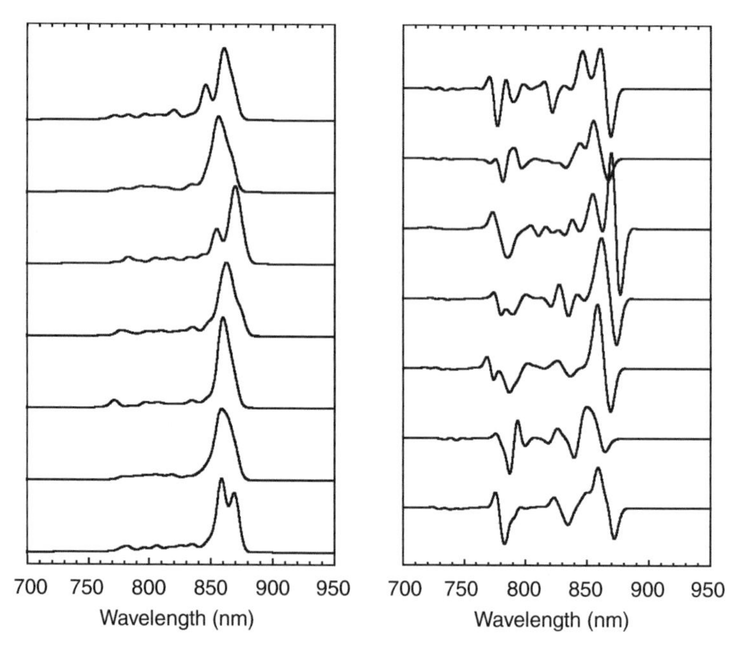

Figure 21. *Left*: Absorption spectra calculated for a random selection of single LH2 complexes (*Rps. acidophila*, 77K). *Right*: The corresponding calculated circular dichroism spectra. Only diagonal disorder is included in the site energies of the monomers, but note the dramatic effects it has on state energies and intensities.

owing to their interpenetration (exchange effects related to the Coulombic interactions make a very minor contribution for molecules). There has been considerable speculation regarding the role of V^{short} in photosynthetic light harvesting since if this coupling becomes significant relative to V^{Coul} at close separations, then the EET rate could increase over that estimated from the Förster rate by a factor $\left|V^{\text{Coul}} + V^{\text{short}}\right|^2 / \left|V^{\text{Coul}}\right|^2$. However, evaluation of the significance of the V^{short} component of the coupling by quantum chemical calculation poses a difficult problem.

We attempted to quantify the total V^{short} for the closely interacting BChl pigments in LH2 of *Rps. acidophila* using CI-singles calculations (6-31G* level) of the excited states of Bchl dimers within the B850 ring, as well as individual Bchl molecules. The dimer calculations provided an estimate of the total electronic coupling $\left(V^{\text{short}} + V^{\text{Coul}}\right)$, but where V^{Coul} was overestimated, just as for the TDC calculations based on CI-singles TDs. We could use TDC calculations, then, to determine the (overestimated) V^{Coul} and hence retrieve V^{short} (which is not similarly overestimated). The usual scaling procedure provided a reasonable estimate of V^{Coul}. We summarize our results in Table III, where we collect the scaled V^{Coul}, calculated by the TDC method, V^{short} derived from the *ab initio* supermolecule calculations once V^{Coul} had been determined, and the total electronic coupling, equal to $V^{\text{short}} + V^{\text{Coul}}$ (where this is the scaled Coulombic interaction).

Combining the calculated electronic couplings with various experimental data, we have simulated the energy transfer dynamics in the wild-type *Rps. acidophila* B800–B850 complex, as well as in four reconstituted complexes in which the B800 band lies at 765, 753, 694, and 670 nm (which we refer to as B765, B753, etc.). There are no adjustable parameters in these calculations, since the mean Bchl site energies in the B800 and B850 rings are set in order to simulate the absorption and circular dichroism spectra. The mutant and

TABLE III
Nearest-Neighbor Electronic Couplings V (cm^{-1}) Calculated Between the Bchl Q_y Transitions in the B850 Ring of the LH2[a] of *Rps. acidophila* Using the CI-singles/6-31G* Method (see Text and Ref. 74), Along with Next-to-Nearest Neighbor Couplings Calculated Using the CI-Singles/3-21G*/TDC Method [73]

	Separation[b] (Å)	Total Coupling	V^{Coul} Part	V^{short} Part	$V^{\text{dipole–dipole}}$
αB850$_A$–βB850$_A$	9	320	265	55	415
βB850$_A$–αB850$_B$	9.5	255	195	60	330
αB850$_A$–αB850$_B$	18	−46	−46	0	−48
βB850$_A$–βB850$_B$	19	−37	−37	0	−37

[a]See Fig. 20 for the labeling convention.
[b]Center-to-center.

TABLE IV
Calculated B800–B850 Energy Transfer Times (ps) in LH2 of *Rps. acidophila* and Reconstituted Complexes (See Text and Refs. 16 and 22)

	GFT[a] (77 K)	GFT[a] (300 K)	Experiment[b]	FT[c] (300 K)
B800–B850	0.96	0.91	0.9	6
B765–B850	0.76	0.75	1.4	9
B753–B850	1.90	1.34	1.8	11
B694–B850	17.3	13.8	4.4	18
B670–B850	49.6	43.7	8.3	37

[a]Generalized Förster Theory for molecular aggregates (see Section VI.A and Ref. 63).
[b]Results reported by Herek et al. [122] (300 K).
[c]Förster Theory predictions [122].

temperature dependence of the B800–B850 EET rates provides a convincing test of the theory for energy transfer in molecular aggregates, Eq. (15).

In Table IV we summarize the results of these calculations for both 77 K and 300 K. Note that here it is assumed that each of the substituted chlorophylls has the same transition moment magnitude and orientation, and therefore coupling to the B850 BChls, as the wild-type B800s. We see from the results collected in Table IV that (i) the calculated energy transfer times for B800–B850 and B753–B850 correspond closely to the experimental values reported by Herek et al. [122]; (ii) the calculated B800-B694 and B800-B670 energy transfer times are much slower than revealed by experiment, suggesting that the carotenoid S_1 state may be mediating the energy transfer for these donors; (iii) While the "B800"-type donor has appreciable overlap with the B850 density of states, which spans 720–870 nm, the 'B800'–B850 energy transfer time is rapid and is sensitive (i.e., can be tuned by a factor of two in magnitude) to the exact location of the donor emission spectrum; (iv) The EET rate is insensitive to temperature, which is a well-known characteristic of the wild type LH2 complex.

In Fig. 22 the spectral overlaps calculated for seven individual LH2 complexes (*Rps. acidophila*) are shown and compared to the corresponding coupling-weighted spectral overlaps, calculated according to the GFT. First we note that the spectral overlaps do not correspond to the Förster spectral overlaps, since we have correctly calculated the overlap between the donor emission and the absorption density of states (not absorption spectrum) of each acceptor eigenstate according to Eq. (16), rather than as the overlap between the donor emission and total B850 absorption spectrum. Second, it is evident that a proper ensemble average over the individual complexes is crucial because of the significant static disorder, as seen in the single complex absorption spectra shown in Fig. 21. However, comparison of the calculated spectral overlaps with

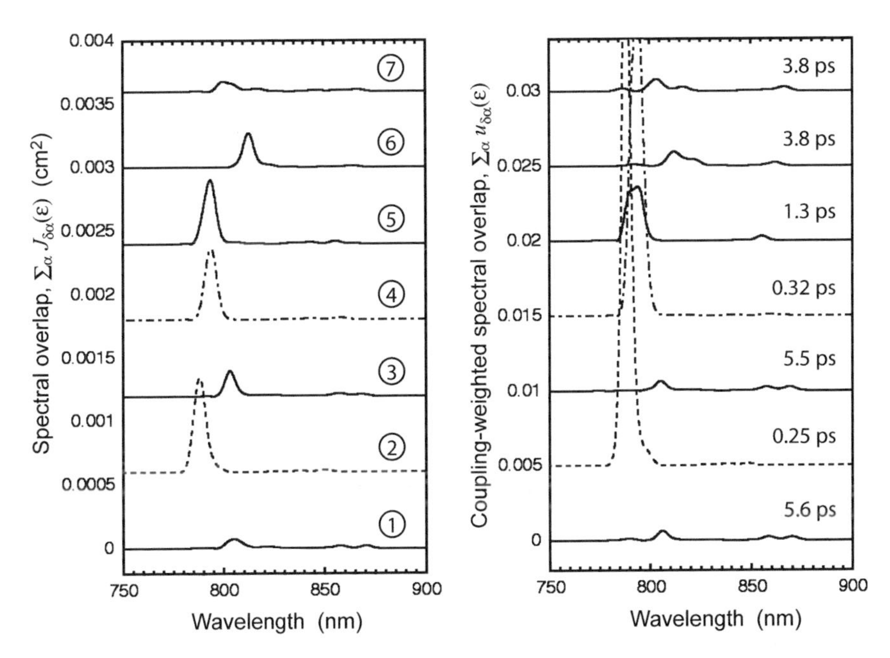

Figure 22. *Left*: Spectral overlaps calculated using the GFT model for B800–B850 energy transfer within each of the single LH2 complexes shown in Fig. 21. The dashed and dashed–dot lines for spectra 2 and 4, respectively, are simply meant to guide the eye in the right-hand panel of the figure. *Right*: Coupling-weighted spectral overlaps and energy transfer times calculated for these systems. See text.

the calculated coupling-weighted spectral overlaps immediately reveals significant differences. For example, from the spectral overlap calculations, the B800–B850 EET times for complex 4 and 5 are expected to be similar. However, a calculation according to the GFT reveals via the coupling-weighted spectral overlaps that the EET times in these two complexes actually differ by a factor of 4! Such observations emphasize the importance of keeping the electronic coupling and corresponding spectral overlap factor associated.

LH2 typifies a complex donor–acceptor system for which EET rates cannot be understood according to Förster theory. To elucidate B800–B50 EET rates, it was crucial to understand the nested averages in a microscopic picture of the dynamics that are important even at the level of single complexes. Thus, the summations over effective donor and acceptor states, and the ensemble average over disorder must be carefully treated. These subtleties are all contained in Eq. (15), the GFT, and cannot be ignored for the sake of an expedient solution. The key quantity for characterizing EET in molecular aggregates is the coupling-weighted spectral overlap, $u_{\delta\alpha}(\varepsilon)$, Eq. (25).

2. *Exciton Dynamics in B850*

In the case of the B850 molecules of LH2 or the B875 molecules of LH1, the situation is more complex. Now the excitonic interactions are at least similar to the site energy disorder, k_BT, and the electron–phonon coupling and the electronic (and perhaps vibrational) state are not localized on individual molecules. This delocalization leads to the phenomenon of exchange narrowing [123] whereby the distribution of site energies is apparently narrowed by the averaging effect of the delocalized states. Thus the intracomplex disorder, σ, and, therefore, the total disorder, becomes dependent on the electronic coupling. The intercomplex disorder Σ, however, remains independent of the coupling.

In such a system the initially prepared state can evolve very rapidly via electron–phonon coupling. This rapid relaxation among exciton levels gives rise to a kind of lifetime broadening, which competes against the exchange narrowing. The two effects cannot be separated by analysis of the linear absorption spectrum, but can be resolved by analysis of photon-echo signals [124]. Thus in order to properly characterize a system such as the B850 ring of LH2 or the B875 ring of LH1, it is necessary to be able to have a formalism with which one is able calculate both linear and nonlinear optical signals from the same approach.

Energy disorder (both diagonal and off-diagonal) plays a major role in determining the electronic structure and consequent dynamics of such a system. As before, averaging over the ensemble must be done after the microscopic dynamics are calculated. Figure 23 shows the exciton levels and associated

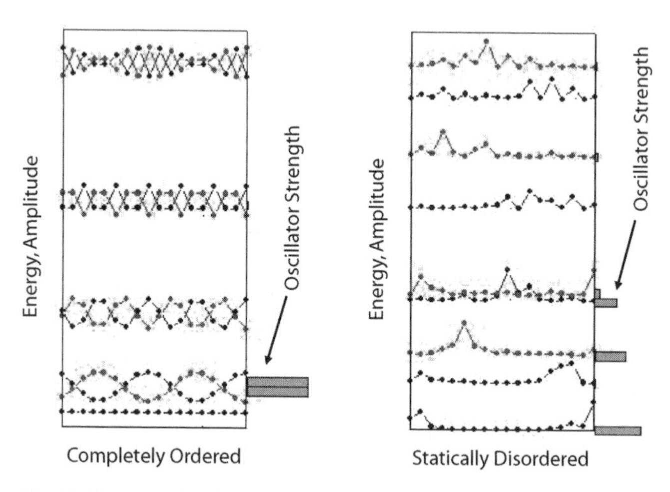

Figure 23. Exciton wavefunctions calculated for the B850 band of LH2. *Left*: For a completely ordered model with $J_{\text{intra}} = J_{\text{inter}} = 320\,\text{cm}^{-1}$. *Right*: For a statically disordered model, with $J_{\text{intra}} = 320\,\text{cm}^{-1}$, $J_{\text{inter}} = 255\,\text{cm}^{-1}$, $\sigma(\text{disorder}) = 150\,\text{cm}^{-1}$ and an energy offset between α and β Bchls (see Fig. 20) of $530\,\text{cm}^{-1}$. See Ref. 40.

wavefunctions for the B850 either as completely ordered or for a particular complex selected from a distribution with $\sigma = 150\,\text{cm}^{-1}$. In the perfectly ordered case, the lowest state is optically dark and all the oscillator strength is concentrated in the next two (degenerate) levels. The wavefunctions are completely delocalized for all levels. In the disordered case, the oscillator strength is distributed over multiple levels; in particular, the lowest state is no longer dark, and the wavefunction is "broken up" such that it has amplitude on typically 2–4 molecules. Figure 23 is useful for visualizing the system but does not allow calculation of the dynamics or nonlinear optical response. For this, we turn to a formalism based on the density matrix and calculate the dynamics using Redfield theory [124]. The formal approach has been described in detail in Refs. 124 and 121. Here we give a brief overview of the approach beginning with theory for the linear absorption and third-order response of a simple model system for molecular aggregates which consist of monomers with two electronic states. The standard description of the electronic states of molecular aggregates is based on the Frenkel-exciton Hamiltonian.

$$H = H^{el} + H^{el-ph} + E^{ph} \tag{26}$$

$$H^{el} = \sum_{n=1}^{N} |n\rangle \varepsilon_n \langle n| + \sum_{\substack{m,n \\ m \neq n}}^{N} J_{mn} |m\rangle\langle n| \tag{27}$$

$$H^{el-ph} = \sum_{n=1}^{N} |n\rangle u_n \langle n| \tag{28}$$

where $|n\rangle$ is the electronic excited state of the monomer n and ε_n is the static energy of the electronic excited state of the nth monomer. u_n describes the fluctuation of the transition energy due to the electron–phonon coupling. E^{ph} is the Hamiltonian of the phonon bath. The interaction between monomers n and m is given by J_{mn}, which is assumed to be homogeneous. We also assume that each monomer is coupled to its own bath and that the baths belonging to different monomers are uncorrelated.

For the one-exciton state, the eigenstates and eigenenergies of the Frenkel excitons are obtained by numerical diagonalization of the electronic part of the exciton Hamiltonian [Eq. (27)]. The exciton wavefunctions and exciton energies for the two exciton band can be constructed from those of the one exciton band by use of Bethe's Ansatz [125]. The electron–phonon coupling Hamiltonian H^{el-ph} is responsible for pure dephasing of the exciton states (diagonal in exciton basis) and population transfer between the exciton states (off-diagonal in exciton basis).

The linear and third-order response functions of the molecular aggregates are described by a density matrix formalism in the exciton basis. The evolution of the density matrix is given by

$$\frac{d}{dt}\rho(t) = -i(L_0 + L')\rho(t) \tag{29}$$

where L_0 is a diagonal Liouville operator governing the exciton dynamics in the absence of any exciton transfer process. The off-diagonal term, L', is responsible for the population transfer process between the exciton states. The usual method to obtain the time evolution of the exciton state is to reduce the full density matrix [Eq. (29)] to the excitonic space by taking an average over the bath. In this case, we lose detailed information of the dynamics of the bath. However, the photon-echo peak shift method [124] is sensitive to the non-Markovian behavior of the bath, and thus we need to keep the dynamics of the bath to accurately describe the experimental data. As a first approximation, the operator of the second term of the right-hand side of Eq. (29) is replaced with a rate equation.

$$\frac{d}{dt}\rho(t) \approx -iL_0\rho(t) - K\rho(t) \tag{30}$$

where K is the Redfield tensor which is based on a second-order approximation with respect to the off-diagonal Hamiltonian in the exciton representation [124]. Equation (30) is a kind of mean-field description of the population transfer since the phonon-dependent operator has been replaced by a phonon-averaged rate equation. We note, however, that the fast phonon dynamics is correctly described by the first term of Eq. (30) in contrast with the usual reduced density matrix approach. As usual, we introduce the so-called secular approximation in which the nonsecular elements of the Redfield tensor are assumed to be zero:

$$K_{\alpha\beta,\alpha'\beta'} = 0 \qquad \text{when} \qquad \left|\omega_{\alpha\beta} - \omega_{\alpha'\beta'}\right| \neq 0 \tag{31}$$

where $\omega_{\alpha\beta}$ is the energy difference between the states α and β. The contributions of the nonsecular terms are averaged out on a time scale of $\left|\omega_{\alpha\beta} - \omega_{\alpha'\beta'}\right|^{-1}$. By this approximation, contributions for the population and coherence elements are not likely to be coupled to each other in the presence of static disorder. The Redfield tensors then consist of three terms: (a) population transfer from α to γ ($\alpha \neq \gamma$), $K_{\alpha\alpha,\gamma\gamma}$, (b) population decay from α, $K_{\alpha\alpha,\alpha\alpha}$, and (c) decay of the coherence (dephasing) due to population transfer, $K_{\alpha\alpha',\alpha\alpha'}$.

For the linear absorption spectrum, the system evolves in a coherence between the ground and one-exciton states after the interaction with the first pulse. Exciton relaxation occurs during the coherence period and influences the broadening of the absorption spectrum. The linear absorption spectrum of an aggregate in the presence of population transfer is given by

$$I(\omega) = \mathrm{Re} \int_0^{\infty} dt \exp(i\omega t) \sum_{k=1}^{N} \left| d_{gk} \right|^2 \exp\left[-iE_k t - g_{kk}(t) - K_{kk,kk} t \right] \tag{32}$$

where $K_{kk,kk}$ is the population decay rate from the kth level of the one exciton state and E_k is the static energy of the kth exciton state and $g_{kk}(t)$ is the exchange narrowed line-shape function which is given by

$$g_{kk}(t) = C_k g(t) \tag{33}$$

For completely localized states, we have $C_k = 1$. Therefore the line broadening function for the aggregate is the same as that for monomer. In the absence of static disorder, we have $C_k \sim \frac{1}{N}$. In other words, the width of absorption spectrum of the aggregate becomes significantly narrower. The narrowing results from the fact that delocalized exciton states average over the disorder in the transition frequency of the individual molecules. Due to the presence of the disorder in the system, the degree of the delocalization of the exciton in the aggregates is smaller than the actual size of the aggregates.

The procedure for calculating both the linear absorption spectrum and the third-order nonlinear signals is shown schematically in Fig. 24. After diagonalizing the Hamiltonian, we construct the exciton wavefunctions and energies and calculate the transition dipole moments. Then we calculate the exciton population transfer rates from the expression for the Redfield tensor with the spectral density of the phonon. Inserting the solution of Eq. (30) into the linear and third-order response functions and taking into account the finite laser pulse duration, we calculate the third-order nonlinear signals. These procedures are repeated over different sets of static energies of the monomer until our calculated result converges.

The width of the absorption spectrum for the aggregate is significantly narrower than that for monomer as a result of the exchange narrowing. As the value of the static disorder decreases, the exciton becomes more delocalized in the aggregates and the width of the absorption spectrum decreases when we only consider the exciton structure and the exchange narrowing mechanism. Now exciton population transfer contributes to the width of the absorption spectrum via lifetime broadening, and the width of the absorption spectrum no longer depends on the degree of the delocalization. The fast nuclear fluctuations of the monomers appear as exciton energy fluctuation and population transfer in the exciton basis. The energy fluctuations are subject to the exchange narrowing, and the population relaxation produces lifetime broadening. Because

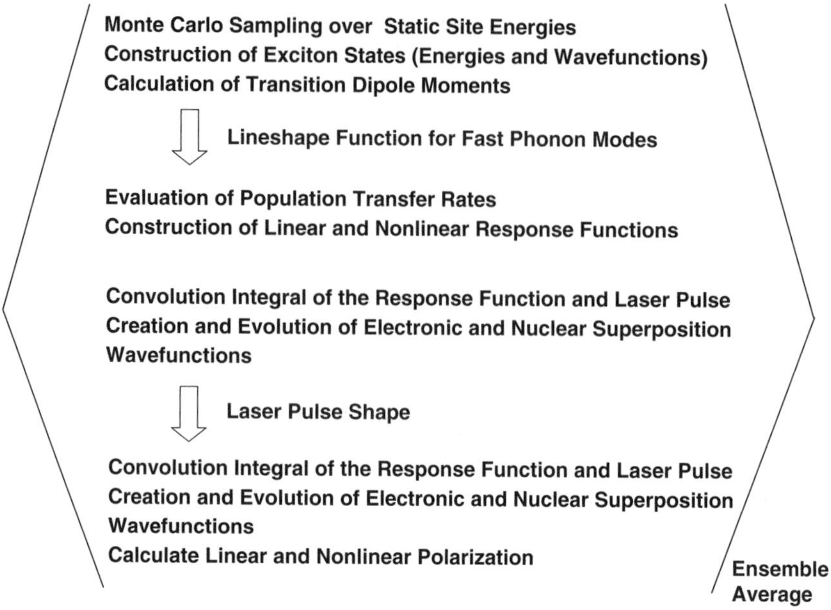

Monte Carlo Sampling over Static Site Energies
Construction of Exciton States (Energies and Wavefunctions)
Calculation of Transition Dipole Moments

Lineshape Function for Fast Phonon Modes

Evaluation of Population Transfer Rates
Construction of Linear and Nonlinear Response Functions

Convolution Integral of the Response Function and Laser Pulse
Creation and Evolution of Electronic and Nuclear Superposition Wavefunctions

Laser Pulse Shape

Convolution Integral of the Response Function and Laser Pulse
Creation and Evolution of Electronic and Nuclear Superposition Wavefunctions
Calculate Linear and Nonlinear Polarization

Ensemble Average

Figure 24. Schematic diagram of the procedure used to calculate the linear absorption and third-order nonlinear signals in molecular exciton systems. See Ref. 124.

of this, in contrast with static disorder (which is completely exchange-narrowed), the dynamical disorder is not completely subject to the exchange narrowing. Knoester and co-workers [126, 127] investigated the effect of the dynamic disorder in the optical line shapes for circular aggregates and arrived at a similar conclusion. For small aggregates, they found that the exchange narrowing factor is equal the number of molecules in the aggregate, while for large aggregates it saturates. The number of molecules at which it saturates depends on the amplitude of the fast fluctuation and the intermolecular coupling.

Figure 25 summarizes the processes included in the calculation in a pictorial fashion. The procedure outlined in Figs. 24 and 25 is combined with the calculated electronic couplings for B850 (Sections III.C and VI.E), a line-shape function obtained by fitting photon-echo peak shift data and static disorder from hole-burning and other experimental methods; the absorption spectrum and photon-echo peak shift decay are calculated for B850, without adjustable parameters [40] (Fig. 26). The agreement with both experimental measures is rather good, including the slow decay of the echo peak shift evident from 200 fs to 1 ps. What do the timescales evident in the peak shift decay of

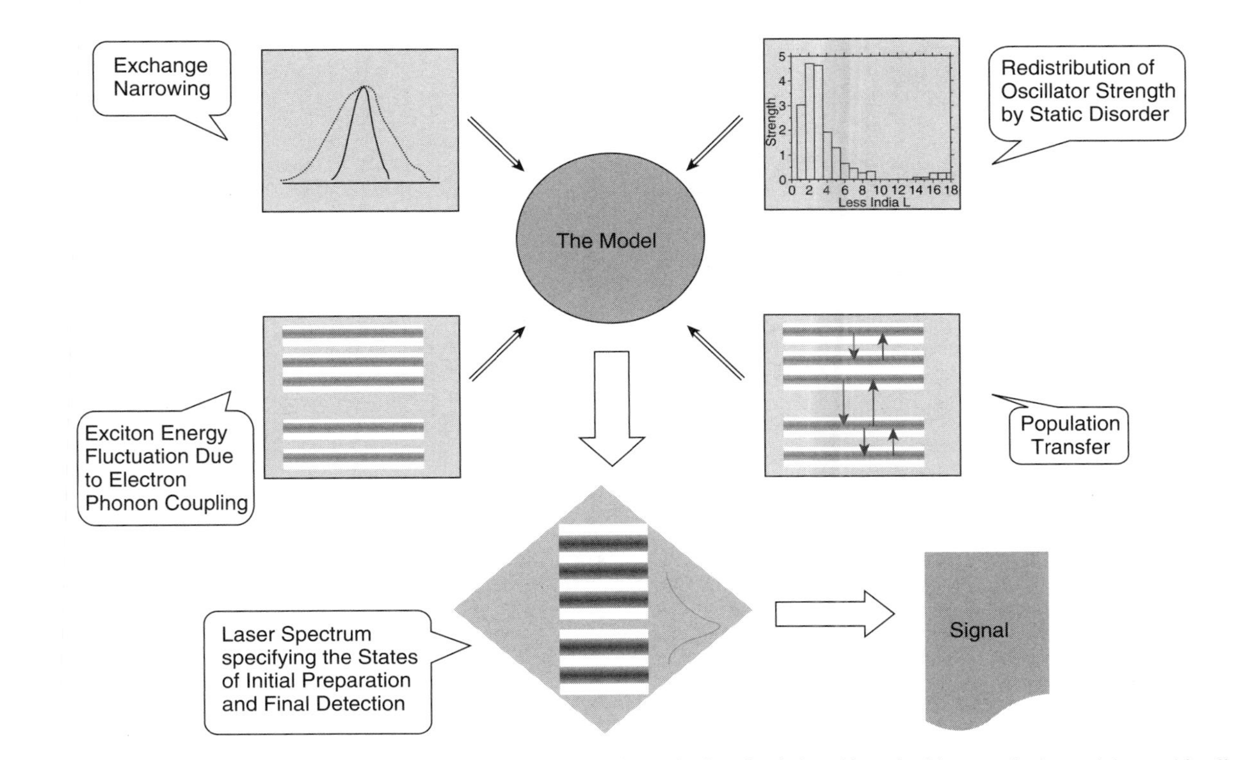

Figure 25. Pictorial representation of the calculation of exciton population and relaxation induced by pulsed laser excitation and the resulting linear and nonlinear optical signals. See text for details.

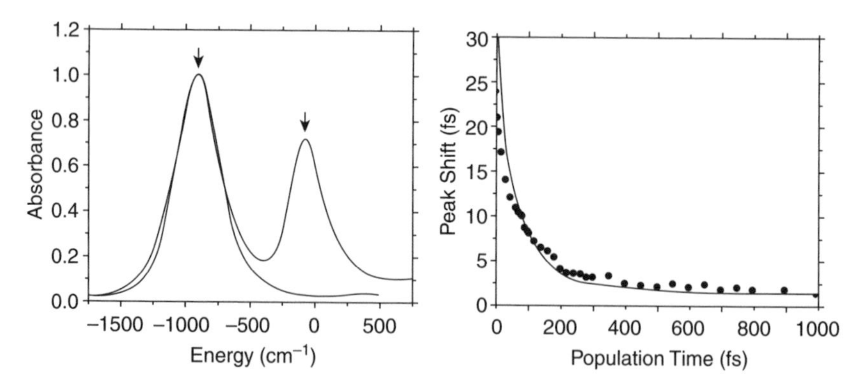

Figure 26. Comparison for the absorption spectrum and three-pulse-echo peak shift determined from experiment for the B850 band of LH2 with that calculated using the model described in the text. Parameters are the same as in Fig. 23 (*right panel*), plus a spectral density (see Ref. 40).

Fig. 26 represent? To investigate this, we consider the following ansatz: The peak shift decay ($\tau^*(\mathrm{T})$ versus T) can be approximated as a product of a term describing the exchange narrowing effect and a term describing population relaxation:

$$\delta\tau^*(T) \approx \frac{\tau^* exchange(T)}{\tau^* exchange(0)} \cdot \delta\xi(T) \tag{34}$$

where

$$\delta A(T) \equiv \frac{A(T) - A(\infty)}{A(0) - A(0)}$$

is a normalized peak shift on population term and

$$\xi(T) \equiv \frac{P_D(T)}{P_D(T) + P_A(T)} \tag{35}$$

where $\tau^*_{exchange}(T)$ is the echo peak shift of the exchange narrowed system when exciton relaxation is turned off, $P_D(T)$ is the weighted (by the oscillator strength and the laser spectrum) sum of population remaining on the initially prepared states, and $P_A(T)$ is the (similarly) weighted sum of the population on states transferred to that are within the laser bandwidth.

The points in Fig. 27 are calculated exactly from the full theory for two different values of the diagonal disorder. The solid lines are calculated according to the ansatz of Eq. (34). For a large value of the disorder, the

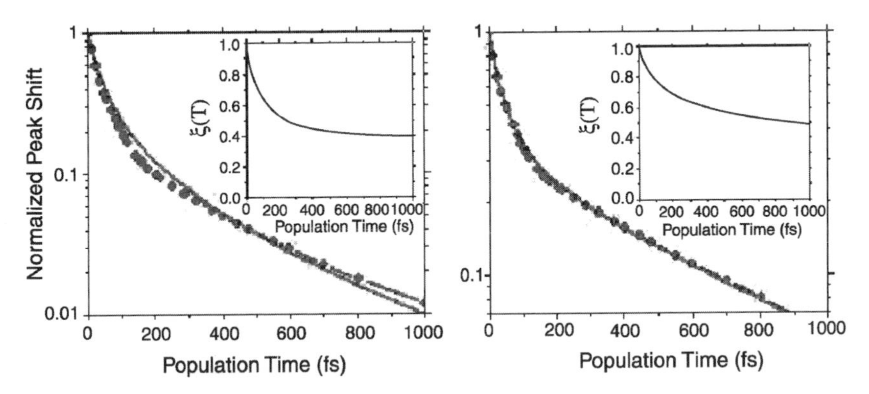

Figure 27. Illustration of the ansatz described by Eq. (34). The points represent the exact calculated photon-echo peak shift, while the solid line is calculated via Eq. (34). The insert shows the population term [Eq. (35)]. The Left panel is for a disorder (σ) of 160 cm^{-1}, and the right panel is for $\sigma = 320\,\text{cm}^{-1}$. Other parameters are as in Figs 23 and 26.

agreement between the exact result and the factored form is quantitative. For the small disorder, the two results deviate slightly, because the disorder is not large enough to entirely decorrelate the energy levels, but still the agreement is very good. The insets show the population relaxation contribution showing that both fact and slow exciton relaxation can be captured by the photon-echo peak shift method.

Finally, it is possible to make a pictorial representation of the exciton dynamics in B850 of LH2. Figure 28 shows the density matrix at 0 fs, 50 fs, and 100 fs in both the exciton basis (upper) and the site basis (lower). In interpreting Fig. 28, it is important to recall that two levels of ensemble averaging are involved in generating the microscopic observable. First, the density matrix approach averages over the fluctuations induced by the phonon modes, and, second, after the phonon-averaged molecular response is calculated, an average over the static disorder is required. The plots in Fig. 28 contain only the first average. Turning to the plots themselves, note that the initial excitation is highly delocalized with two nodes evident in the site basis at the position of monomers whose transition dipoles lie perpendicular to the excitation polarization. Within 50 fs, the excitation becomes localized on groups of 2–4 molecules (site basis), and the coherence evident in the off-diagonal ($k \neq k'$) amplitudes in the exciton basis has almost disappeared. By 100 fs, the exciton representation is almost fully diagonal and the exciton populations (diagonal terms) have redistributed. The clear localization to 2–4 molecules in the site representation suggests that a reasonable physical image of the dynamics can be visualized via Fig. 23. Noting that the Redfield equations for the population relaxation contain the overlap of the wavefunctions of the initial and final states, the picture emerges of excitation

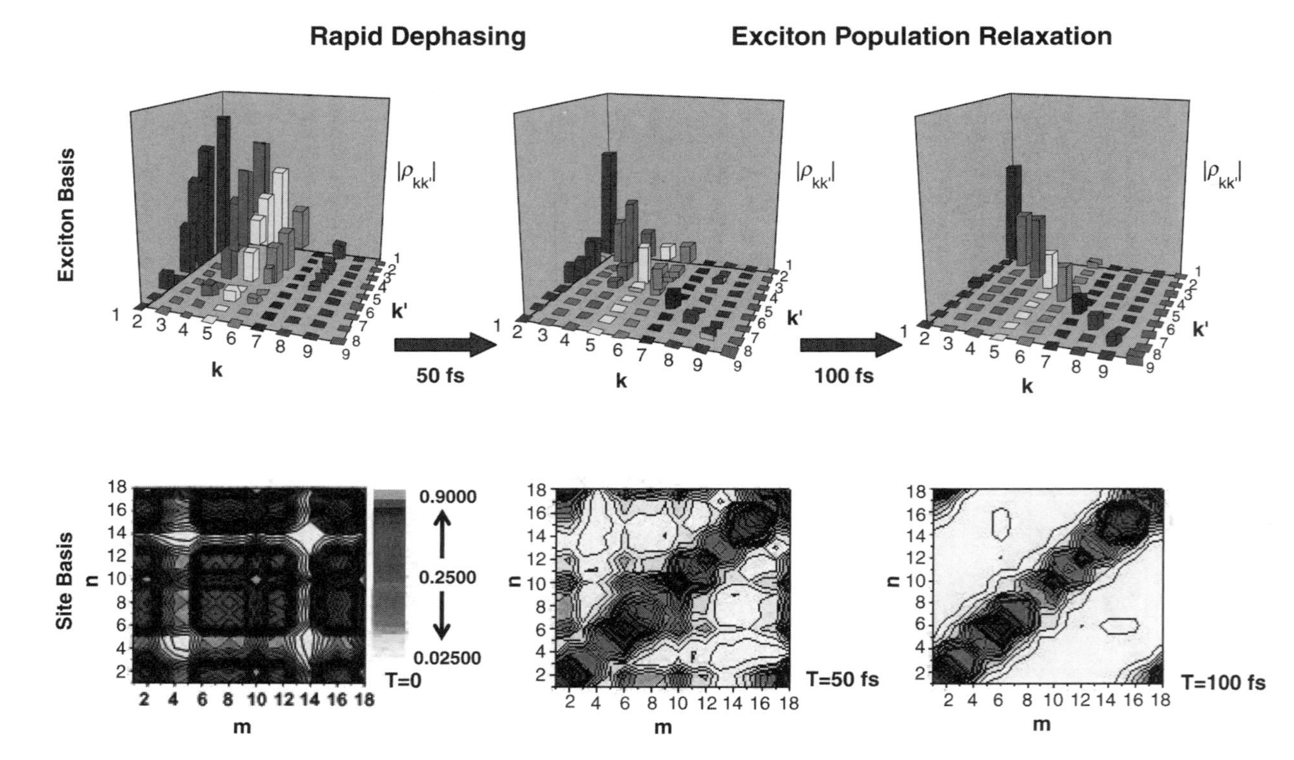

Figure 28. The density matrix as a function of time for a particular realization of the static disorder for the B850 band of LH2 in the exciton basis (*top panels*) and site basis (*bottom panels*). See text. Two nodes initially appear at the position of monomers with perpendicular transition dipole moments to the excitation laser polarization. See color insert.

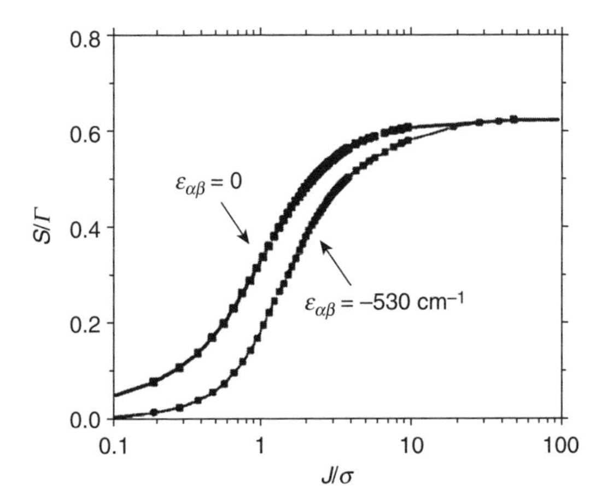

Figure 29. Ratio of exciton population relaxation rate (S) to the exciton pure dephasing rate (Γ) as a function of J/σ, where J is the electronic coupling and σ is the disorder. $\sigma = 150\,\mathrm{cm}^{-1}$ for B850 of LH2. The upper curve is for $\varepsilon_{\alpha\beta} = 0$, and the lower curve is for $\varepsilon_{\alpha\beta} = 550\,\mathrm{cm}^{-1}$. As S/Γ increases, energy transfer becomes more coherent.

"hopping" from one set of 2–4 molecules to an adjacent set of similar size. To what an extent is the hopping description justified. This can be quantified by considering the ratio of the hopping and dephasing rates. In the delocalized exciton picture, electron–phonon coupling is responsible for both the exciton hopping and dephasing processes. While the hopping rate between a pair of exciton states is proportional to spatial overlap of the states, the dephasing rate of an exciton state is proportional to the self-overlap of the spatial distribution of the state. Figure 29 shows the ratios of the two quantities as a function of J/s for a 18-Chl ring when s is fixed at $150\,\mathrm{cm}^{-1}$. For a given value of J/s, we take an average of the ratio over many realizations of static energies which determine spatial distributions of the partially delocalized exciton states. The black symbols are for the case when the mean energy of 18 Chls is identical, and the red symbols are the case when two Chls with an energy difference of $530\,\mathrm{cm}^{-1}$ are arrayed alternately. In both cases, the hopping rate is less than half of the dephasing rate when $J \leq 300\,\mathrm{cm}^{-1}$, which corresponds to the case of LH2. In other words, in the partially delocalized exciton picture, we can ignore coherence transfer, and the incoherent hopping process over the partially delocalized exciton states seems to be a reasonable description of the energy transfer dynamics in LH2.

F. Energy Transfer in PS-I

Figure 30 shows the results of global fits to the fluorescence decay of Photosystem I obtained by Kennis et al. [128] using the upconversion technique.

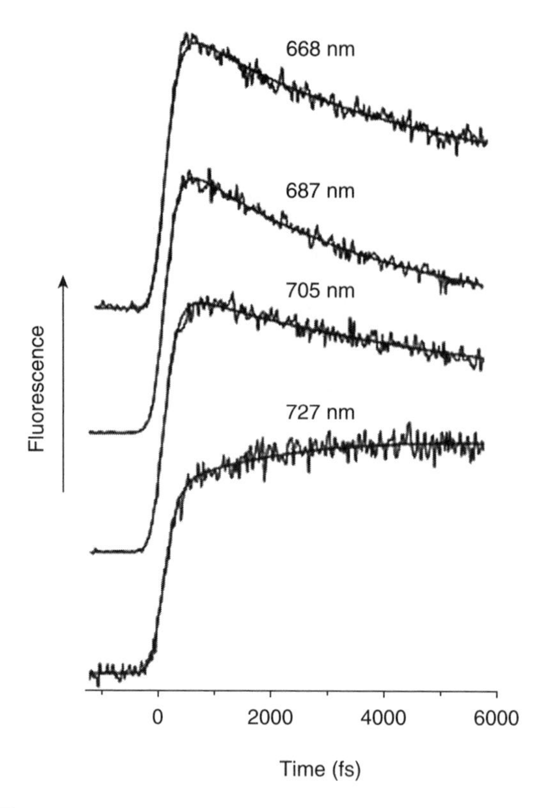

Figure 30. Fluorescence upconversion data from PS-I as a function of emission wavelength. Global fits to the data are also shown. See Ref. 139 for details.

The global fit of data recorded throughout the PSI fluorescence spectrum yielded four exponential components: 360 fs, 3.6 ps, 9.8 ps, and 38 ps. The longest time scale corresponds to the overall trapping time by the reaction center in PSI. A major difficulty of such fitting attempts is knowing whether the other time scales can be ascribed to specific physical processes or result from complex averages of microscopic time scales. In Ref. 128 the 360 fs was assigned to equilibration among Chla pigments in the bulk antenna, while the 3.6-ps component was associated with equilibration between bulk Chla and the red-shifted Chls, which seem unique to PSI. The 9.8 ps may relate to interactions between monomeric PSI units in the naturally occurring trimer.

Using calculated transition frequencies for all 96 Chls, spectral densities from experiment that reflect the inhomogeneity in electron–phonon coupling discussed in Section V.B, and the theoretical formalism described in Section VI.C, Yang et al. calculated the time scales of fluorescence decay in a PSI

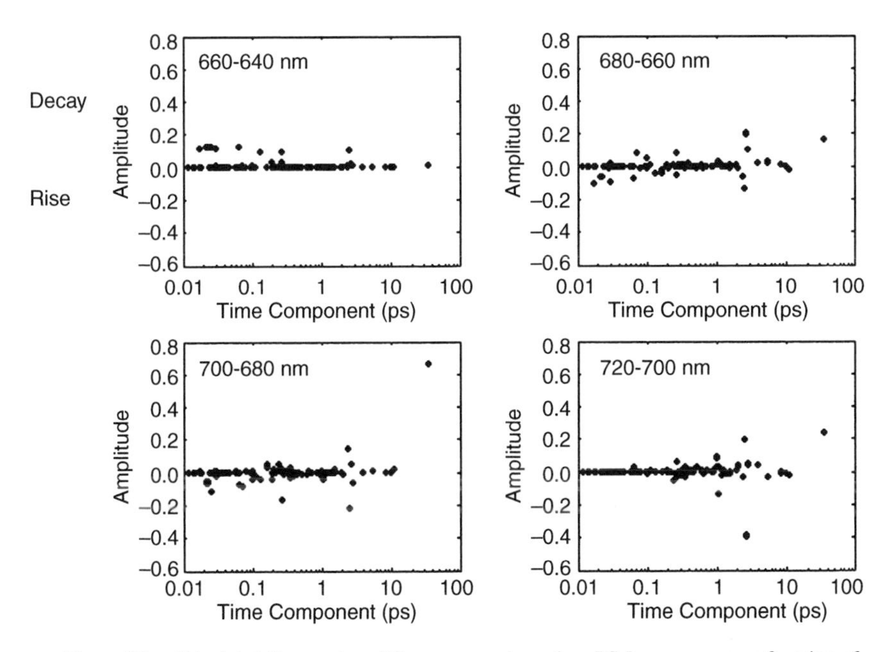

Figure 31. Calculated time scales of fluorescence decay in a PS-I monomer as a function of emission wavelength [97]. Excitation is at 640–660 nm, and the panels show the amplitudes of eigenvalues of the rate matrix for four different detection wavelengths. The amplitudes clearly cluster into four groups: < 100 fs, ~300 fs, 2–3 ps, and 38 ps, with the latter representing the overall trapping time.

monomer following excitation of a subset of Chls on the blue side of the spectrum (640–660 nm). Figure 31 shows the amplitudes and time scales (inverse of the rate matrix eigenvalue) for four different detection windows. Negative amplitudes correspond to a rising, and positive to a decaying, component. The plots in Fig. 31 reveal a small number of clusters of time constants which we divide into four groups: sub-100 fs, 0.3 ps, 2–3 ps, and 35–40 ps. Given that the fluorescence up-conversion study of Kennis et al. most likely did not have the time resolution to obtain the sub-100-fs components, the correspondence with the experimental data is striking. The 9.8-ps component does not appear strongly in the calculated result, suggesting that it is indeed associated with trimer formation.

Analysis of the decay associated spectra (DAS) with 10-nm resolution confirms the physical picture of the various time scales. The shortest time scales correspond to energy flow out of highest-energy Chls. The 0.3-ps component appears as a decay in the 650- to 670-nm windows and as a major rise at 680–700 nm (680 nm is the maximum of the absorption spectrum). The 2- to 3-ps component appears as a rise only in the 690- to 710-nm window and corresponds to a steady flow of excitation from the blue to red via the 660- to 690-nm region during this time period. Finally, the 35- to 40-ps component is obtained as a major decay component at all wavelengths above 670 nm. This demonstrates that a steady state in the spectral distribution is reached before the longest time scale; in this steady state, excitation energy has been depleted in the blue region of the spectrum.

The remarkable consistency of the calculated and experimental time scales suggests that the calculations can be utilized to explore the microscopic details of the energy transfer processes. Such an analysis leads to the following conclusions:

1. The overall trapping time scale in PSI ($\sim$40 ps) has two main contributions: (a) Excitation energy diffusion in the antenna and transfer from the antenna to the RC Chls for the first time, which we refer to as the primary rate-determining step (RDS). This process contributes about 54% of the total time scale. (b) Subsequent processes that lead to the arrival of the excitation at P700 after the excitation has arrived at the RC. We call this the secondary RDS, and it includes energy transfer back to the antenna. The secondary RDS contributes the remaining 46% of the total trapping timescale, and it distinguishes PSI from the LHI/purple bacterial RC system where the equivalent of the primary RDS accounts for essentially the entire trapping time scale. This difference arises from (a) the energetic difference of the primary electron donor (P860) from the remaining RC components, which means that the antenna transfers only to the P860 and that P860 cannot transfer to the other RC components; (b) the lower dimensionality of the purple bacterial system; and (c) the absence of linker Chls in the bacterial system making the final step from LHI to P860 by far the slowest ($\sim$35 ps) in the overall trapping time scale of $\sim$50 ps.
2. Spectral equilibration occurs within the antenna in less than 5 ps and leads to a state characterized as a transfer equilibrium state, rather than a thermodynamic equilibrium state. By this we mean that single exponential fluorescence decay kinetics are observed at all detection wavelength on timescales longer than 5 ps.
3. As described in Section V.A, the energy configuration of the six RC Chls and two "linker" Chls is highly optimized for efficient trapping at P700 by forming a quasi-funnel structure.
4. The energy configuration of the remaining 88 Chls of PSI does not (at room temperature) influence the overall trapping time greatly. This arises from the high connectivity (dimensionality) of the PSI antenna, which mitigates against trapping of excitation on energetically unfavorable sites.

This makes the system very robust with respect to energetic disorder, again in contrast to the purple bacterial system which is quasi-one-dimensional.

5. The orientations of the antenna pigments (via their influence on the Coulombic couplings) do influence the efficiency of trapping to a moderate extent. The model suggests an electron transfer time scale in the range 0.87–1.7 ps from P700 to the primary electron acceptor, and this time scale does not have a strong influence on the overall trapping time scale.

VII PROTECTION AGAINST PHOTOCHEMICAL DAMAGE

Highly reactive, photo-oxidative species are inevitable byproducts of photosynthesis, and plants, cyanobacteria, and photosynthetic bacteria have evolved various mechanisms to deal with this problem. By far the most sophisticated mechanisms exist in green plants. An excess photon flux can exacerbate the damage caused by these intermediates, leading to problems ranging from reversible decreases in photosynthetic efficiency, to, in the worst case, death of the plant. Carotenoid molecules (Cars) constitute a key component of the protection system in all photosynthetic systems. In addition, carotenoids also act as light-harvesting pigments, providing spectral coverage between the Chl Q_y/Q_x and Soret bands. In some species, at least 95% of the excitation absorbed by the carotenoids is transferred to the B(Chls), while in other species significantly lower efficiencies are reported.

A crucial aspect of the photoprotective role of Cars is their ability to efficiently quench chlorophyll triplet states, thereby preventing the formation of excited, singlet oxygen by triplet–triplet energy transfer from its ${}^3\sum_g^-$ ground state:

$$\begin{aligned} {}^3\left({}^1Car\ {}^3Chl^*\right) &\rightarrow {}^3\left({}^3Car^*\ {}^1Chl\right) \\ {}^1\left({}^3O_2\ {}^3Chl^*\right) &\rightarrow {}^1\left({}^1O_2^*\ {}^1Chl\right) \end{aligned} \tag{36}$$

The mechanism of the efficient Chl-Car TT-EET has been investigated by Damjanovic et al. [129], but quantitatively accurate calculations of the electronic coupling have not yet been possible. Both triplet–triplet energy transfer and sensitization of singlet oxygen are mediated by interactions that depend on orbital overlap. Calculations are therefore highly sensitive to the accuracy of the wavefunctions. We note that a purely exchange-mediated interaction only operates when the wavefunctions are orthogonal and do not interpenetrate. The most significant orbital overlap-dependent coupling involves exchange of electrons by coupled, screened one-electron matrix elements [75, 130]. It is straightforward to write down and compare the matrix elements corresponding to the EET processes of scheme (36), but this seems to be of limited practical use until it is possible to quantify such electronic couplings for realistic systems.

The mechanism of the efficient Chl-Car triplet–triplet transfer has been investigated by Damjanovic et al. [131], but quantitatively accurate calculations of the electronic couplings have yet not been possible.

In the singlet manifold, carotenoids have, like all polyenes, an unusual electronic structure: The first excited state (S_1) has the same symmetry, A_g^-, as the ground state, and thus one-photon transitions from S_0 to S_1 are forbidden. In other words, the S_1 state does not appear in the absorption (or emission) spectrum of carotenoids (with more than 9 double bonds), which is dominated by the very strong $S_0 \rightarrow S_2\ (B_u^+)$ transition. Carotenoids also possess a state of B_u^- symmetry, which may lie near S_2, though evidence for the spectroscopic observation of this state remains controversial [132–135]. Finally, some unusual carotenoids with polar substituents, such as peridinin, may also have low-lying charge transfer states [42, 136, 137].

A decade ago it was considered that all carotenoid (B) Chl energy transfer was mediated through the S_1 state of the carotenoid because $S_2 \rightarrow S_1$ internal conversion would be too fast (100–200 fs) to allow significant transfer from S_2. More recently, it was concluded that a significant fraction, up to 100%, of the Car-Chl energy transfer does, in fact, take place from S_2. Although neither the Dexter nor Forster theories provide satisfactory predictions, we showed in Section III.D that Car S_1 states can have significant Coulombic coupling with Chl molecules despite the forbidden nature of the $S_0 \rightarrow S_1$ transition. Thus whether or not Car S_1 states play a role in energy transfer depends on the location of the S_1 state with respect to the (B) Chl Q_x and Q_y states. In the next section, we briefly describe recent experiments to determine the energies of S_1 states of photosynthetic carotenoids.

A. Carotenoids: Energy of the S_1 State

Two different methods have been used recently to determine the S_0–S_1 spectrum of carotenoids. Our group has utilized the two-photon allowed character of the S_0–S_1 transition to populate S_1, followed by subsequent detection of either S_1–S_n absorption or (B) Chl fluorescence resulting from S_1–Q_y energy transfer to determine the spectrum. Figure 32 shows both the energy level scheme and the excitation spectrum result for sphaeroidene, the carotenoid in the LH2 complex of *Rb. sphaeroides*. Note that the S_1 state lies in the region of negligible one-photon absorption between the strongly allowed B800/B850 (Q_y) bands and the Car S_0–S_2 band (Fig. 32). The fact that the S_1 lifetime of sphaeroidene is reduced from a solution value of 9 ps to 1.9 ps in LH2 is clear evidence for efficient $S_1 \rightarrow$ BChl energy transfer. In contrast, the S_1 state of rhodopin glucoside the carotenoid present in LH2 of *Rps. Acidophila* lies at too low an energy for

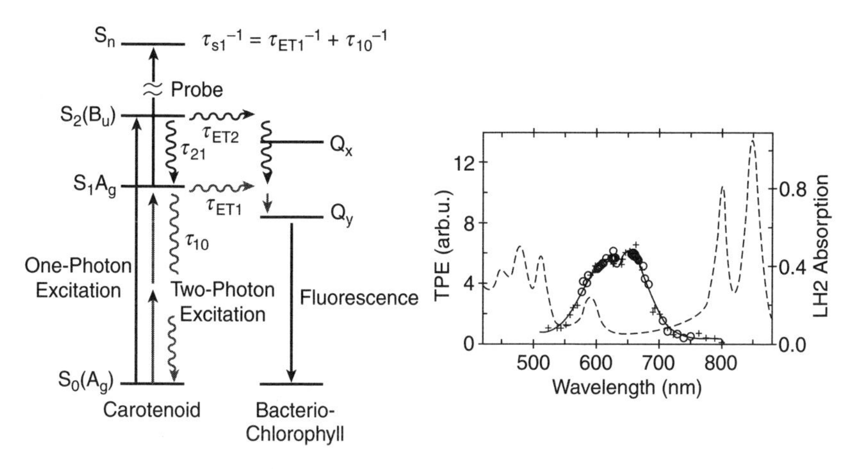

Figure 32. *Left*: Level scheme, excitation, and probing steps for carotenoid–chlorophyll interactions. *Right:* Two-photon excitation spectrum of sphaeroidene obtained by detecting fluorescence from Bchl in LH2 of *Rb. Sphaeroides*. See Ref. 156.

efficient $S_1 \rightarrow Q_y$ transfer and the S_1 lifetime in the complex is essentially unchanged from solution. Table V summarizes the energy transfer time constants and efficiencies for these two species. Such variations in the effectiveness of Car to Chl transfer seem quite common even for the same carotenoid in different contents. For example, the efficiency of β-carotene to Chla transfer is significantly higher in Photosystem I of plants and cyanobacteria than it is in the various light-harvesting complexes associated with Photosystem II in the same species [138–142].

The second method of obtaining the energy of the S_1 state was developed by Polivka, Zigmantis, Sundstrom and co-workers. It involves populating S_2 with an ultrashort laser pulse, allowing S_2–S_1 internal conversion to proceed and then scanning the S_1–S_2 absorption in their near-infrared [143]. The S_1 state energy is then obtained by subtraction. For reasons that are not entirely clear, the two-photon and near-IR probe methods do not always agree precisely on the S_1 state energies, although differences are quite small in most cases. For some

TABLE V
Calculated Car S_1–BChl Energy Transfer Time Constants and Efficiencies

Species	τ_{S1}/ps	τ_{ET1}/ps	ϕ_{ET1}	$\phi_{OA(9,11)}$	ϕ_{ET2}	$\phi_{ET1}\phi_{21}$	ϕ_{21}
Rb. sphaeroides	1.9 ± 0.5	2.4 ± 0.5	80%	>95%	>75%	<20%	<25%
Rps. acidophila	6.5 ± 0.5	>25	<28%	~70%	>60%	<10%	<40%

$\phi_{OA} = \phi_{ET2} + \phi_{ET1}\phi_{21}$

carotenoids, fluorescence and resonance Raman methods have been used to obtain S_0–S_1 energies, but these are extremely difficult to apply in intact photosynthetic complexes.

B. Regulation of Energy Transfer Efficiency

Photosystem II of plants [144] (specifically the D1 protein) is damaged sufficiently to require dismantling and repair in about 30 minutes in bright sunlight. To achieve even this degree of robustness, light harvesting in Photosystem II is highly regulated on both long and short time scales, by which in this context we mean hours versus minutes. The short-term regulation process is called *nonphotochemical quenching* (NPQ), which involves thermal dissipation of excitation energy-absorbed in PSII that exceeds a plant's capacity for CO_2 fixation [145]. Feedback de-excitation or energy-dependent quenching (qE) [146, 147] is the major rapidly reversible component of NPQ in a variety of plants. qE is characterized by a light-induced absorbance change at 535 nm [148], the shortening of the overall chlorophyll fluorescence lifetime (or equivalent reduction in fluorescence yield) [149]. It requires the buildup of a pH gradient across the thylakoid membrane, under conditions of excess light. The pH gradient, in turn, triggers the enzymatic conversion of the carotenoid vioaxanthin (Vio) to zeaxanthin (Zea) via the xanthophylls cycle [150]. In addition, the presence of a specific pigment-binding protein, Psbs (CP22) is essential for *qE* [151].

Currently, two hypotheses concerning the mechanism of *qE* exist, one in which the effect of Zea is solely structural (called *indirect quenching*) and the other in which Zea acts as an energy acceptor for excitation transfer from the Chl Q_y state (called *direct quenching*). Clearly, the direct quenching mechanism depends strongly on the relative Q_y–S_1 energies levels of the Chl and Car molecules. Using the near-IR probing method, Polivka et al. found that in solution both Vio and Zea S_1 states lie below the Chl Q_y energy and could both act as quenchers in principle, although Vio does not [143]. Very recently, Ma et al. [152] found strong evidence for the formation of the Zea S_1 state following Chl excitation, only under conditions of maximum *qE*. This result appears to strongly support the direct quenching mechanism, but much remains to be clarified before a molecular mechanism for NPQ is at hand.

VIII. SUMMARY AND CONCLUSIONS

Early studies—for example, the work of Arnold and co-workers [153, 154] and Duysens [155]—exposed the role of energy transfer in the capture of light by chlorophyll pigments and subsequent transfer to a trap. Such studies were considerably aided by Förster theory, which provided a means to predict energy transfer rates based on simple experimental observables.

In recent years, ultrafast spectroscopies revealed the most rapid energy transfer events in photosynthetic proteins, which could not be readily explained by predictions from Förster theory. As high-resolution structural models became available and more realistic models for light harvesting were explored, it became clear that conventional Förster theory was missing some essential element of the problem. A number of possibilities were discussed, including the possibility that orbital overlap effects were important. Finally, we realized that (i) we needed to think about molecular aggregates differently than molecular pairs—we needed to retain structural information in the model for EET as in the GFT model; (ii) we had to learn how to calculate Coulombic couplings between molecules which led to the development of the TDC method; (iii) we needed to incorporate such ideas into dynamical models for large, complex aggregates in order to simulate various ultrafast spectroscopies, including photon echoes.

At this point, we can suggest some open questions:

1. *The Exact Nature of the Protein as a Phonon Bath and Dielectric Environment.* At this point there exists no satisfactory quantitative description of dielectric screening and local field effects in energy transfer, except in the limit of large donor–acceptor separations (the $1/n^4$ factor). This appears to be an important point to resolve, since the effect on the rate can amount to a factor of ~ 4. A particular challenge will be to calculate or measure medium effects in a protein as opposed to a dielectric continuum.
2. *The Time Scales and Mechanism of "Quasi-coherent" Excitation Hopping Within B850/B875 Rings.* This appears to be an area where simple theory cannot apply. It will be a challenge for experimentalists and theorists to address this issue collaboratively. For example, it is not clear whether linear coupling to a harmonic bath is adequate to describe such systems. For example, it may be necessary to include multiphonon and Duschinsky effects on the dynamics in order to describe the influence of temperature on such systems.
3. *The Signatures of Interactions Between Carotenoids and Chlorophylls in Ultrafast Experiments.* These interactions are not well-characterized at present. The influence of orbital mixing [156] and the potential formation of low-lying charge transfer states between carotenoid and chlorophyll molecules [157] should be detectable spectroscopically. An understanding of these interactions will help to elucidate the role of carotenoids in mediating long-range Chl–Chl energy transfer and in the poorly understood phenomenon of nonphotochemical quenching described in Section VII.B.
4. *The Mechanism of Energy Flow in Photosystem II.* Partly because of the lack of atomic-level structural information and partly because the energy

landscape of Photosystem II is nearly flat, giving no time-scale separation between energy and electron transfer dynamics, the overall energy flow within the Photosystem II supercomplex is not understood in detail. Because of the size and complexity of the entire system—in particular, the need to incorporate regulatory systems—some kind of coarse-graining will almost certainly be necessary. Yang and Fleming [158] have developed a "domain" model that enables the identification of bottlenecks and key time scales in any disordered antenna system. The model requires a rate matrix for the system, but given this, the method provides a systematic way to define compartment models of the type often used intuitively to describe energy transfer in multicomponent systems. Energy flow in PSII will require modeling of EET between pigment–protein complexes. Here questions of excitation delocalization become critical since they define "short"- and "long"-range interactions. It will be interesting to ascertain the role, if any, of molecular aggregates (i.e., Generalized Forster Theory) in this process.

5. *The Experimental Characterization of Spatially and Energetically Disordered (in Both the Diagonal and Off-Diagonal Senses) Energy Transfer Systems.* Despite the great advances in ultrafast spectroscopy, characterizing multicomponent systems where both electronic and electron–phonon couplings are distributed remains a challenge. Multidimensional spectroscopy [159]—in particular, the two-color photon echo [160]—holds promise for significantly more incisive studies of such systems, but considerable development work, both experimentally and theoretically, remains to be done.

Acknowledgments

The work at Berkeley was supported in its entirety by the Director, Office of Science, Office of Basic Energy Sciences, Chemical Sciences Division, of the U.S. Department of Energy under Contract DE-AC03-76SF00098. We are grateful to Dr. Mary Gress for her support of our work. The ideas and results described here represent contributions from many colleagues and co-workers. Our debt to the following people is great: Brent Krueger, Peter Walla, Chao-Ping (Cherri) Hsu, Jenny Yom, Xanthipe Jordanides, Ana Damjanovic, Harsha Vaswani, Nancy Holt, Ying-Zhong Ma, Patricia Linden, John Kennis, Ritesh Agarwal, Bradley Prall, Abbas Rizvi, Mino Yang, Martin Head-Gordon, Petra Fromme, Krishna Niyogi, and Jeffrey Reimers.

References

1. C.-P. Hsu, P. J. Walla, M. Head-Martin, and G. R. Fleming, *J. Phys. Chem. B* **105**, 11016 (2001).
2. R. v. Grondelle, J. P. Dekker, T. Gillbro, and V. Sundstrom, *Biochim. Biophys. Acta* **1187**, 1 (1994).
3. G. R. Fleming, S. A. Passino, and Y. Nagasawa, *Philos. Trans. R. Soc. London A* **356**, 389 (1998).
4. D. L. Dexter, *J. Chem. Phys.* **21**, 836 (1953).

5. G. R. Fleming and R. von Grondelle, *Curr. Opin. Struct. Biol.* **7**, 738 (1997).
6. T. Pullerits and V. Sundstrom, *Acc.Chem. Res.* **29**, 381 (1996).
7. V. Sundstrom, T. Pullerits, and R. von Grondelle, *J. Phys. Chem. B* **103**, 2327 (1998).
8. X. Hu, A. Damjanovic, T. Ritz, and K. Schulten, *Proc. Natl. Acad. Sci. USA* **95**, 5935 (1998).
9. R. G. Alden, E. Johnson, V. Nagarajan, W. W. Parson, C. J. Law, and R. G. Cogdell, *J. Phys. Chem. B* **101**, 4667 (1997).
10. R. Jimenez, S. N. Dikshit, S. E. Bradforth, and G. R. Fleming, **100**, 6825 (1996).
11. S. Tretiak, C. Middleton, V. Chernyak, and S. Mukamel, *J. Phys. Chem. B* **104**, 9540 (2000).
12. W. W. Parson and A. Warshel, *J. Am. Chem. Soc.* **109**, 6152 (1986).
13. Y. Won and R. A. Friesner, *J. Phys. Chem.* **92**, 2208 (1988).
14. E. J. P. Lathrop and R. A. Friesner, *J. Phys. Chem.* **98**, 3056 (1994).
15. M. A. Thompson and M. C. Zerner, *J. Am. Chem. Soc.* **113**, 8210 (1991).
16. G. D. Scholes and G. R. Fleming, *J. Phys. Chem. B* **104**, 1854 (2000).
17. X. J. Jordanides, G. D. Scholes, and G. R. Fleming, *J Phys. Chem. B* **105**, 1652 (2001).
18. G. McDermott, S. M. Prince, A. A. Freer, A. M. Hawthornthwaite-Lawless, M. Z. Papiz, and R. J. Cogdell, *Nature* **374**, 517 (1995).
19. S. Karrasch, P. Bullough, and R. Ghosh, *EMBOJ* **14**, 631 (1995).
20. A. Freer, S. Prince, K. Sauer, M. Papiz, A. Hawthornthwaite-Lawless, G. McDermott, R. Cogdell, and N. W. Isaacs, *Structure* **4**, 449 (1996).
21. J. Koepke, X. Hu, C. Muenke, K. Schulten, and H. Michel, *Structure* **4**, 581 (1996).
22. X. Hu, T. Ritz, A. Damjanovic, and K. Schulten, *J. Phys. Chem. B* **101**, 3854 (1997).
23. X. Hu and K. Schulten, *Biophys. J.* **75**, 683 (1998).
24. K. J. Visscher, H. Bergstrom, V. Sundstrom, C. N. Hunter, and R. von Grondelle, *Photosynth. Res.* **3**, 211 (1989).
25. H. Bergstrom, R. von Grondelle, and V. Sundstrom, *FEBS Lett.* **2**, 503 (1989).
26. A. Freiberg, J. P. Allen, J. C. Williams, and N. W. Woodbury, *Photosynth. Res.* **1–2**, 309 (1996).
27. T. Pullerits, K. J. Visscher, S. Hess, V. Sundstrom, A. Freiberg, K. Timpmann, and R. von Grondelle, *Biophys. J.* **66**, 236 (1994).
28. S. Hess, M. Chachisvilis, K. Timpmann, M. R. Jones, G. J. S. Fowler, C. N. Hunter, and V. Sundstrom, *Proc. Natl. Acad. Sci. USA* **26**, 12333 (1995).
29. V. Nagarajan and W. W. Parson, *Biochemistry* **36**, 2300 (1997).
30. H. Bergstrom, V. Sundstrom, R. von Grondelle, T. Gillbro, and R. Cogdell, *Biochim. Biophys. Acta* **936**, 90 (1988).
31. Y.-Z. Ma, R. J. Cogdell, and T. Gillbro, *J Phys. Chem. B* **101**, 1087 (1997).
32. J. T. M. Kennis, A. M. Streltsov, T. J. Aartsma, T. Nozawa, and J. Amesz, *J. Phys. Chem.* **100**, 2438 (1996).
33. T. Pullerits, S. Hess, J. L. Herek, and V. Sundstrom, *J Phys. Chem. B* **101**, 10560 (1997).
34. A. P. Shreve, J. K. Trautman, H. A. Frank, T. G. Owens, and A. C. Albrecht, *Biochim. Biophys. Acta* **1058**, 280 (1991).
35. S. Hess, F. Feldchtein, A. Babin, I. Nurgaleev, T. Pullerits, A. Sergeev, and V. Sundstrom, *Chem. Phys. Lett.* **216**, 247 (1993).
36. H. J. Kramer, R. V. Grondelle, C. N. Hunter, W. H. J. Westerhaus, and J. Amesz, *Biochim. Biophys. Acta* **765**, 156 (1984).

37. J. M. Salverda, F. von Mourik, G. van der Zwan, and R. von Grondelle, *J Phys. Chem. B* **104** (2000).
38. R. Agarwal, M. Yang, Q.-H. Xu, and G. R. Fleming, *J Phys. Chem. B* **105**, 1187 (2001).
39. H. A. Frank and R. J. Cogdell, *Photochem. Photobiol.* **63**, 257 (1996).
40. M. Yang, R. Agarwal, and G. R. Fleming, *Photochem. Photobiol. Part A* **142**, 107 (2001).
41. Y. Koyama, M. Kuki, P.-O. Andersson, and T. Gillbro, *Photochem. Photobiol.* **63**, 243 (1996).
42. D. Zigmantas, R. G. Hiller, V. Sundstrom, and T. Polivka, *Proc. Natl. Acad. Sci. USA* **99**, 16760 (2002).
43. J. Breton, J.-L. Martin, A. Migus, A. Antonetti, and A. Orszag, *Proc. Natl. Acad. Sci. USA* **83**, 5121 (1986).
44. J. Breton, J.-L. Martin, G. R. Fleming, and J.-C. Lambry, *Biochemistry* **27**, 8276 (1988).
45. Y. W. Jia, D. M. Jonas, T. H. Joo, Y. Nagasawa, M. J. Lang, and G. R. Fleming, *J. Phys. Chem.* **99**, 6263 (1995).
46. D. M. Jonas, M. J. Lang, Y. Nagasawa, S. E. Bradforth, S. N. Dikshit, R. Jiminez, T. Joo, and G. R. Fleming, in *Proceedings of the Feldafing III Workshop, Munich* (1995).
47. R. J. Stanley, B. King, and S. G. Boxer, *J. Phys. Chem.* **100**, 12052 (1996).
48. S. Lin, A. K. W. Taguchi, and N. W. Woodbury, *J. Phys. Chem.* **100**, 17067 (1996).
49. G. Haran, K. Wynne, C. C. Moser, P. L. Dutton, and R. M. Hochstrasser, *J. Phys. Chem.* **100**, 5562 (1996).
50. M. H. Vos, J. Breton, and J. L. Martin, *J. Phys. Chem. B* **101**, 9820 (1997).
51. B. A. King, R. J. Stanley, and S. G. Boxer, *J. Phys. Chem. B* **101**, 3644 (1997).
52. S. I. E. Vulto, A. M. Streltsov, A. Y. Shkuropatov, V. A. Shuvalov, and T. J. Aartsma, *J. Phys. Chem. B* **101**, 7249 (1997).
53. D. C. Arnett, C. C. Moser, P. L. Dutton, and N. F. Scherer, *J. Phys. Chem. B* **103**, 2014 (1999).
54. B. A. King, T. McAnaney, A. deWinter, and S. G. Boxer, *J. Phys. Chem. B* **104**, 8895 (2000).
55. H. Michel, *J. Mol. Biol.* **158**, 567 (1982).
56. J. Deisenhofer, O. Epp, I. Sinning, and H. Michel, *J. Mol. Biol.* **246**, 429 (1995).
57. C. H. Chang, O. Elkabbani, D. Tiede, J. Norris, and M. Schiffer, *Biochemistry* **30**, 5352 (1991).
58. U. Ermler, G. Fritzsch, S. K. Buchanan, and H. Michel, *Structure* **2**, 925 (1994).
59. L. Slooten, *Biophys. Biochim. Acta* **256**, 452 (1972).
60. T. Förster, *Ann. Phys.* **6**, 55 (1948).
61. S. Mukamel, *Principles of Nonlinear Optical Spectroscopy*, Oxford University Press, New York, 1995.
62. B. W. van der Meer, G. I. Coker, and S.-Y. Chen, *Resonance Energy Transfer, Theory and Data*, VCH Publishers, New York, 1994.
63. G. D. Scholes, X. J. Jordanides, and G. R. Fleming, *J. Phys. Chem. B.* **105**, 1640 (2001).
64. G. D. Scholes, *Annu. Rev. Phys. Chem.* **54**, 57 (2003).
65. G. J. S. Fowler, W. Crielaard, R. W. Visschers, R. von Grondelle, and C. N. Hunter, *Photochem. Photobiol.* **57**, 2 (1993).
66. A. Gall, G. J. S. Fowler, C. N. Hunter, and B. Robert, *Photochem. Photobiol.* **36**, 16282 (1997).
67. G. J. S. Fowler, R. W. Visschers, G. G. Grief, R. von Grondelle, and C. N. Hunter, *Nature* **355**, 848 (1992).
68. R. K. and R. M. Hochstrasser, *J Chem. Phys.* **109**, 855 (1998).
69. T. P. and A. Freiberg, *Chem. Phys.* **149**, 409 (1991).

70. M. Beauregard, I. Martin, and A. R. Holzwarth, *Biochim. Biophys. Acta* **1060**, 271 (1991).
71. S. Hess, E. Akesson, R. J. Cogdell, T. Pullerits, and V. Sundstrom, *Biophys. J.* **69**, 2211 (1995).
72. S. Jang, S. E. Dempster, and R. J. Silbey, *J Phys. Chem. B* **105**, 6655 (2001).
73. B. P. Krueger, G. D. Scholes, and G. R. Fleming, *J. Phys. Chem. B* **102**, 5378 (1998).
74. G. D. Scholes, I. R. Gould, R. J. Cogdell, and G. R. Fleming, *Phys. Chem. B* **103**, 2543 (1999).
75. G. D. Scholes and K. P. Ghiggino, *J. Phys. Chem.* **98**, 4580 (1994).
76. R. McWeeny, *Methods of Molecular Quantum Mechanics*, 2nd ed., Academic Press, London, 1992.
77. G. D. Scholes, R. D. Harcourt, and G. R. Fleming, *J. Phys. Chem. B* **101**, 7302 (1997).
78. D. P. Craig and T. Thirunamachandran, *Molecular Quantum Electrodynamics*, Academic Press, New York, 1984.
79. R. J. Buehler and J. O. Hirschfelder, *Phys. Rev.* **83**, 628 (1951).
80. A. D. Buckingham, *Intermolecular Forces—From Diatomics to Biopolymers*, Wiley, New York, 1978.
81. F. London, *J. Phys. Chem.* **46**, 305 (1942).
82. D. Beljonne, G. Pourtois, C. Silva, E. Hennebicq, L. M. Herz, R. H. Friend, G. D. Scholes, S. Setayesh, K. Mullen, and J. L. Bredas, *Proc. Natl. Acad. Sci. USA* **99**, 10982 (2002).
83. P. J. Walla, P. A. Linden, C.-P. Hsu, G. D. Scholes, and G. R. Fleming, *Proc. Natl. Acad. Sci. USA* **97**, 10808 (2000).
84. R. Mulliken, *J. Chem. Phys.* **23**, 1833 (1955).
85. R. Mulliken, *J. Chem. Phys.* **23**, 2343 (1955).
86. S. Tretiak and S. Mukamel, *Chem. Rev.* **102**, 3171 (2002).
87. S. Mukamel, S. Tretiak, T. Wagersreiter, and V. Chernyak, *Science* **277**, 781 (1997).
88. A. J. Hoff and J. Deinsenhofer, *Phys. Rep.* **287**, 1 (1997).
89. C. A. Wraight and R. K. Clayton, *Biochim. Biophys. Acta* **333**, 246 (1974).
90. R. K. Clayton, *Photochem. Photobiol.* **1**, 201 (1962).
91. J. A. Jackson, S. Lin, A. K. W. Taguchi, J. C. Williams, J. P. Allen, and N. W. Woodbury, *J. Phys. Chem. B* **101**, 5747 (1997).
92. X. Jordanides, G. D. Scholes, W. A. Shapley, J. R. Remers, and G. R. Fleming, *J. Phys. Chem. B* **108**, 1753 (2004).
93. J. R. Reimers and N. S. Hush, *J. Am. Chem. Soc.* **117**, 1302 (1995).
94. R. S. Knox and H. von Amerongen, *J Phys. Chem. B* **106**, 5289 (2002).
95. J. D. Dow, *Phys. Rev.* **174**, 962 (1968).
96. C.-P. Hsu, M. Head-Gordon, T. Head-Gordon, and G. R. Fleming, *J. Chem. Phys.* **114**, 3065 (2001).
97. M. Yang, A. Damjanovic, H. Vaswani, and G. R. Fleming, *Biophys. J.* **85**, 1 (2003).
98. R. Agarwal, A. H. Rizvi, B. S. Prall, J. D. Olsen, C. N. Hunter, and G. R. Fleming, *J. Phys. Chem. A.* **106**, 7573 (2002).
99. A. Damjanovic, H. M. Vaswani, P. Fromme, and G. R. Fleming, *J. Phys. Chem. B.* **106**, 10251 (2002).
100. V. Zazubovich, S. Matsuzuki, T. W. Johnson, J. M. Hayes, P. R. Chitnis, and G. J. Small, *Chem. Phys.* **275**, 47 (2002).
101. I. R. Mercer, I. R. Gould, and D. R. Klug, *J. Phys. Chem. B.* **103**, 7720 (1999).
102. G. Zucchelli, R. C. Jennings, F. M. Garlaschi, G. Cinque, R. Bassi, and O. Cremonesi, *Biophysical Journal* **82**, 378 (2002).

103. T. Pullerits and A. Freiberg, *Chem. Phys.* **149**, 409 (1991).

104. T. Pullerits and A. Freiberg, *Biophys. J.* **63**, 879 (1992).

105. S. E. Bradforth, R. Jimenez, F. von Mourik, R. von Grondelle, and G. R. Fleming, *J. Phys. Chem.* **99**, 16179 (1995).

106. J. M. Jean, C. K. Chan, and G. R. Fleming, *Isr. J. Chem.* **28**, 169 (1988).

107. R. Agarwal, B. P. Krueger, G. D. Scholes, M. Yang, J. Yom, L. Mets, and G. R. Fleming, *J. Phys. Chem. B* **104**, 2908 (2000).

108. O. Kuhn and V. Sundstrom, *J Chem. Phys.* **107**, 4154 (1997).

109. A. G. Redfield, *Adv. Magn. Reson.* **1**, 1 (1965).

110. W. T. Pollard, A. K. Felts, and R. A. Friesner, *Adv. Chem. Phys.* **93**, 77 (1996).

111. J. M. Jean, R. A. Friesner, and G. R. Fleming, *J Chem. Phys.* **96**, 5827 (1992).

112. J. M. Jean and G. R. Fleming, *J Chem. Phys.* **103**, 2092 (1995).

113. A. K. Felts, W. T. Pollard, and R. A. Friesner, *J. Phys. Chem.* **99**, 2929 (1995).

114. M. Morillo, C. Denk, and R. I. Cukier, *Chem. Phys.* **212**, 157 (1996).

115. J. M. Jean, *J. Chem. Phys.* **104**, 5638 (1996).

116. J. M. Jean, *J. Phys. Chem. A* **102**, 7549 (1998).

117. C. F. Jen and A. Warshel, *J. Phys. Chem. A* **103**, 11378 (1999).

118. P. Herman and I. Barvik, *J. Phys. Chem. B* **103**, 10892 (1999).

119. P. Herman and I. Barvik, *J. Lumin.* **83-84**, 247 (1999).

120. D. Kilin, U. Kleinekathofer, and M. Schreiber, *J. Phys. Chem. A* **104** (2000).

121. M. Yang and G. R. Fleming, *Chem. Phys.* **282**, 161 (2002).

122. J. L. Herek, N. J. Fraser, T. Pullerits, P. Martinsson, T. Polivka, H. Scheer, R. J. Cogdell, and V. Sundstrom, *Biophys. J.* **78**, 2590 (2000).

123. E. W. Knapp, *Chem. Phys.* **85**, 73 (1984).

124. K. Ohta, M. Yang, and G. R. Fleming, *Chem. Phys.* **115**, 7609 (2001).

125. H. B. Thacker, *Rev. Mod. Physics* **53**, 253 (1981).

126. M. Wubs and J. Knoester, *Chem. Phys. Lett.* **284**, 63 (1998).

127. L. D. Bakalis, M. Coca, and J. Knoester, *J. Chem. Phys.* **110**, 2208 (1999).

128. J. T. M. Kennis, B. Gobets, I. H. M. von Stokkum, J. P. Dekker, R. von Grondelle, and G. R. Fleming, *J. Phys. Chem.* **105**, 4485 (2001).

129. A. Damjanovic, I. Kosztin, U. Kleinekathoefer, and K. Schulten, *Phys. Rev. E.* **65**, 031919 (2002).

130. G. D. Scholes, R. D. Harcourt, and K. P. Ghiggino, *J. Chem. Phys.* **101**, 10521 (1994).

131. A. Damjanovic, T. Ritz, and K. Schulten, *Phys. Rev.* **59**, 3293 (1999).

132. T. Sashima, H. Nagae, M. Kuki, and Y. Koyama, *Chem. Phys. Lett.* **299**, 187 (1999).

133. T. Sashima, Y. Koyama, T. Yamada, and H. Hashimoto, *J. Phys. Chem. B* **104**, 5011 (2000).

134. M. Yoshizawa, H. Aoki, and H. Hashimoto, *Phys. Rev. B* **63**, 180301 (2001).

135. G. Cerullo, D. Polli, G. Lanzani, S. D. Silvestri, H. Hashimoto, and R. J. Cogdell, *Science* **298**, 2395 (2002).

136. D. Zigmantas, T. Polivka, R. G. Hiller, A. Yartsev, and V. Sundstrom, *J. Phys. Chem. A.* **105**, 10296 (2001).

137. J. A. Bautista, R. E. Connors, B. B. Raju, R. G. Hiller, F. P. Sharples, D. Gosztola, M. R. Wasielewski, and H. A. Frank, *J. Phys. Chem. B* **103**, 8751 (1999).

138. J. T. M. Kennis, *Unpublished results.*
139. B. Gobets, J. T. M. Kennis, J. A. Ihalainen, M. Brazzoli, R. Croce, I. H. M. von Stokkum, R. Bassi, J. P. Dekker, H. von Amerongen, G. R. Fleming, and R. von Grondelle, *J. Phys. Chem. B* **105**, 10132 (2001).
140. J. van der Lee, D. Bald, S. L. S. Kwa, R. von Grondelle, M. Rogner, and J. P. Dekker, *Photosynth. Res.* **34**, 311 (1993).
141. F. L. der Weerd, I. H. M. von Stokkum, H. von Amerongen, J. P. Dekker, and R. von Grondelle, *Biophys. J.* **82**, 1586 (2002).
142. R. J. Vandorssen, J. Breton, and J. J. P. et al., *Biochim. Biophys. Acta* **893**, 267 (1987).
143. T. Polivka, J. L. Herek, D. Zigmantas, H.-E. Akerlund, and V. Sundstrom, *P. Natl. Acad. Sci. USA* **96**, 4914 (1999).
144. B. Andersson, A. H. Salter, I. Virgin, I. Vass, and S. Styring, *J. Photochem. Photobiol. B* **15**, 15 (1992).
145. K. K. Niyogi, *Annu. Rev. Plant Phys.* **50**, 333 (1999).
146. P. Horton, A. V. Ruban, and R. G. Walters, *Annu. Rev. Plant Phys.* **47**, 655 (1996).
147. P. Muller, X. P. Li, and K. K. Niyogi, *Plant Physiol.* **125**, 1558 (2001).
148. A. M. Gilmore, *Physiol. Plant.* **99**, 197 (1997).
149. A. M. Gilmore, T. L. Hazlett, and Govindjee, *Proc. Natl. Acad. Sci. USA* **92**, 2273 (1995).
150. A. M. Gilmore and H. Y. Yamamoto, *P. Natl. Acad. Sci. USA* **5**, 1899 (1992).
151. X. P. Li, O. Bjorkman, and C. Shih, *Nature* **403**, 391 (2000).
152. Y.-Z. Ma, N. E. Holt, X. P. Li, K. K. Niyogi, and G. R. Fleming, *Proc. Natl. Acad. Sci. USA* **100**, 4377 (2003).
153. R. Emerson and W. Arnold, *J. Gen. Physiol.* **16**, 191 (1932).
154. W. Arnold and J. K. Oppenheimer, *J. Gen. Physiol.* **33**, 423 (1949).
155. L. N. M. Duysens, *Thesis. Leiden.* 1952.
156. B. P. Krueger, J. Yom, P. J. Walla, and G. R. Fleming, *Chem. Phys. Lett.* **310**, 57 (1999).
157. A. Dreuw, G. R. Fleming, and M. Head-Gordon, *Phys. Chem. Chem. Phys.* **5**, 3247 (2003).
158. M. Yang and G. R. Fleming, *J. Chem. Phys.* **119**, 5614 (2003).
159. H. M. Vaswani, C.-P. Hsu, M. Head-Gordon, and G. R. Fleming, *J. Phys. Chem. B* **107**, 7940 (2003).
160. R. Agarwal, B. S. Prall, A. H. Rizvi, M. Yang, and G. R. Fleming, *J Chem. Phys.* **116**, 6243 (2002).
161. L. J. Kaufman, D. A. Blank, and G. R. Fleming, *J. Chem. Phys.* **114**, 2312 (2001).
162. P. Jordan, P. Fromme, H. T. Witt, O. Klukas, W. Saenger, and N. Krauss, *Nature* **411**, 909 (2001).
163. M. Yang and G. R. Fleming, *Chem. Phys.* **282**, 161 (2002).

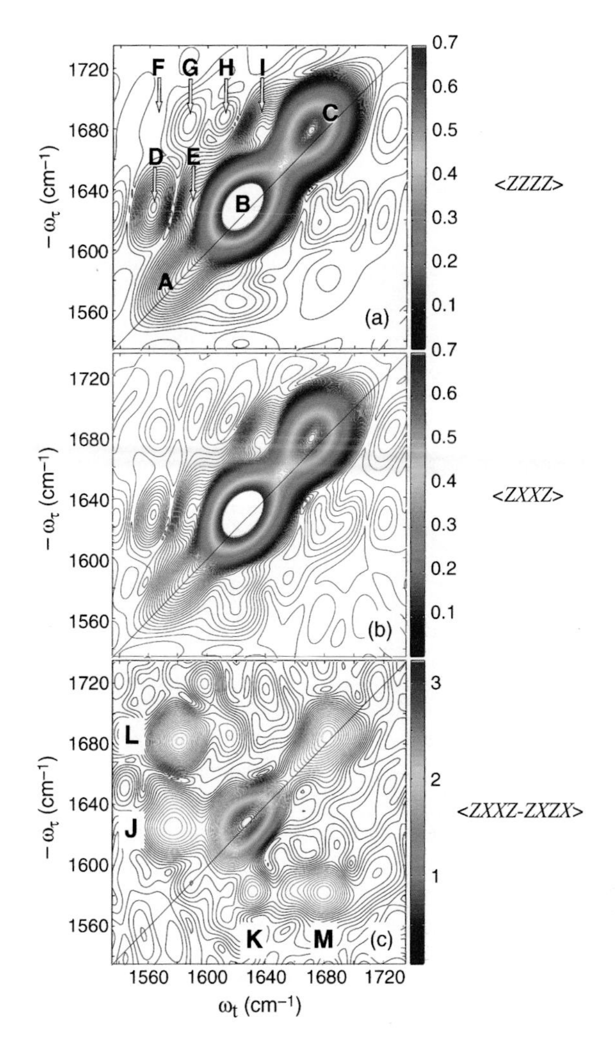

Figure 6. The 2D-IR spectrum of acylproline for three polarization conditions. These three measurements represent all possible orientational information on this isotropic system. This figure is adapted from published data (see References. 7 and 43).

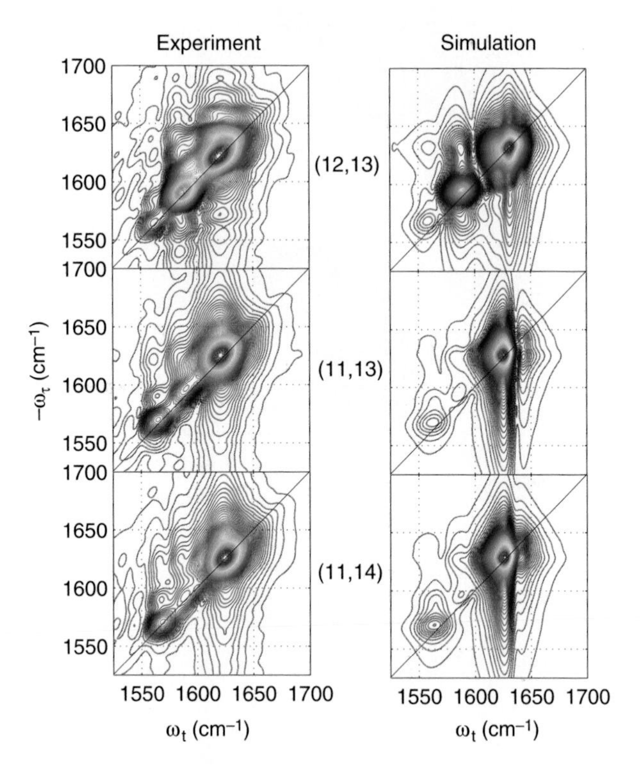

Figure 10. The absolute magnitudes of the experimental 2D-IR spectra of a 25-residue α-helix with carbonyl isotopic substitution at residue positions numbers [$^{13}C{=}^{16}O$, $^{13}C{=}^{18}O$]. Simulated spectra are shown to the right of each spectrum. (After Refs. 112 and 113.)

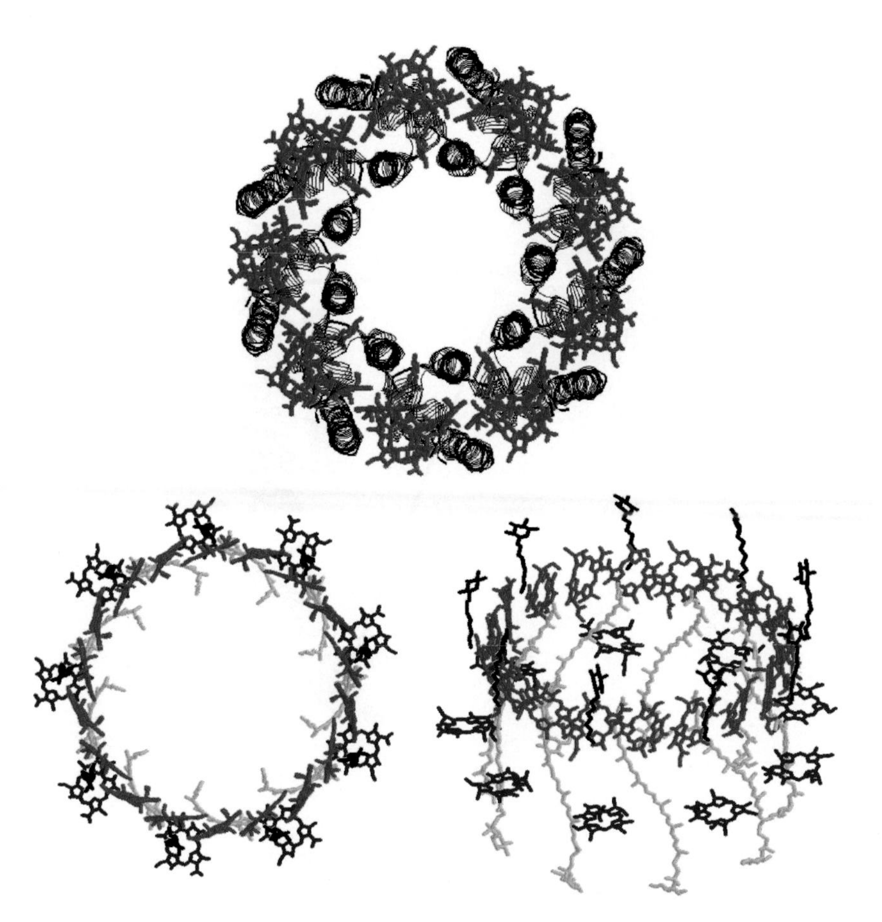

Figure 1. Illustration of the structure of the peripheral light-harvesting complex LH2 of the purple bacterium *Rps. acidiphila* strain 10050 [18]. The top view with α-helices represented as ribbons is shown at the top of the figure. The same view, but without the protein, leaving just the bacteriochlorophyll and carotenoid pigments, is shown at the lower left. On the lower right, this structure is shown tilted on its side, revealing the upper B850 ring of 18 Bchl pigments, the lower B800 ring of 9 Bchl pigments, and the carotenoids that weave their way between these rings.

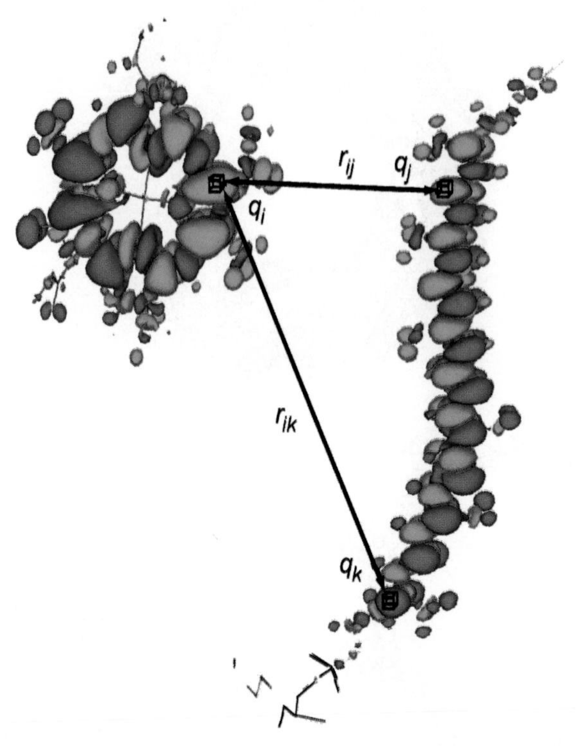

Figure 7. Transition densities calculated for a Bchl molecule and a carotenoid. Density elements, containing charge q_i, q_j, and so on, are depicted together with their corresponding separation r_{ij}. Summing the Coulombic interaction between all such elements gives the total Coulombic interaction, which, according to the TDC method, promotes energy transfer.

Figure 9. A comparison of transition densities for rhodopin glucoside calculated using TDDFT (6-31 + + g** basis set). On the right the $S_0 \rightarrow S_2$ transition is shown, with its large dipole transition moment being evidenced by the change in sign of this TD from one end of the molecule to the other. On the left the $S_0 \rightarrow S_1$ transition is shown. The symmetry of the TD causes the transition to be optically forbidden.

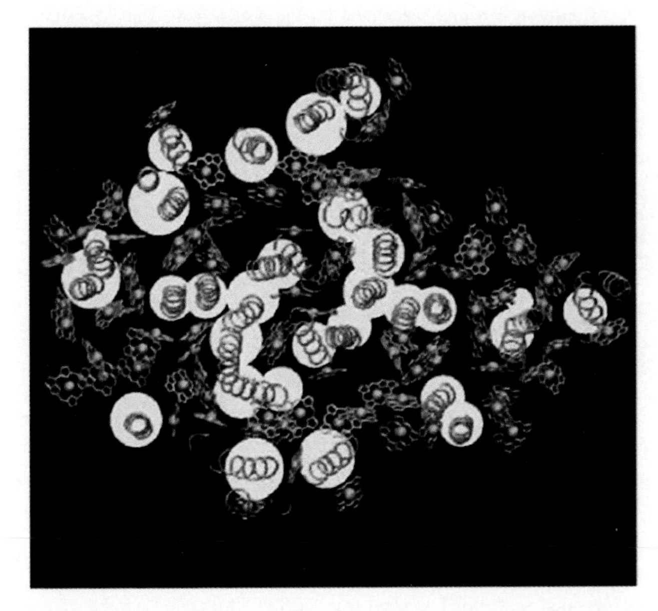

Figure 10. Dielectric model of the protein. Within this model, the protein medium (i.e., the medium with the refractive index of $n = 1.2$) is represented with a set of cylinders. The cross section of these cylinders is shown with white circles. The real location of the transmembrane part of α-helices in PSI are indicated by coiled structures. Chlorophylls are presented as Mg-chlorin rings, lacking the phytyl tail. Chlorophyll Mg atoms are shown in van der Waals representation.

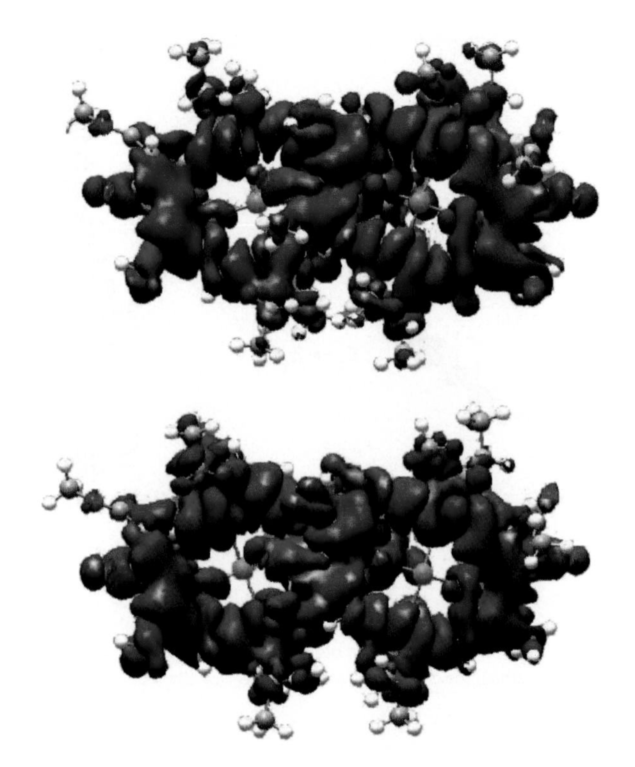

Figure 17. Transition densities calculated for the special pair. *Top*: Transition densities for P_-, *Bottom*: Transition density for P_+.

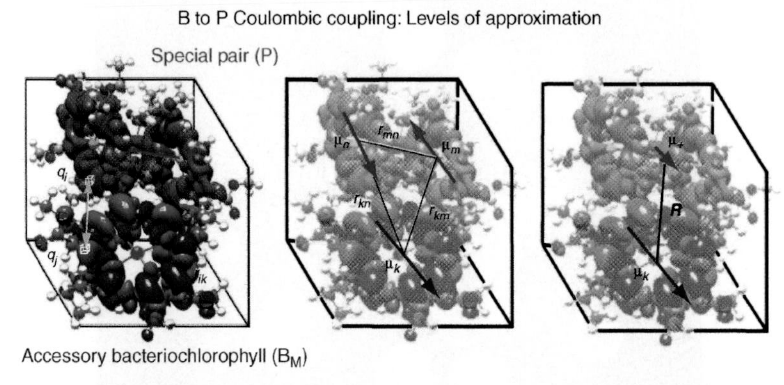

Figure 18. An illustration of the levels of approximation used in estimating the B to P_+ electronic coupling. (*a*) An essentially "exact" calculation can be made using the TDC method. (*b*) Distributed dipoles used in the GFT method (see Section VI.B) represent the minimal acceptable approximation. (*c*) The harsh dipole approximation, in which the correct physical picture of the system is completely washed away.

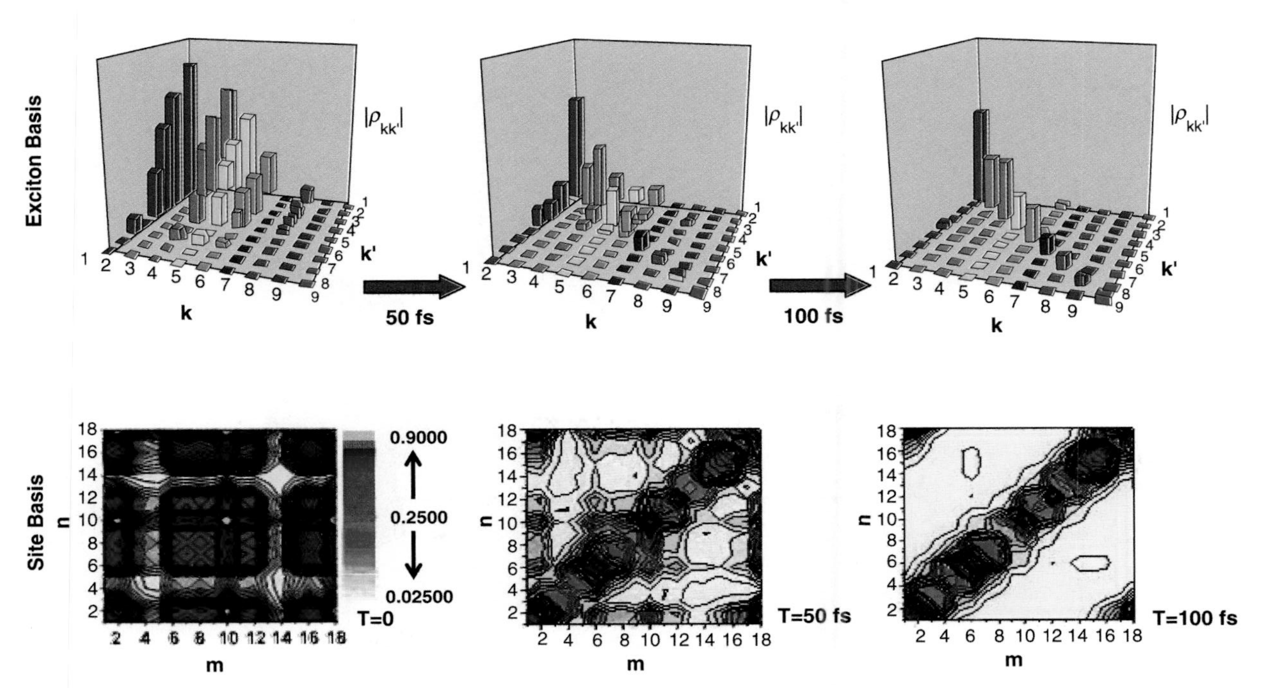

Figure 28. The density matrix as a function of time for a particular realization of the static disorder for the B850 band of LH2 in the exciton basis (*top panels*) and site basis (*bottom panels*). See text. Two nodes initially appear at the position of monomers with perpendicular transition dipole moments to the excitation laser polarization.

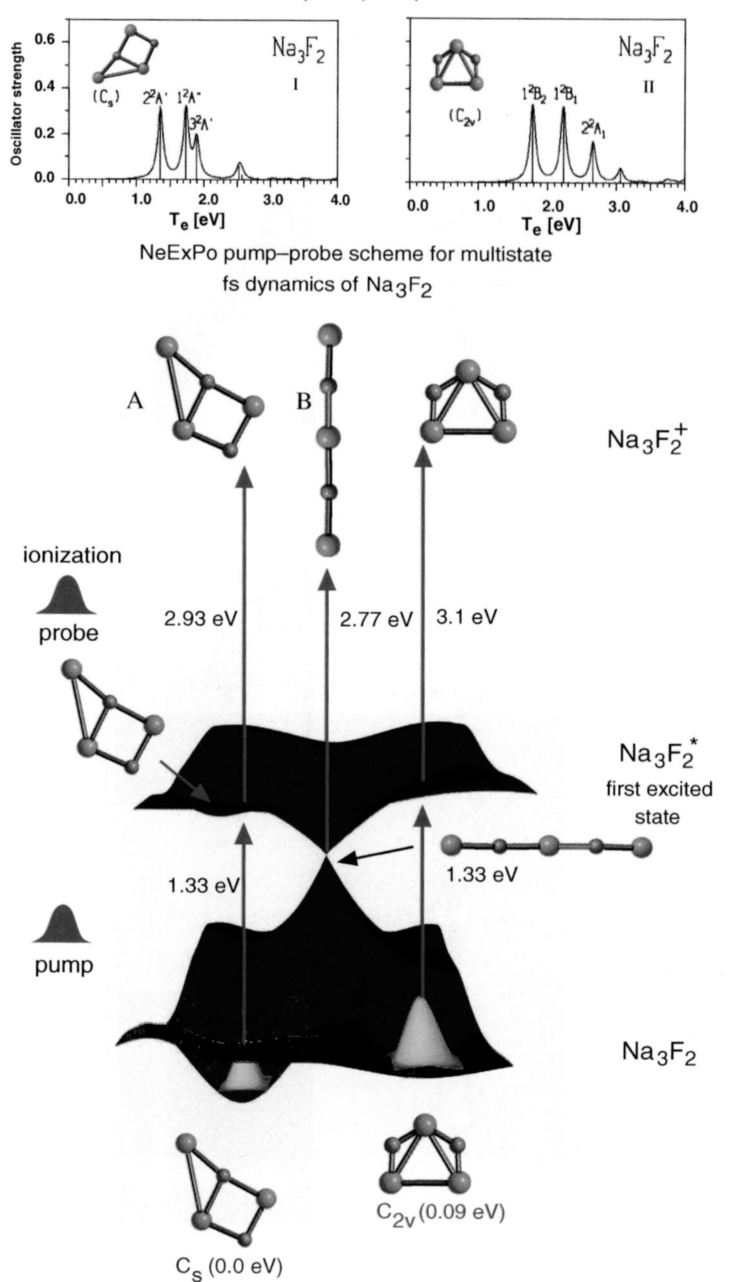

Figure 8. Absorption spectra for two isomers I and II of Na_3F_2 obtained from one electron "frozen ionic bonds" approximation [46] (upper part). Scheme of the multistate fs dynamics for NeExPo pump-probe spectroscopy of Na_3F_2 including conical intersection with structures and energy intervals for the pump and probe steps [46].

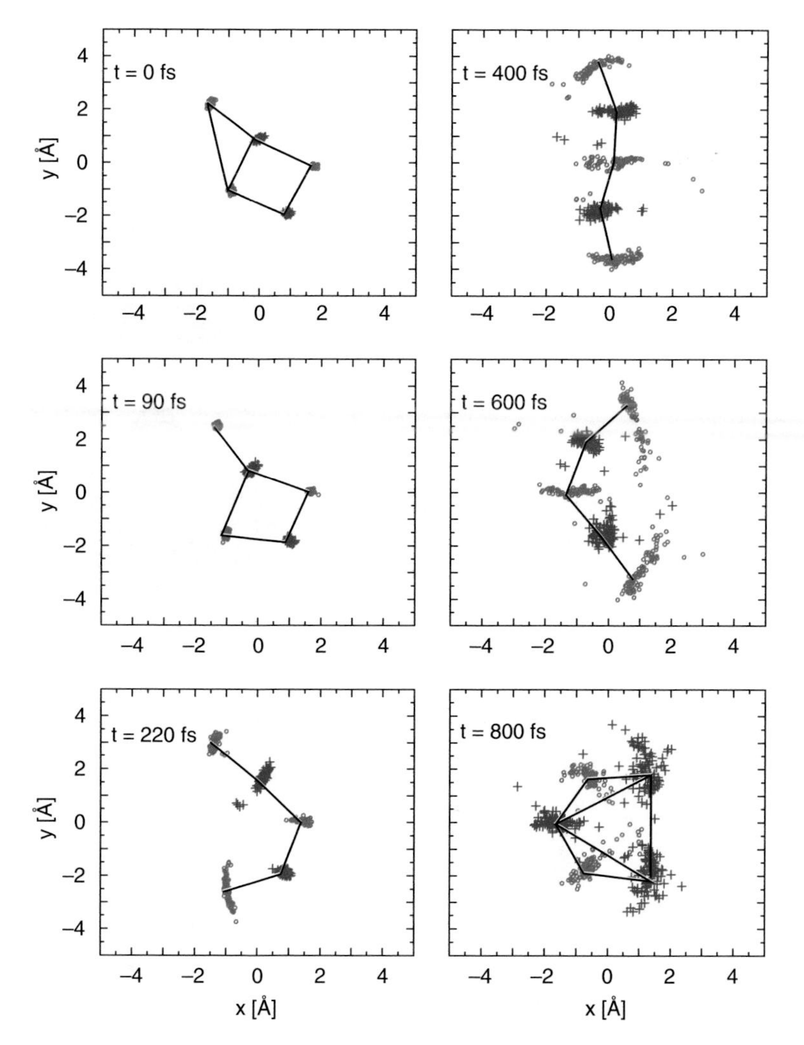

Figure 9. Snapshots of the phase space distribution (PSD) obtained from classical trajectory simulations based on the fewest-switches surface-hopping algorithm of a 50 K initial canonical ensemble [46]. Na atoms are indicated by black circles, and F atoms are indicated by gray crosses. Dynamics on the first excited state starting at the C_s structure ($t = 0$ fs) over the structure with broken Na–Na bond ($t = 90$ fs) and subsequently over broken ionic Na–F bond ($t = 220$ fs) toward the conical intersection region ($t = 400$ fs), Dynamics on the ground state after branching of the PSD from the first excited state leads to strong spatial delocalization ($t = 600$ fs). The C_{2v} isomer can be identified at $\sim$800 fs in the center-of-mass distribution.

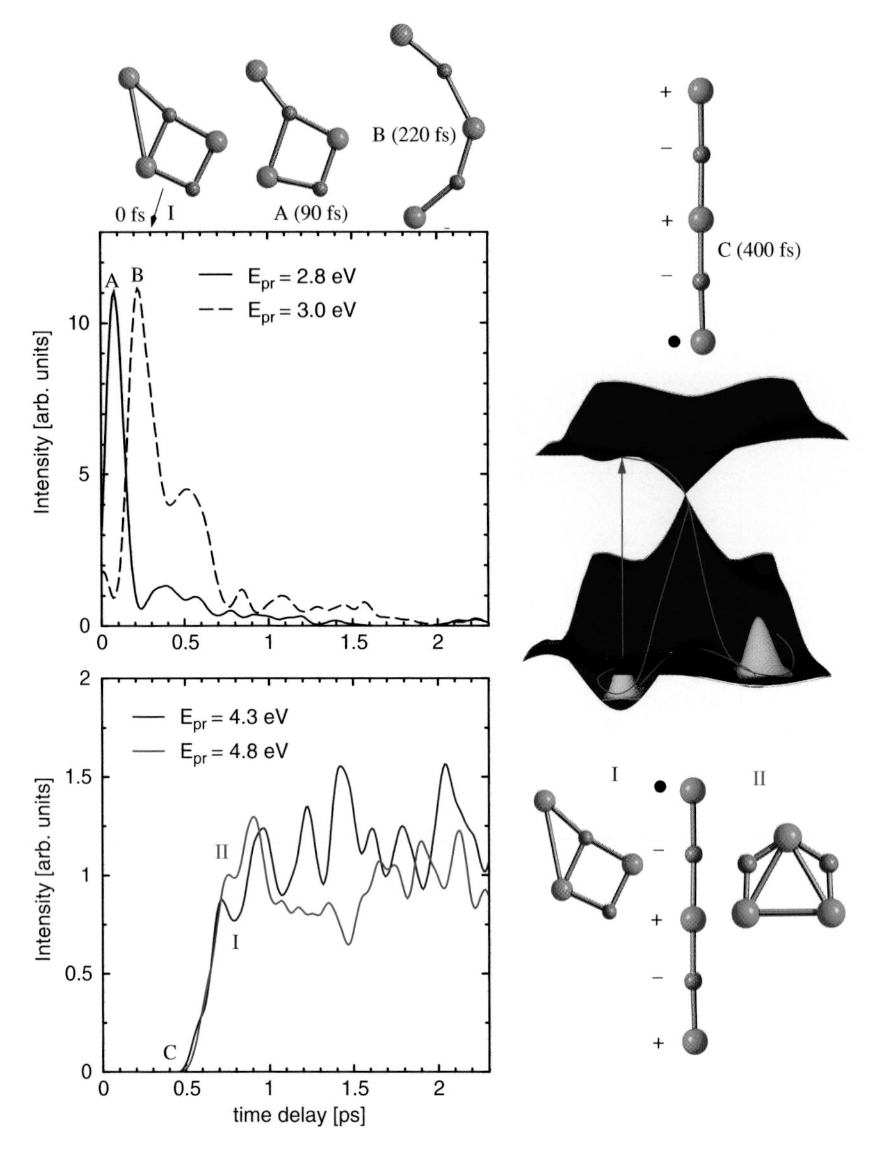

Figure 10. Simulated NeExPo pump-probe signals for the 50 K initial temperature Na_3F_2 ensemble at different excitation energies of the probe laser monitoring the geometric relaxation on the first excited state involving bond-breaking processes and passage through the conical intersection as well as geometric relaxation and IVR processes on the ground state after the passage (left-hand side). The isomerization through the conical intersection is schematically illustrated on the right-hand side [46].

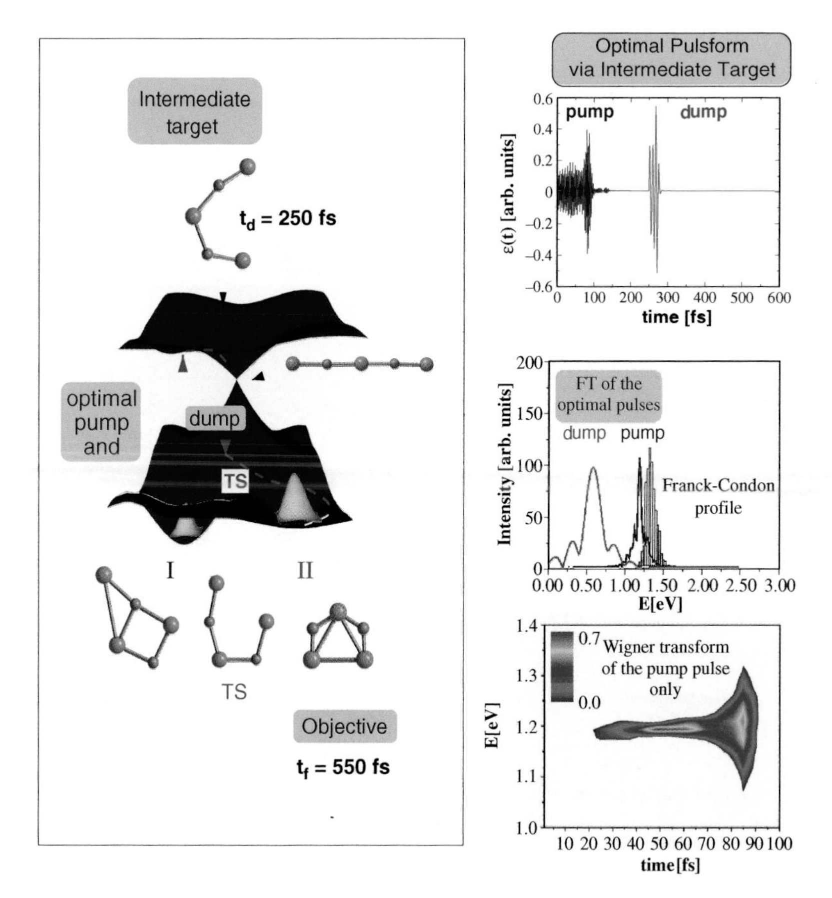

Figure 11. *Left-hand side*: Scheme for pump-dump optimal control in the Na_3F_2 cluster with geometries of the two ground-state isomers and of the transition state separating them, the conical intersection, and the intermediate target. *Upper panel, right-hand side*: The optimal electric field corresponding to the pump and dump pulses [51]. The mean energy of the pump pulse is 1.20 eV and the mean energy of the dump pulse is 0.6 eV. *Middle panel, right-hand side*: Fourier transforms of the optimal pump and dump pulses and the Franck–Condon profile for the first excited state corresponding to the excitation energy $T_e = 1.33$ eV. *Bottom panel, right-hand side:* Wigner transform of the optimal pump pulse.

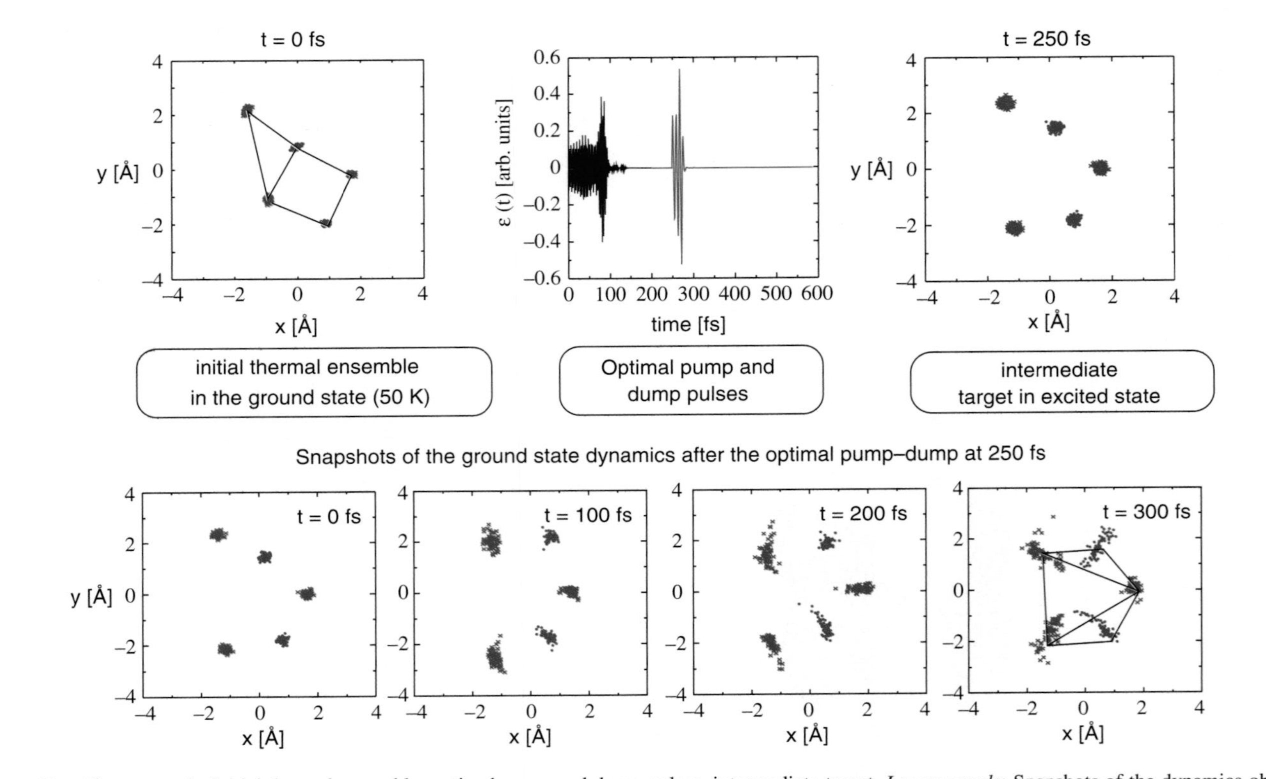

Figure 12. *Upper panels*: Initial thermal ensemble, optimal pump and dump pulses, intermediate target. *Lower panels*: Snapshots of the dynamics obtained by propagating the ensemble corresponding to the intermediate target after the optimized pump-dump at 250 fs on the ground state showing the localization of the phase space density in the basin corresponding to isomer II [51].

SECOND- AND FIRST-ORDER PHASE TRANSITIONS IN MOLECULAR NANOCLUSTERS

A. PROYKOVA

Atomic Physics, University of Sofia, Sofia 1126, Bulgaria

R. STEPHEN BERRY

Department of Chemistry, The University of Chicago, Chicago, Illinois 60637, USA

I. P. DAYKOV

Department of Physics, Cornell University, Ithaca, New York 14850, USA

CONTENTS

I. INTRODUCTION

Small finite systems exhibit solid- and liquid-like behavior much like that of bulk matter and, on the other hand, exhibit specific, quite interesting properties that distinguish them from the bulk [1–3]. "Small" is defined here with respect to the

Adventures in Chemical Physics: A Special Volume in Advances in Chemical Physics, Volume 132, edited by R. Stephen Berry and Joshua Jortner. Series editor Stuart A. Rice

range of the interaction potential: A system is small if its linear size is of the order of the potential range. In molecular clusters with Coulomb interaction included, most of the sizes amenable to computer simulations are small.

To define a phase in a cluster is not a straightforward task [4]. We speak about phase-like forms, with specific pair-distribution functions [5], that help to distinguish between different thermodynamic states in small systems. In what follows we use terms "phase change" and "phase transformations" for small systems, preserving the term "phase transitions" for bulk matter.

A first-order transition is characterized in the mean field theory or the Ehrenfest scheme by a discontinuity in a suitable order parameter and the specific heat has a δ-function singularity. However, the mean field theory is inaccurate in the vicinity of phase transitions, because it neglects the role of thermodynamic fluctuations. For instance, it predicts a finite discontinuity in the heat capacity at the ferromagnetic transition, which is implied by Ehrenfest's definition of "second-order" transitions. In real ferromagnets, the heat capacity diverges to infinity at the transition. In the modern classification scheme, which we think is correct for finite-size systems as well, the first-order phase transitions are those that involve a *latent heat.* During such a transition, a system either absorbs or releases a fixed (and typically large) amount of energy. Because energy cannot be instantaneously transferred between the system and its environment, first-order transitions are associated with "mixed-phase regimes" in which some parts of the system have completed the transition and others have not. Hence, an important indication of a first-order transition is the presence of metastable states in the transition region where the free energy has two local minima and the system may be trapped in the upper metastable state for a time shorter than the relaxation time of that metastable state. When the system is perturbed in a metastable state (e.g., by a temperature change), it may exhibit a two-step relaxation, a feature typical of a first-order transition. It is important that the relaxation time of the new metastable state, at a new temperature, could be so long that the state may mistakenly be considered as a stable one (the true equilibrium state). The macroscopic manifestation of trapping in metastable states is hysteresis in the cooling and heating branches of the internal energy [6], that is, the system's behavior in the transition region depends on its thermal history. Some systems may have several local minima in their free energy [7], which can give rise to the stable coexistence of three or even more phases-for small systems, of course.

For most systems, the mean temperature of a phase change shifts toward low temperatures with the reduction of the cluster size [8]. (Recently, exceptions to this general behavior have been observed [9–11].) The phase changes of finite-size systems are rounded-off and occur smoothly through the points of equal chemical potentials, even though they are sharp and effectively discontinuous in the bulk. (It has been predicted that, instead of discontinuities at the points of equal chemical potentials, clusters should show discontinuities in the equilibrium constants for two-phase equilibrium, at the temperatures at which local minima for a phase disappears [1], analogous to the limits of spinodal curves, but such discontinuities have not yet been observed.) Hence distinguishing between a discontinuous transition and a continuous (no latent heat) transition in a small system is a challenging task for both simulations and real experiments. In fact, some phase changes of small systems that have the characteristics that would identify them as first-order become second-order if the system is made large enough.

In a number of publications [12], classification of phase transitions in small systems has been presented. This scheme is based on the distribution of zeroes of the canonical partition function in the complex temperature plane. Among others, Gross has suggested a microcanonical treatment [13], where phase transitions of different order are distinguished by the curvature of the entropy $S = k_B \ln \Omega(E)$. According to this scheme, a back-bending in the microcanonical caloric curve $T(E) = 1/d_E \ln (\Omega(E))$ (i.e., the appearance of negative heat capacities) is a mandatory criterion for a first-order transition. Caloric curves without back-bending, where the associated specific heat shows a hump, are classified as higher-order transitions.

In simulations, however, the observation or nonobservation of back-bending effect might be an artifact of the computations. Hence, analytical theories are needed to confirm any statements. Some light has been shed on the origin and nature of the order–disorder structural phase changes on the basis of symmetry considerations [14]. We have shown that the near-neighbor intermolecular interactions of a cluster can be cast in terms of local molecular and site symmetry in a manner that accounts for the multistep phase changes that these clusters exhibit. In particular, we have shown how translation–rotation and rotation–rotation interactions enter into the O_h–D_{4h}–$[D_{3d}]$ transition of TeF_6 to yield a phase change with two local free-energy minima for the small system, but only rotation–rotation interactions enter into the lower-temperature phase change from partial to complete orientational ordering of the molecules on a monoclinic lattice, a change that, according to all indications, involves only a single local free-energy minimum.

Clusters, with their relatively short time scales, exhibit *dynamic* equilibrium between different phases, with passage between phases typically in the gigahertz range. This is how they exhibit phase coexistence [6, 7] within a temperature interval [15] rather than at a unique temperature (for a given pressure), typical of bulk phase transitions. Here we study the effect that width has on the detection of coexisting phases in the case of a discontinuous transition.

Most MD studies [6,16–20] of molecular clusters have been performed at constant energy. The current results have been obtained in canonical MD simulations. A question then arises: What can be learned about the microcanonical behavior of a system that has a first-order phase transition as revealed by its canonical solution? A general answer is well known: If a system is infinite and extensive (i.e., the interaction is not long-range), then microcanonical and canonical behavior are the same. However, clusters of AF_6 molecules can be considered as having infinitely long-range interaction (due to Coulomb interaction), and of course the infinite system size limit can hardly be applied here. Thus, it is instructive to compare the similarities and differences of results obtained from different ensembles in the case of plastic clusters, because the equivalence of the thermodynamic ensembles is under question.

Section II briefly reviews some arguments about the applicability of one or another ensemble in studying various aspects of small systems. For example, negative heat capacities can be detected in microcanonical ensembles [21–24] if the entropy has a convex dip. The canonical ensemble of the same system does not show any negative heat capacity [25], which is consistent with the general theory; for example, the heat capacity is proportional to the energy variance in the canonical ensemble and can never be negative.

Here we address as well the question of the ergodicity of small systems undergoing phase transformations. It can be addressed by a comparison of the present results with those obtained from canonical Monte Carlo simulations [26].

In Section III we present the potential used to simulate the thermal behavior and evolution of free molecular clusters. The thermostat is described with the Nosé–Hoover algorithm [5, 27]. Section IV describes first-order changes of the clusters found by detection of coexisting phases for both TeF_6 and SF_6 clusters of various sizes; predictions are made regarding the bulk transition temperature. In Section V, the continuous transformation associated with the molecular orientation is illustrated with TeF_6 clusters. Finally, we draw conclusions about how to relate the transformations observed in small systems to their bulk counterparts.

II. ENSEMBLES

By definition, the microcanonical ensemble contains all possible configurations in the 6N-dimensional phase space with the same energy and a constant probability of being in each configuration; N is the number of particles in the system under consideration. This ensemble describes an isolated system with constant N and V, or constant N and zero external pressure [28]. Constant-energy simulations are not recommended for equilibration because, without the energy flow facilitated by the temperature control methods, the desired temperature cannot be achieved. However, during the data collection phase, if one is interested in exploring the constant-energy surface of the conformational space, or, for other reasons, does not want the perturbation introduced by temperature- and pressure-bath coupling, this is a useful ensemble. This is the best ensemble for free-surface clusters.

In contrast, a system in contact with a thermal bath (constant-temperature, constant-volume ensemble) can be in a state of all energies, from zero to arbitrary large energies; however, the state probability is different. The distribution of the probabilities is obtained under the assumption that the system plus the bath constitute a closed system. The imposed temperature varies linearly from start-temp to end-temp. The main techniques used to keep the system at a given temperature are: velocity rescaling, Nosé, and Nosé–Hoover-based thermostats. In general, the Nosé–Hoover-based thermostat is known to perform better than other temperature control schemes and produces accurate canonical distributions. The Nosé–Hoover chain thermostat has been found to perform better than the single thermostat, since the former provides a more flexible and broader frequency domain for the thermostat [29]. The canonical ensemble is the appropriate choice when conformational searches of molecules are carried out in vacuum without periodic boundary conditions.

The constant-temperature, constant-pressure ensemble (*NPT*) allows control over both the temperature and pressure. The unit cell vectors are allowed to change, and the pressure is adjusted by adjusting the volume. This is the ensemble of choice when the correct pressure, volume, and densities are important in the simulation. This ensemble can also be used during equilibration to achieve the desired temperature and pressure before changing to the constant-volume or constant-energy ensemble when data collection starts.

The constant-temperature, constant-stress ensemble (*NST*) is an extension of the constant-pressure ensemble. In addition to the hydrostatic pressure that is applied isotropically, constant-stress ensemble allows you to control the *xx*, *yy*, *zz*, *xy*, *yz*, and *zx* components of the stress tensor (sometimes also known as the pressure tensor). This ensemble is particularly useful if one wants to study the stress–strain relationship in polymeric or metallic materials.

The constant-pressure, constant-enthalpy ensemble (*NPH*) is the analogue of constant-volume, constant-energy ensemble. Enthalpy $H = E + PV$ is constant when the pressure is kept fixed without any temperature control.

Conventional molecular dynamics (MD) and Monte Carlo (MC) simulations explore only parts of the entire phase space (MD) or configuration space (MC) of complex systems with rugged potential energy surfaces due to the finite time of the computations. Thus MD trajectories are usually trapped in one of the local potential energy minima for a long time or MC samples a part of the configuration space, eventually leading to inaccurate thermodynamic quantities. Such a problem should be carefully solved in simulation studies of protein folding and phase transitions.

Several novel simulation strategies have been proposed to address such quasi-ergodic behavior. The multicanonical ensemble, developed by Berg and Neuhaus [30] and Lee [31] is one of the most powerful methods for overcoming the problem of quasi-ergodicity. In the multicanonical ensemble, the potential energy distribution covers a wide energy range. As a result, a broad potential energy landscape can be explored by the scheme and, more importantly, correct canonical distributions can be reproduced. In actual implementations, the multicanonical distribution is obtained by introducing a weighting function resulting from the conventional canonical sampling at high temperature [32]. Usually, preparation of the weighting function is a tedious and time-consuming process. However, once the multicanonical distribution is established, the canonical distributions at various temperatures can be easily obtained by a simple reweighting technique [33]. Recently, several attempts have been made to simplify the generation of the multicanonical ensemble [34] and to study structural transitions in biomolecules [35]. Furthermore, it has been shown that the multicanonical ensemble could be combined with other novel simulation methods to accelerate overall sampling efficiency [36].

III. EVOLUTION OF FREE CLUSTERS FROM CANONICAL MOLECULAR DYNAMICS

We represent the cluster as composed of rigid molecules interacting through a sum of pairwise atom–atom potentials of Lennard-Jones and Coulomb types [37]:

$$U(i,j) = \sum_{\alpha,\beta=1}^{7} \left[4\epsilon_{\alpha\beta} \left[\left(\frac{\sigma_{\alpha\beta}}{r_{ij}^{\alpha\beta}} \right)^{12} - \left(\frac{\sigma_{\alpha\beta}}{r_{ij}^{\alpha\beta}} \right)^{6} \right] + \frac{q_{i\alpha} q_{j\beta}}{4\pi\epsilon_0 r_{ij}^{\alpha\beta}} \right] \tag{1}$$

$$U = \sum_{i,j=1(i<j)}^{n} U(i,j) \tag{2}$$

The Lennard-Jones potential parameters α, β, $\epsilon_{\alpha\beta}$ are chosen [20] to agree with diffraction experiments [17]. The charge q has been computed with the help of a linear combination of plane orbitals (LCPO) [37] or of Gaussian orbitals (LCGO). In both cases, q is significantly less than $1e$; $q_F = 0.1e$ in TeF_6 (LCPO); $q_F = 0.25e$ in TeF_6 (LCGO) [38].

The constant temperature (canonical) ensemble is realized following Hoover's approach [5] to the Nosé algorithm [27]. An additional degree of freedom is introduced in the system to describe the heat reservoir (bath). The temperature is controlled by a "friction," ζ:

$$\begin{aligned}
\frac{d\mathbf{q}_i}{dt} &= \frac{\mathbf{p}_i}{m_i} \\
\frac{d\mathbf{p}_i}{dt} &= -\frac{\partial U}{\partial \mathbf{p}_i} - \zeta \mathbf{p}_i \\
\frac{d\zeta}{dt} &= \left(\sum_i \frac{\mathbf{p}_i^2}{2m_i} - g k_b T \right) \Big/ M
\end{aligned} \tag{3}$$

where g is the number of degrees of freedom, q_i and p_i are the positions and momenta respectively, m_i is the molecular mass, and k_b is the Boltzmann constant.

The parameter M, which has the meaning of a bath size, governs the heat transfer rate between the system and the reservoir. Too small M means decoupling of the new degree of freedom from the physical system (Fig. 1), resulting in a bimodal distribution of the kinetic energy (much like a set of very

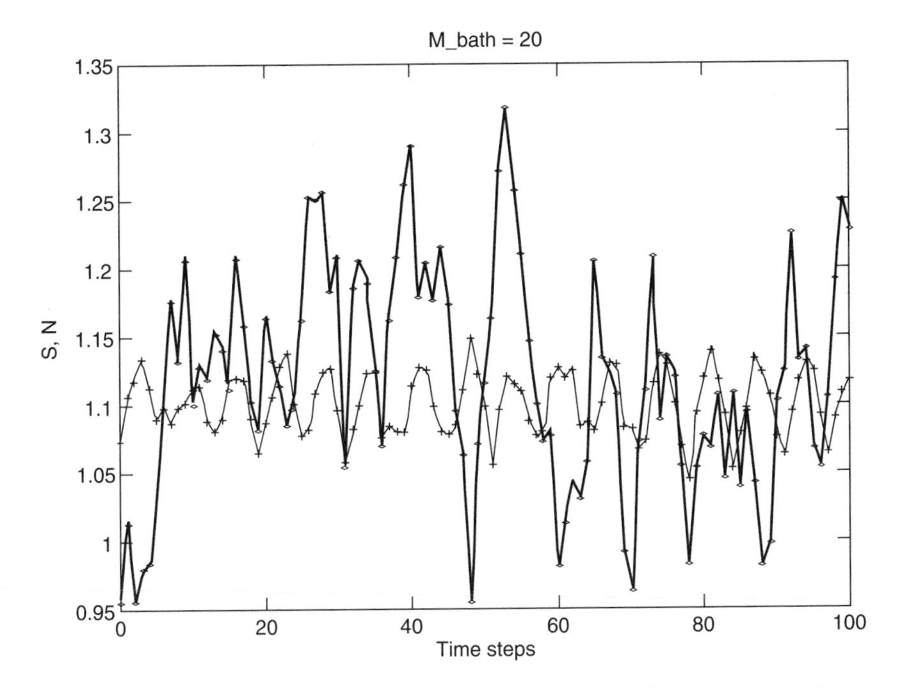

Figure 1. The new degree of freedom decouples from the system if its mass is too small.

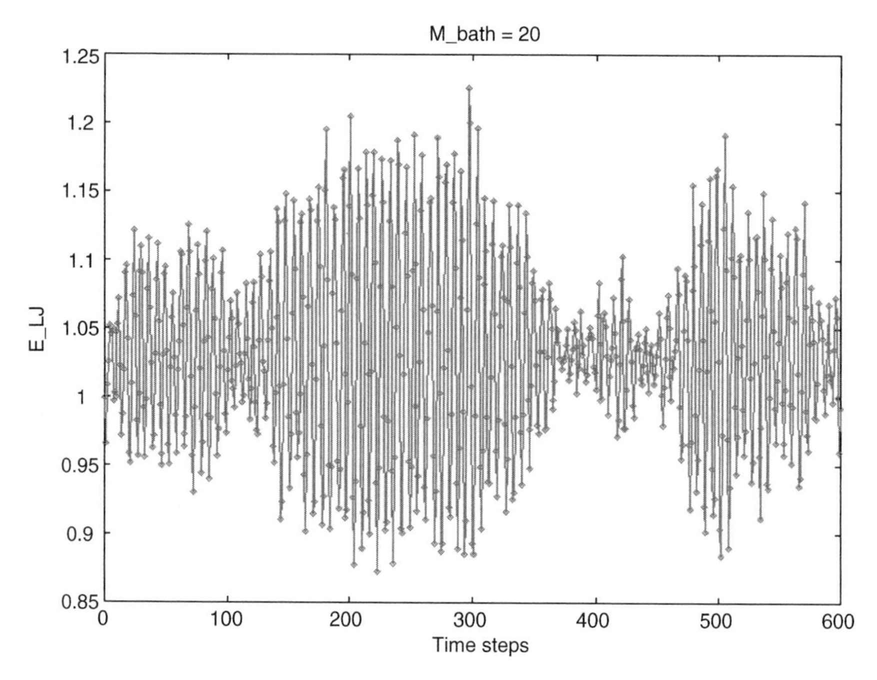

Figure 2. The system and the bath behave like coupled harmonic oscillators if the bath mass is rather small.

weakly coupled harmonic oscillators; Fig. 2) rather than the desired Gaussian distribution. Likewise, it should not be too large; if it is, the heat reservoir is "too heavy" and one has to wait too long to establish thermal equilibrium. The optimal bath size depends both on the system size and the interaction potential. For the potentials studied here, a suitable bath size is $M = 200$ for a 59-molecule TeF_6 or SF_6 cluster, while $M = 300$ is optimal for a 137-molecule cluster. Figure 3 illustrates the mean kinetic energy of the cluster immersed in a bath with a suitable mass. (We use mean kinetic energy to measure temperature; in a canonical system, this choice is not controversial, as it would be for a microcanonical system. There, the mean kinetic energy and the derivative of energy with respect to entropy are not necessarily equivalent and the two choices may lead to different conclusions.) The proper choice of the bath mass becomes very important in the transition region. Our systematic study showed that the mass plays a role similar to that of the total energy in the microcanonical ensemble for placing the system into a state that is entropically unfavorable.

A more extensive sampling of the phase space can be achieved by introducing a chain of thermostats [39], rather than a single bath. Such an

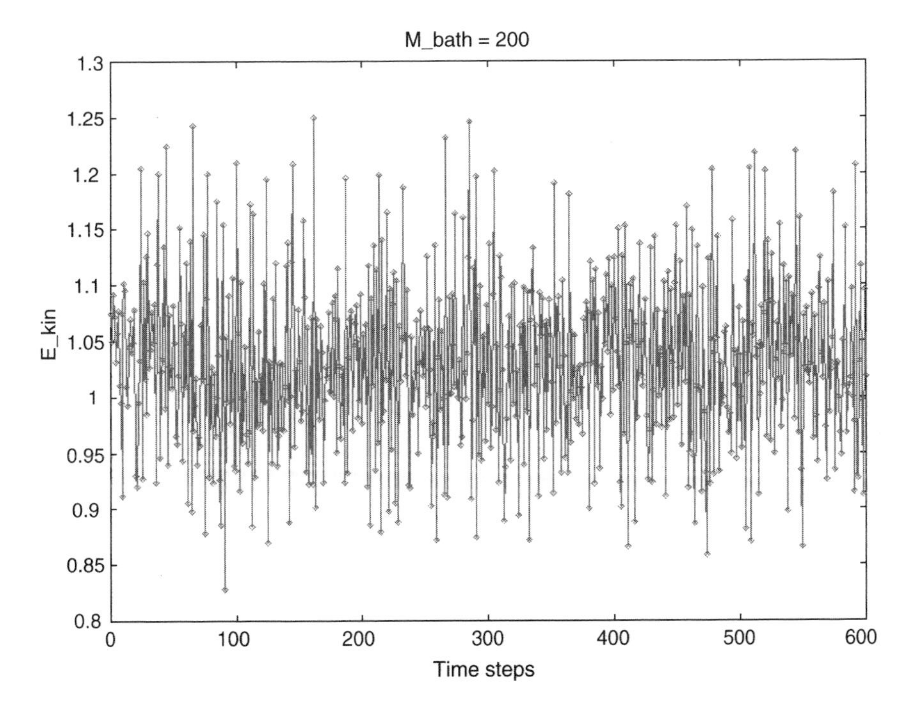

Figure 3. The proper bath size ($M = 200$ for a 59-molecule cluster) ensures the desired mean temperature.

approach increases the computational time about 10–15% for the smallest clusters, without essentially improving the temperature control. However, a chain of thermostats is extremely important for proper simulations of large, flexible biomolecules.

IV. PHASE COEXISTENCE IN FIRST-ORDER SOLID–SOLID TRANSFORMATION OF CLUSTERS

The clusters we study are known from constant energy simulations [6, 16, 18, 40] and experimental data [41] to undergo at least two temperature-driven orientational order–disorder transformations below the freezing point. The current study confirms the finding [16] that there are two successive transformations: First, when the cluster is cooled below its range of solidification, it exhibits a discontinuous solid–solid transformation, involving partial molecular orientation and lattice reconstruction. A further cooling induces

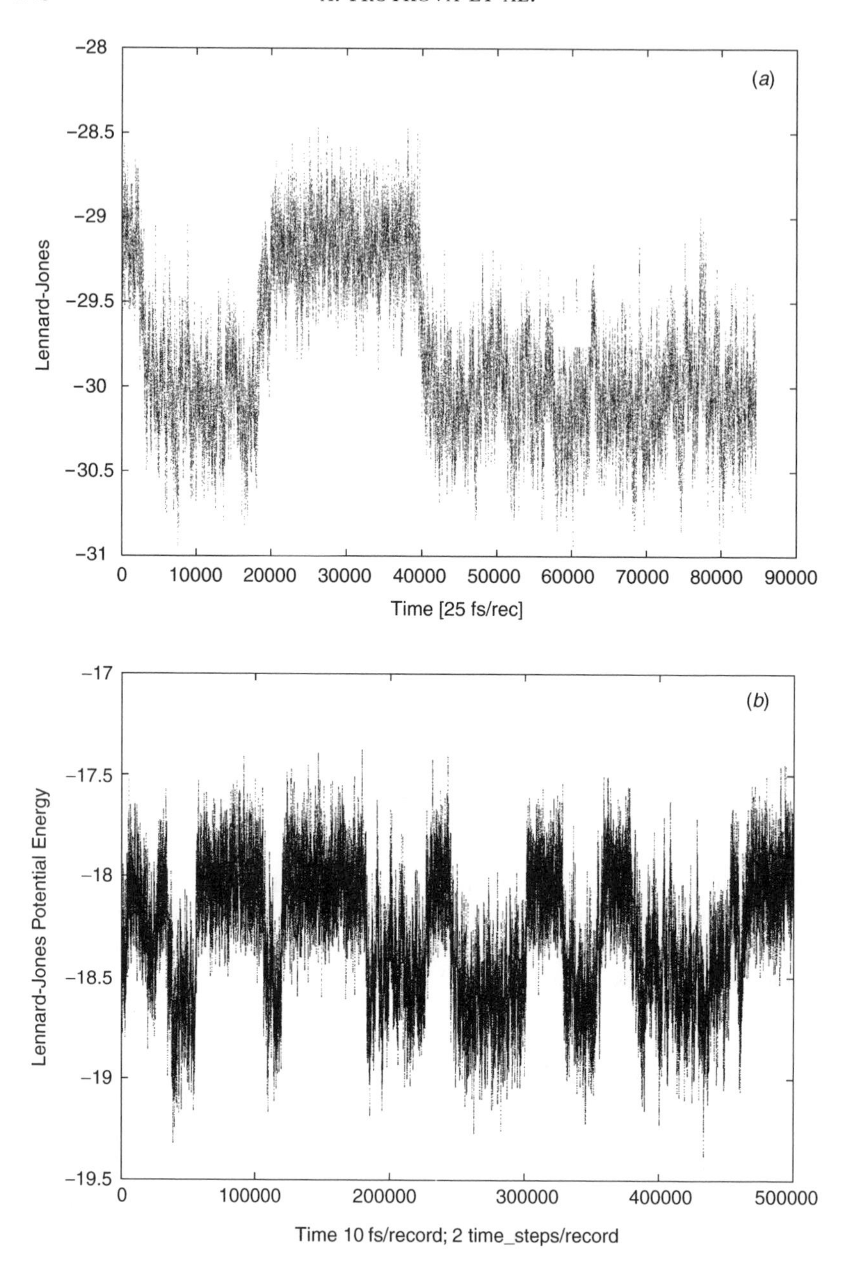

Figure 4. Phase coexistence of orientation-disordered and orientation-ordered phase in TeF_6 clusters of various sizes: (*a*) 137 molecules ($Q = 300$, $T = 93.3\,K$, $T_0 = 92.5\,K$), (*b*) 89 molecules (Lennard-Jones potential energy, $Q = 300$, $T = 88\,K$, 5-ns run), (*c*) 59 molecules ($Q = 200$, $T = 76\,K$, $T_0 = 76\,K$).

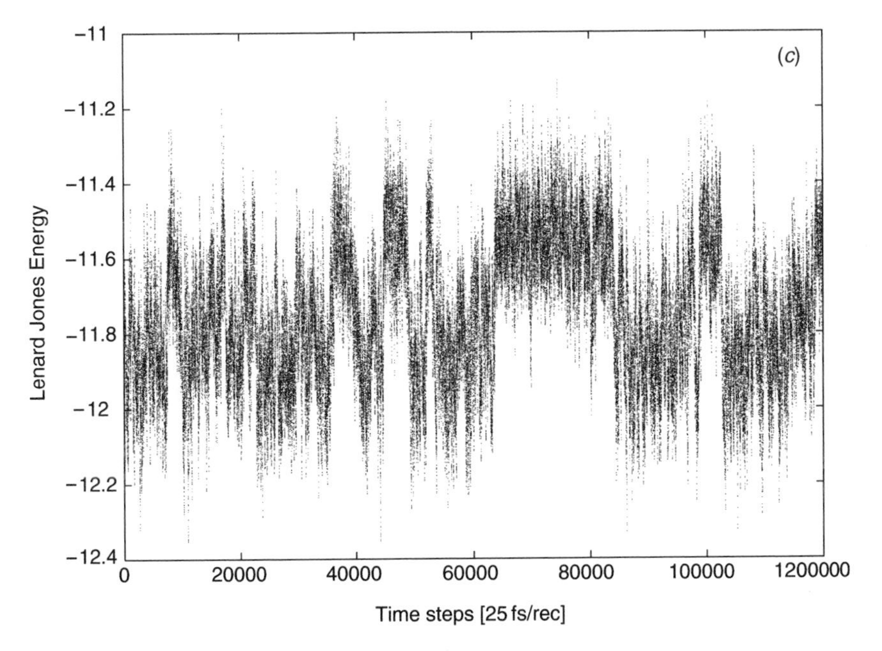

Figure 4. (*Continued*)

another change of the cluster—a continuous solid–solid transformation to a completely oriented structure.

One way to see that a transition is discontinuous is to detect a coexistence of two phases, in this case the orientationally ordered and disordered phases, in a temperature interval. This is revealed by time variation of the potential energy of the cluster. In the temperature region of phase coexistence, each cluster dynamically transforms between the phases, and its potential energy fluctuates around two different mean values (Fig. 4). In an ensemble of clusters, the coexistence of different phases is observable insofar as a fraction of the clusters (e.g., in a beam [17]) can exhibit the structure of one phase, while another fraction takes on the structure of another phase.

Figure 4 shows the dynamical coexistence of two structures as a function of cluster size: The examples are 137-, 89-, and 59-molecule TeF_6 clusters. In panels a–c the orientation-disordered phase has a higher potential energy (and higher entropy) than the orientation-ordered phase. The free energy is the same of course for both when the fractions are equal. (One more difference between phase coexistence of bulk and small systems is that the latter may exhibit coexistence of *unequal* fractions of two phases—that is, of phases with somewhat different free energies. Likewise, ensembles of small

Table I

L(Å)	T_c	δT (K)
$\sim 59^{1/3}d = 3.89d$	76	3.2
$\sim 89^{1/3}d = 4.46d$	88	1.7
$\sim 137^{1/3}d = 5.16d$	93.3	0.7
$\sim 227^{1/3}d = 6.10d$	97.8	0.2

homogeneous systems may exhibit coexistence of more than two phases in equilibrium [1].)

The temperature range of dynamical coexistence is narrower for the larger clusters than for the smaller, as seen in Table I. This compilation shows the dependence of the mean transformation temperature and the width of the temperature interval as a function of the cluster's linear size $L \sim N^{1/3}\, d$, where N is the number of molecules and d is the average distance between them in a crystalline state.

The relative increase of the linear size is ~0.86, from one to the next of the reported clusters. The transformation temperature increases in a more complicated way and reaches a plateau at about $50d$. These results allow us to estimate the transition temperature for bulk to be about 215 K, which is 5% less than the observed bulk transition temperature [17].

The shrinkage of the coexistence temperature interval is one factor that makes it difficult to detect dynamical coexistence in large clusters. The other is that the fraction of the unfavored phase decreases with the number of molecules in the cluster. This dependence appears in the exponent of the ratio of the amounts of the two phases—that is, in the exponent of the equilibrium ratio of concentrations [1]. One necessarily must compute an extremely long phase trajectory in order to distinguish between two different phases and especially to establish the equilibrium ratio of the amounts of the two phases. This explains why absence of coexistence was reported [42] in simulations of clusters containing more than 500 molecules.

We have succeeded in detecting coexisting phases in clusters of SF_6 molecules both in constant-energy simulations [6] and in the current, constant temperature calculations (Fig. 5).

Just below the freezing point the solid clusters have continuous rotational symmetry, which breaks at lower temperatures because of the spontaneous alignment of the molecules along preferred orientations, as can be seen in Fig. 6. This is a plot of the mutual orientations of the molecular axes of symmetry in the disordered and ordered solid state.

We check whether the cluster is solid by computing a modification of the Lindemann index [43] and by plotting the radial distributions of the molecular

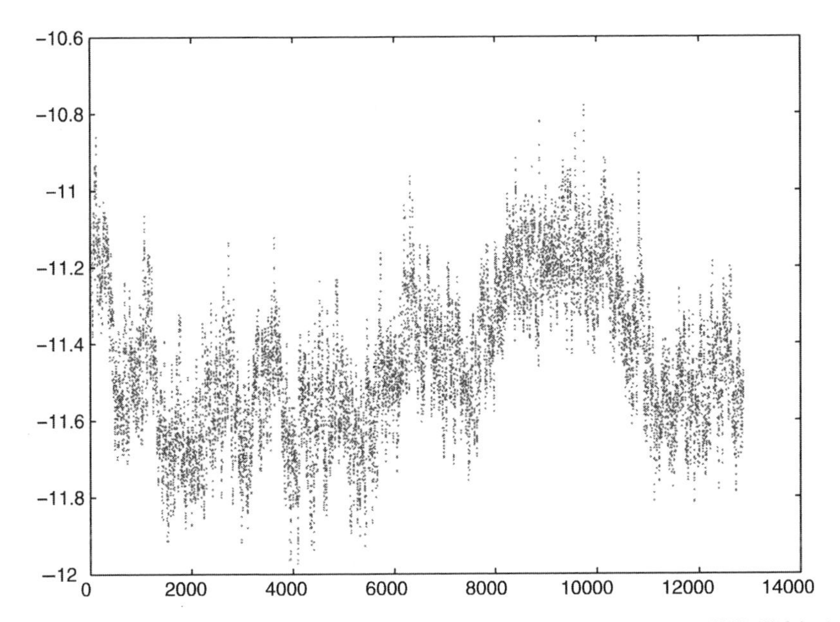

Figure 5. Dynamically coexisting phases in an 89-molecule SF_6 cluster ($T = 58$ K, 50 fs/rec). The potential energy surface of a sulfur cluster is shallower than that of a tellurium cluster with the same size, and the SF_6 cluster spends less time in any phase.

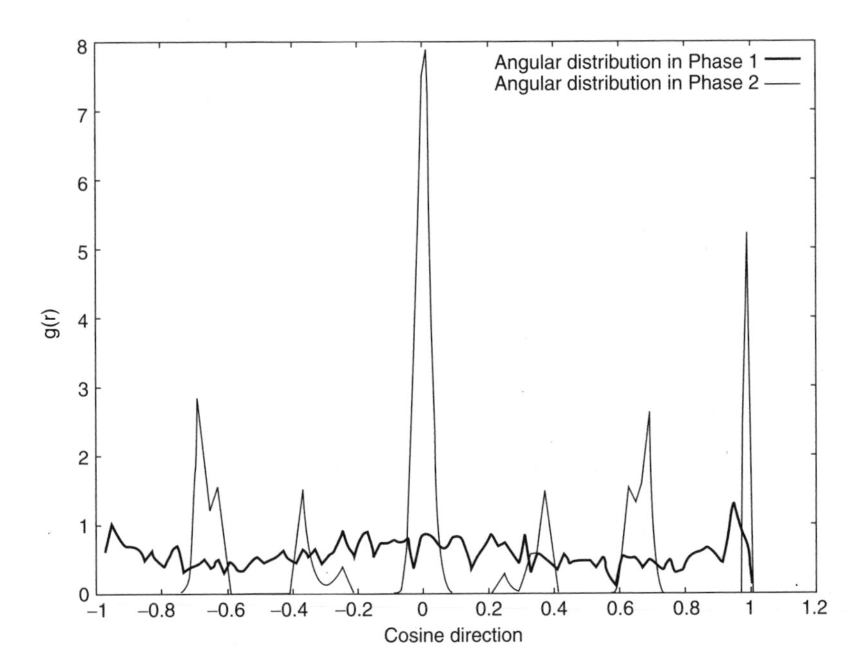

Figure 6. Orientational distribution of the molecular axes in a liquid and an oriented solid state of a tellurium cluster.

centers of mass. The radial distribution of a liquid-like system should have a typical periodic pattern [5]. To characterize a liquid-like state of a cluster and differentiate it from a solid, we use the quantity $\delta_{lin} \geq 0.08$. Here δ_{lin} is computed from

$$\delta_{lin} = \frac{2}{N(N-1)} \sum_{i,j(>i)=1}^{N} \frac{\sqrt{\langle r_{ij}^2 \rangle - \langle r_{ij} \rangle^2}}{\langle r_{ij} \rangle} \tag{4}$$

where $|\mathbf{r}_{ij}| = |\mathbf{r}_i(t)| - |\mathbf{r}_j(t)|$ and $\langle . \rangle$ denotes time averaging. For bulk, $\delta_{lin} \geq 0.1$ corresponds to a melted phase. (The original Lindemann index used displacements from equilibrium, rather than interparticle distances, as the fluctuating quantities.)

The radial distribution of the molecular centers of mass is computed from $g(r) = \text{Norm}\langle \sum_{i=1}^{N} \sum_{j\neq i}^{N} (r - r_{ij}) \rangle$. The normalization factor Norm ensures that the integral of $g(r)$ over r gives N. The plot in Fig. 7 shows the appearance of structures when the cluster is cooled down.

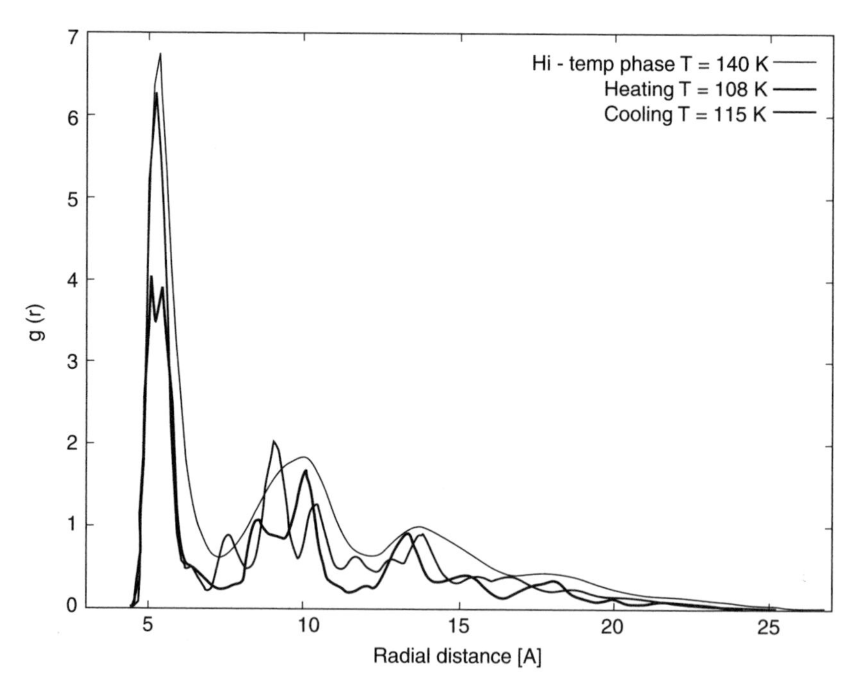

Figure 7. The radial distribution of a 59 TeF_6 molecule cluster in coexisting region. Two structures (up-configuration, down-configuration) are mixed and the peaks are split.

To understand the difficulties in detecting coexisting phases in large clusters, let us discus the ergodicity of a system. One signal for the phase transition is a change of time scale necessary to establish global, rather than just regional ergodicity. One can ask about a complex system, "How long does it take to establish ergodicity?" The answer to this question was studied for atomic clusters by looking at the distribution of sample values of Liapunov exponents as a function of the length or duration of the trajectory of the system used to evaluate the exponents [44]. It was found that in the solid–liquid coexistence region, the distribution of Liapunov exponents was bimodal if they were estimated from short trajectories, but the distribution became unimodal if the trajectories were made longer. In the case of molecular clusters we observe that long computations, due to truncation error in the number presentation, could bring the cluster in a region of the multidimensional phase space where coexistence cannot be observed [6]. (It is also possible that a cluster may pass between phases so rapidly that the time spent in one phase is too brief for thermal equilibration. This phenomenon, known for some small Lennard-Jones clusters [45], makes it literally impossible to distinguish phase coexistence. Instead, one can only observe a sort of dynamic "slush.")

The potential surface of our system is extremely rugged, with high barriers and bottlenecks [6]. Very long trajectories develop nonzero probabilities or rates to pass through these bottlenecks. Therefore the evolution of the system depends on its history. The time to establish ergodicity may be too long for practical simulations which are usually limited to the scale of nanoseconds. Even if the system is technically ergodic, we may not be able to observe such "inhibited" passages, and the probability that any given system visits specific regions of its phase space may depend sensitively on the starting configuration. This is especially important in the vicinity of a phase change, where the time to establish ergodicity increases.

What is important in the observation of nanosize systems is that phases that are metastable in large systems are actually thermodynamically stable, but are minority species, in ensembles of small systems. The metastability of a thermodynamically unfavored bulk phase is indeed a result of a nonequilibrium population of that phase. However, in an ensemble of small systems, an unfavored phase can easily be not so very unfavored and, in fact, may have an observable minority population. The smoothness of the phase change of small systems is reflected in the continuous change—with temperature, for example—of the ratio of (observable) amounts of the two species. Moreover, more than two phases can be present simultaneously, in equilibrium, if the systems are small like Ar_{55} [7].

The mean transition temperatures obtained in molecular dynamics simulations of a canonical (Can) and a microcanonical (Mic) ensembles differ slightly but systematically:

Size, N	T_c(Can), K	T_c(Mic), K
59	76	78
89	88	90
137	93.3	95
227	97.8	101

In these small systems the different ensembles do not give precisely the same results, which is understandable: The systems are quasi-ergodic, and thermodynamic limit is far away. The results obtained in Can and Mic simulations converge as $1/N$ for clusters containing more than 100,000 molecules. The surfaces of these larger clusters contain only phase transitions and can be neglected [46]. An illustration: In the solid–liquid coexistence range, a microcanonical ensemble may exhibit a negative slope in the dependence of mean kinetic energy (here used to define mean temperature) on energy, if the entropy and potential energy of the liquid region of the potential surface are great enough, compared with those of the solid region. A canonical ensemble does not show this "negative heat capacity."

V. CONTINUOUS SOLID–SOLID TRANSFORMATION OF CLUSTERS

In our simulations we detect the continuous solid-solid transformation for all cluster sizes greater than 27 molecules; this corresponds to a cluster diameter greater than about 1.5–2 nm. The caloric curves of 89-molecule clusters of TeF_6 and SF_6 below 50 K and 35 K, respectively, show changes of their slopes. One example for a cluster of 89 TeF_6 molecules is shown in Fig. 8. The heat capacity computed from the energy fluctuations, a modified Lebowitz formula [6], shows that the slope of the caloric curve changes at $\sim$44 K and $\sim$37 K (Fig. 9).

The orientational ordering of the molecules at low temperatures breaks the isotropic rotational symmetry [14]. The fact that the lowest energy state has broken orientational and translational symmetry means that the system is stiff in both ways: Modulating the order parameter must cost some energy. The broken rotational symmetry introduces orientational elastic stiffness and low-frequency rotational librational waves. The higher-temperature structural transitions of clusters of interest here break both the translational order and orientational order. The translational aspect introduces a rigidity to shear deformations, and low-frequency phonons (Goldstone modes) appear. Hence sound measurements may distinguish between the two solid structures coexisting in the first-order-like phase change.

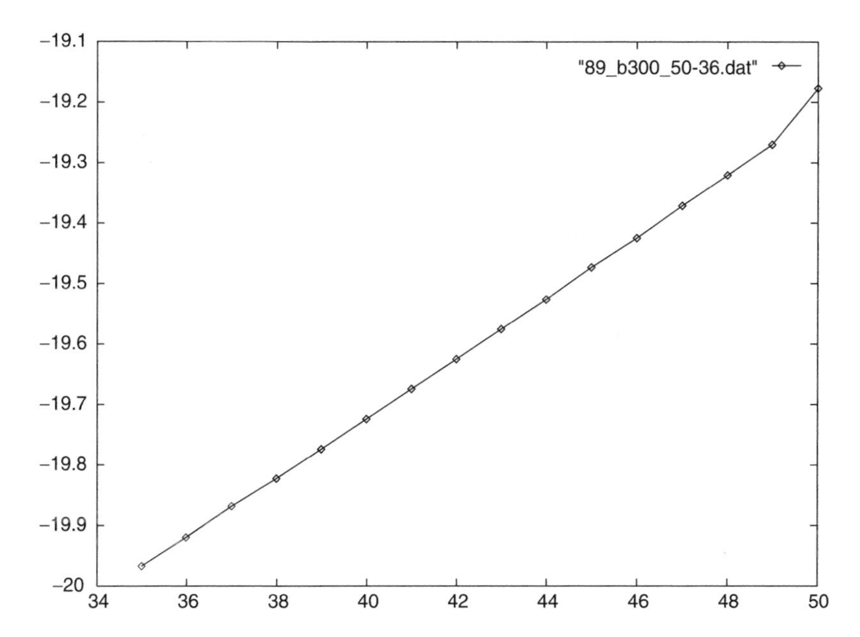

Figure 8. The caloric curve of an 89 TeF_6 cluster in the vicinity of the continuous transition ($Q = 300$, 0.5 ns/K): Changes of the slope can be seen in the derivative of the curve at ~44 K, and another one at ~37 K. The latter corresponds to a complete orientational order.

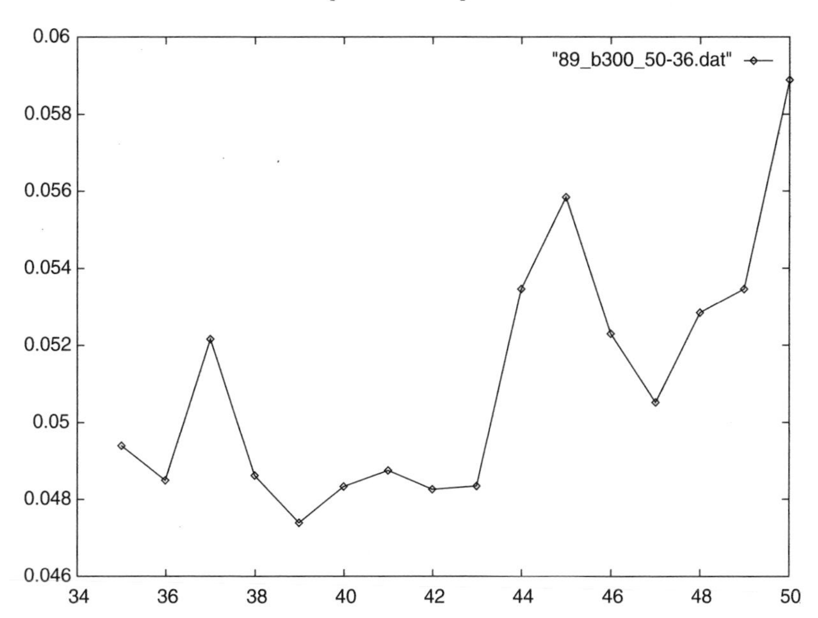

Figure 9. The heat capacity computed from the energy fluctuations of an 89 TeF_6 cluster in the vicinity of the continuous transition ($Q = 300$, 0.5 ns/k): Changes are detected at ~44 K, and another one at ~37 K.

The rotational symmetry breaking cannot be detected in such a way; there are no orientational Goldstone modes. One could look at Raman spectra or neutron diffraction experiments that are sensitive to the molecular orientations. The order parameter field for an orientation order in molecular systems can be chosen to be a three-component field of the cosine distribution of the mutual orientations of molecular axes. This index reveals the continuous, low-temperature transition.

VI. CONCLUSIONS

Simulations of octahedral molecular clusters at constant temperature show two kinds of structural phase changes, a high-temperature discontinuous transformation analogous to a first-order bulk phase transition, and a lower-temperature continuous transformation, analogous to a second-order bulk phase transition. The former shows a band of temperatures within which the two phases coexist and hysteresis is likely to appear in cooling and heating cycles Fig. 10; the latter shows no evidence of coexistence of two phases. The width of the coexistence band depends on cluster size; an empirical relation for that dependence has been inferred from the simulations.

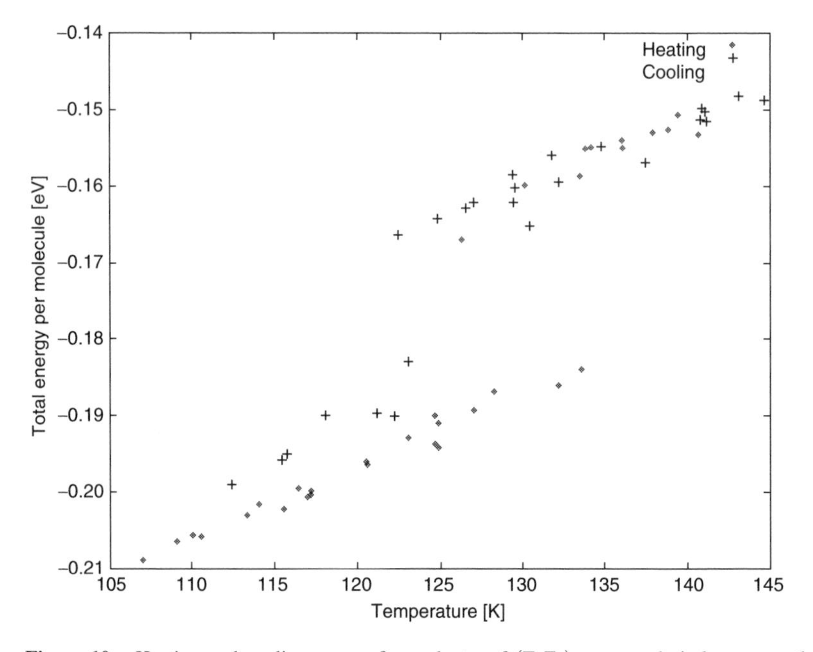

Figure 10. Heating and cooling curves for a cluster of $(TeF_6)_{89}$; crossed circles are on the heating curve, and crosses are on the cooling curve.

Acknowledgments

The research was supported by CL Grant-NATO. A.P. acknowledges the support of a grant (NSF-BG, F-003/2003) for supporting scientific research at the universities. I.P.D. thanks the Department of Physics, Cornell University for the CPU time allocated for the project. R.S.B. acknowledges the support of a grant from the National Science Foundation. The authors thank Mr. S. Pisov for computing the radial distributions and for quenching the structures to obtain the most stable configurations.

References

1. R. S. Berry, *Phase Transitions* **24–26**, 269–270 (1990); R. S. Berry, in *Theory of Atomic and Molecular Clusters*, J. Jellinek, ed., Springer-Verlag, Berlin, 1999, p. 1.
2. D. J. Wales and R. S. Berry, *J. Chem. Phys.* **92**, 4473 (1990); D. J. Wales and R. S. Berry, *Phys. Rev. Lett.* **73**, 2875 (1994); J. P. K. Doye and D. J. Wales, *J. Chem. Phys.* **111**, 11070–11079 (1999).
3. M. Bixon and J. Jortner, *J. Chem. Phys.* **91**, 1631 (1989).
4. R. S. Berry and B. M. Smirnov, *J. Chem. Phys.* **113**, 728 (2000); R. S. Berry and B. M. Smirnov, *J. Chem. Phys.* **114**, 6816 (2001).
5. M. P. Allen and D. J. Tildesley, *Computer Simulation of Liquids,* Clarendon Press, Oxford, 1994.
6. A. Proykova, S. Pisov, and R. S. Berry, *J. Phys. Chem.* **115**, 8583 (2001).
7. R. E. Kunz and R. S. Berry, *Phys. Rev.* E **49**, 1985 (1994).
8. *Applications of the Monte Carlo Method in Statistical Physics*, K. Binder, ed., SpringerVerlag, Berlin, 1984; K. Binder, *Rep. Prog. Phys.* **50**, 783–859 (1987).
9. A. A. Shvartsburg and M. F. Jarrold, *Phys. Rev. Lett.* **85**, 2530 (2000).
10. G. A. Breaux, R. C. Benirschke, T. Sugai, B. S. Kinnear, and M. F. Jarrold, *Phys. Rev. Lett.* **91**, 215508 (2003).
11. K. Joshi, D. G. Kanhere, and S. A. Blundell, *Phys. Rev. B* **66**, 155329 (2002); K. Joshi, D. G. Kanhere and S. A. Blundell, *Phys. Rev. B* **67**, 235413 (2002).
12. O. Mulken, H. Stamerjohanns, and P. Borrmann, *Phys. Rev. E* **64**, 047105 (2001); O. Mulken, P. Borrmann, J. Harting, and H. Stamerjohanns, *Phys. Rev. A* **64**, 013611 (2001); P. Borrmann, O. Mulken, and J. Harting, *Phys. Rev. Lett.* **84**, 3511 (2000).
13. D. H. E. Gross, *Rep. Prog. Phys.* **53**, 605 (1990).
14. A. Proykova, D. Nikolova, and R. S. Berry, *Phy. Rev. B* **65**, 085411 (2002).
15. I. Daykov and A. Proykova, *Meetings in Physics at the University of Sofia*, Vol. 2, A. Proykova, ed., Heron Press Science Series, Sofia, 2001, p. 31.
16. A. Proykova, R. Radev, Feng-Yin Li, and R. S. Berry, *J. Chem. Phys.* **110**(8), 3887–3896 (1999).
17. L. S. Bartell, E. J. Valente, and J. Caillat, *Phys. Chem.* **91**, 2498 (1987).
18. A. Proykova and R. S. Berry, *Eur. J. Phys.* D **9**, 445 (1999).
19. J. B. Maillet, A. Boutin, S. Buttefey, F. Calvo, and A. H. Fuchs, *J. Chem. Phys.* **109**(1), 329–337 (1998).
20. K. E. Kinney, S. Xu, and L. S. Bartell, *J. Phys. Chem.* **100**, 6935 (1996).
21. M. Schmidt and H. Haberland, *C. R. Phys.* **3**, 327 (2002).
22. D. H. E. Gross, *Microcanonical Thermodynamics: Phase Transitions in Small Systems*, Lecture Notes in Physics, Vol. 66, World Scientific, Singapore, 2001.
23. F. Gobet, B. Farizon, M. Farizon, M. J. Gaillard, J. P. Buchet, M. Carr, P. Scheier, and T. D. Märk, *Phys. Rev. Lett.* **89**, 183403 (2002).

24. Juan A. Reyes-Nava, Ignacio L. Garzon, Karo Michaelian, *Phys. Rev. B* **67**, 165401 (2003). http://arxiv.org/abs/physics/0302078
25. D. J. Wales and J. P. K. Doye, *J. Chem. Phys.* **103**, 3061 (1995).
26. A. Proykova, S. Pisov, R. Radev, I. Daykov, and R. S. Berry, *Nanoscience and Nanotechnology*, E. Balabanova and I. Dragieva, ed., **2**, 18, Heron Press Science Series, Sofia, 2001; A. Proykova, S. Pisov, R. Radev, P. Mihailov, I. Daykov, and R. S. Berry, *Vacuum* **68**(1), 87 (2002).
27. S. Nosé, *Mol. Phys.* **52**, 255 (1984).
28. T. L. Hill, *Thermodynamics of Small Systems*, Parts I and II, Two Volumes in One, Dover, New York, 1994.
29. S. Jang, Y. Pak, and S. Shin, *J. Chem. Phys.* **116**, 4782 (2002).
30. B. A. Berg and T. Neuhaus, *Phys. Rev. Lett.* **68**, 9 (1991).
31. J. Lee, *Phys. Rev. Lett.* **71**, 211 (1993).
32. Y. Okamoto and U. H. E. Hansman, *J. Phys. Chem.* **99**, 11276 (1995).
33. A. M. Ferrenberg and R. H. Swedensen, *Phys. Rev. Lett.* **61**, 2635 (1988).
34. U. H. E. Hansman and Y. Okamoto, *Chem. Phys. Lett.* **329**, 261 (2000).
35. N. A. Alves, U. H. E. Hansmann, and Y. Peng, *Int. J. Mol. Sci.* **3**, 17–29 (2002).
36. F. Calvo and J. P. K. Doye, *Phys. Rev. E* **63**, 010902 (2001); H. Xu and B. J. Berne, *J. Chem. Phys.* **112**, 2701 (2000); C. Bartels and M. Karplus, *J. Phys. Chem. B* **102**, 865 (1998); R. Radev and A. Proykova, *Comp. Phys. Commun.* **147**/1–2, 242 (2002).
37. A. Proykova and R. S. Berry, *Z. Phys. D* **40**, 215 (1997).
38. S. Pisov and A. Proykova, *Comp. Phys. Commun.* **147/1–2**, 238 (2002).
39. G. H. Martina, M. L. Klein, *J. Chem. Phys.* **97**, 2635 (1992).
40. R. A. Radev, A. Proykova, Feng-Yin Li, and R. S. Berry, *J. Chem. Phys.* **109**, 3596 (1998).
41. T. S. Dibble and L. S. Bartell, *J. Phys. Chem.* **96**, 8603 (1992).
42. A. Boutain, B. Rousseau, and A. H. Fuchs, *Chem. Phys. Lett.* **218**, 122 (1994); A. Boutin, J. B. Maillet, and A. H. Fuchs, *J. Chem. Phys.* **99**, 9944 (1993).
43. J. Jellinek, T. L. Beck, and R. S. Berry. *J. Chem. Phys.* **84**(5), 2783 (1986); Y. Zhou, M. Karplus, K. D. Ball, and R. S. Berry, *J. Chem. Phys.* **116**, 2323 (2002).
44. C. Amitrano and R. S. Berry, *Phys. Rev. Lett.* **68**, 729 (1992); C. Amitrano and R. S. Berry, *Phys. Rev. E* **47**, 3158 (1993).
45. T. L. Beck, J. Jellinek, and R. S. Berry, *J. Chem. Phys.* **87**, 545 (1987).
46. Y. Kondo, Q. Ru, and K. Takayanagi, *Phys. Rev. Lett.* **82**, 751 (1999).

A CALCULUS FOR RELATING THE DYNAMICS AND STRUCTURE OF COMPLEX BIOLOGICAL NETWORKS

R. EDWARDS

Department of Mathematics and Statistics, University of Victoria, Victoria, BC, Canada V8W 3P4

L. GLASS[†]

Centre for Nonlinear Dynamics in Physiology and Medicine, Department of Physiology, McIntyre Medical Sciences Building, McGill University, Montréal, Québec, Canada H3G 1Y6

CONTENTS

[†]We dedicate this to Stuart Rice in honor of his 70th birthday.

Adventures in Chemical Physics: A Special Volume in Advances in Chemical Physics, Volume 132, edited by R. Stephen Berry and Joshua Jortner. Series editor Stuart A. Rice

I. INTRODUCTION

Biological systems at the cellular level display a bewildering complexity. We are all familiar with the charts displaying various biochemical processes such as the citric acid cycle [1, p. 106]. These classical diagrams are now being enriched and enhanced by results obtained using new technologies that demonstrate enormous complexity in the organization of metabolic processes [2–4]. Scientists and industrial concerns are interested in developing methods that can be used to simulate such systems—including in some cases equations with hundreds, if not thousands, of variables representing chemical concentrations of chemical species in different subcellular compartments of the cell. Although one of the unifying themes of science is that the real world, when understood, will be simple, there is a growing perception that biology is complex. It is possible that biology is truly as complex as it now appears. But we should not discount the possibility that we are still at an early stage of understanding and that unifying simplifications will emerge as the depth of our knowledge increases.

The current survey is based on a scheme that was conceived as a way to relate qualitative aspects of the structure and dynamics of complex biological sytems.

Our approach originates with an article by Jacob and Monod from the early 1960s that argued that genetic regulatory elements, which had just been discovered as controllers of transcription of proteins in bacteria, could be strung together to generate regulatory modules that could underly such processes as cellular differentiation or oscillation [5]. The nonmathematical models of Jacob and Monod were cast into a mathematical framework by several researchers including Sugita [6], Kauffman [7, 8], and Thomas [9–11]. The main idea of this early work was to represent genetic control circuits by Boolean switching circuits in which time is discrete and the states of each gene at each time is also discrete—either 1 or 0 ("active" or "inactive"). Kauffman proposed that attractors of the discrete network corresponded to cell types in organisms [7]. Mathematical analysis of the properties of randomly generated Boolean networks, often called Kauffman networks, has been an area of intense theoretical interest [12–18].

However, since clocking devices that lead to synchronous updating of variables are unknown in cellular biological systems, it is more natural to develop differential equations to model the dynamics. Yet, it seems reasonable to preserve the connection and logic of the underlying circuit. Indeed, it is just these interactions that are typically presented in the schematic network diagrams of biochemists. Thus, the basic idea of the current work is to embed the logical structure of a genetic network into a differential equation. Over the years, differential equations in which logical or sigmoidal functions have been embedded as a key control component have been proposed for a wide range of different biochemical and gene networks, including networks describing feedback inhibition [19, 20], the immune system [21], bacteriophage [22, 23], drosophila [24, 25], and synthetic genetic regulatory networks [26, 27]. We do not attempt a complete listing of all the relevant articles, but refer readers to recent excellent reviews [28–31].

In the current review, we do not describe models for individual systems, but rather we develop mathematical methods that can be used to predict and compute the qualitative features of dynamics based on the structure of the network. By qualitative features we mean the asymptotic dynamics of the network including its fixed points, oscillations, or chaotic dynamics. Questions of a qualitative nature include the patterns of change of the variables and the ways in which the dynamics can change as the parameters describing the equations vary. Less central to the qualitative approach is a detailed knowledge of the parameter values describing the equations. The basic assumption, which is unproven in general, but which has been elegantly demonstrated in a particular genetic control network [25], is that the dynamical behavior will emerge robustly for large sets of parameters. Indeed, it is impossible that the detailed kinetic equations in any two cells are the same in a single individual or species, yet cell types may be readily distinguished and classified. Moreover, control circuits between different but related species are surely different, yet preserve important qualitative dynamical and morphological features. Although much current research is being driven by the vast amounts of computer power combined with increased knowledge of biological parameters that may make detailed quantitative models feasible, there is still the danger that the detailed models will be fragile to small parameter changes and will in any case not yield transparent insights into observed dynamics.

In Section II, we demonstrate this approach with an equation that incorporates a nonlinear Hill function to model genetic control representing a mutually inhibitory network of two elements [26], and an inhibitory loop of three elements [27]. Although theoretical models of these types of networks have been known for at least 30 years [20, 32, 33], they took on new life in 2000 with the construction of genetic regulatory circuits in bacteria that were well described by the equations.

In Section III, we show that in the limit when the Hill function control becomes infinitely steep, complex functional control can be represented by logical functions. In this limit the continuous nonlinear differential equation becomes piecewise linear, and the equation can be explicitly integrated by piecing together the linear trajectory segments. The simple two- and three-dimensional networks can be extended to N dimensions, where N is the number of variables. These piecewise linear equations were initially presented in Refs. 33 and 34.

In the remainder of this review we focus on the mathematical analysis of the piecewise linear equations largely based on earlier studies [34–48]. In Section IV we show that the logical structure can be mapped onto a hypercube in N dimensions, where each vertex of the hypercube represents the state of each of the N variables, and the dynamics and logical structure in the network are represented as directed edges on the hypercube.

We then use the hypercube representation to carry out a nonlinear dynamical analysis of these networks. The key insight is that quantitative aspects of flows in phase space can be computed from linear fractional maps that represent the flows between boundaries on the hypercube. Analysis is possible because the composition of two linear fractional maps is a linear fractional map. This analysis is useful for analyzing steady states, limit cycles, and chaotic dynamics in these networks.

The hypercube representation is also useful for examining problems in discrete mathematics suggested by this work. In Section VI we count the number of distinct networks under the symmetry of the hypercube, and we also show how to classify chaotic dynamics.

Finally, in Section VII we discuss the applicability of these methods to study real biochemical networks. There are many simplifications in these equations, and there are still few rigorous results that demonstrate the applicability of these methods to more realistic equations.

II. NONLINEAR DIFFERENTIAL EQUATIONS FOR GENE CONTROL

A gene is a sequence of DNA bases that codes for a sequence of amino acids that constitute a protein (see Ref. 1 for more details on molecular biology). Even though all cells have the potential to make any protein, there are a variety of regulatory mechanisms (still not completely understood) that determine which proteins will be synthesized in each cell. There is also a sequence of steps involved in translating the information contained in the DNA into the actual synthesis of a protein. One mode of gene regulation is through a class of proteins called transcription factors. Transcription factors bind to the DNA turning "on" and "off" the synthesis of specific mRNA molecules that carry the sequence of target genes from the nucleus, where they are stored, to the cytoplasm, where the mRNA sequence is translated into an amino acid sequence. We simplify this complex process in a schematic way by assuming that x_i is the concentration of a chemical species in a cell. Furthermore, the time rate of change of x_i is

$$\frac{dx_i}{dt} = \beta_i h_i(\mathbf{x}) - \gamma_i x_i, \qquad i = 1, \ldots, N \tag{1}$$

where there are N chemical species, $\mathbf{x}$ is a vector giving their concentrations, h_i is a function giving the control of the synthesis of the *ith* chemical species by the others, and λ_i and γ_i are production and decay constants. Differential equation models of biochemical and genetic networks are often of this form (for example, see Refs. 19–34, 37).

Recent work has constructed genetic circuits in bacteria in which there was a loop of either two [26] or three [27] interacting genes. The transcription factor coded by each of the genes, inhibited the synthesis of the next in the loop. The behavior in this type of network can be modeled by the differential equation

$$\frac{dx_1}{dt} = \beta_1 \frac{\theta_1^{n_1}}{\theta_1^{n_1} + x_N^{n_1}} - \gamma_1 x_1, \qquad \frac{dx_i}{dt} = \beta_i \frac{\theta_i^{n_i}}{\theta_i^{n_i} + .x_{i-1}^{n_i}} - \gamma_i x_i, \quad i = 2, \ldots, N \tag{2}$$

where x_i is a concentration of a transcription factor, and θ_i and n_i are constants that can be determined from experimental data. The nonlinear function regulating control is called the Hill function, and n_i is called the Hill coefficient. By including additional equations for the mRNA molecules, these equations can be made more realistic [27]. However, this modification does not affect the qualitative aspects of the dynamics, and therefore in this chapter we do not include explicit equations for mRNA. These and other closely related equations have also been considered and analyzed much earlier [32, 34, 35] as abstract models of nonlinear chemical networks and their properties are well known in some special cases. We summarize this analysis.

To simplify matters we assume that all the Hill coefficients are equal and are given by n, that all the decay and production coefficients are equal to 1, and that $\theta_i = 1/2$ for all values of i.

Figure 1 gives the phase plane portrait of Eq. (2) for $N = 2$, $n = 6$. In this two-dimensional model there are two stable fixed points and an unstable saddle point. At the stable fixed points, one of the transcription factors is at a high concentration and inhibits the synthesis of the other factor. This forms the basis for the synthesis of the toggle switch in Ref. 26. Of course, in the real systems, there were major difficulties in constructing a system that had appropriately set parameters. To appreciate this, notice that the eigenvalues of the fixed point at $x_1 = x_2 = 1/2$ are given by $\lambda_{1,2} = -1 \pm n/2$. Consequently for values of $n < 2$, there is only a single stable fixed point and a toggle behavior would not be predicted.

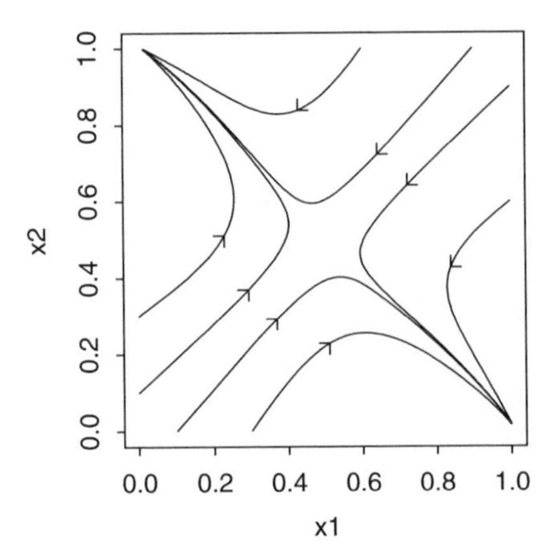

Figure 1. Phase portrait for the bistable toggle switch of Eq. (2) with $N = 2$, and for $i = 1, 2$ using $n_i = 6$, $\theta_i = \frac{1}{2}$, $\beta_i = \gamma_i = 1$.

Figure 2 shows a time trace of the dynamics for $N = 3$, $n = 6$. Now the dynamics follow a stable limit cycle oscillation. This forms the basis for the synthesis of the repressilator [27]. In this case the eigenvalues of the fixed point at $x_1 = x_2 = x_3 = 1/2$ are $\lambda_1 = -1 - n/2$ and $\lambda_{2,3} = -1 + n/4 \pm \sqrt{3}ni/4$ [34]. In this case there is a Hopf bifurcation when $n = 4$ so that for values of $n > 4$ there is a stable limit cycle oscillation corresponding to the repressilator. If the equations for the dynamics of mRNA are included, then oscillations are still found, but now oscillations can be found for smaller values of the Hill

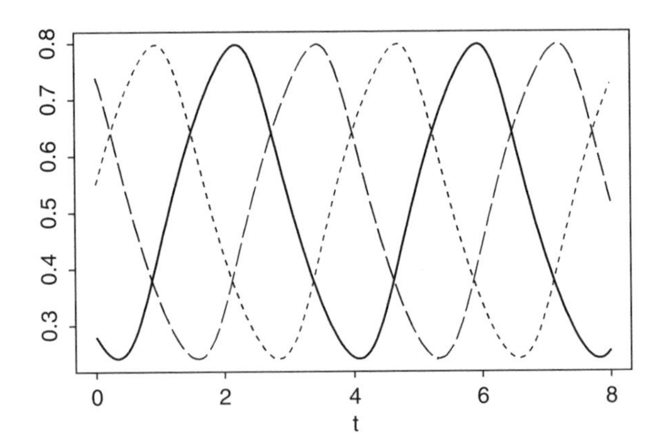

Figure 2. Time trace for the "repressilator" of Eq. (2) with $N = 3$ and other parameters as in Fig. 1 for $i = 1, 2, 3$.

coefficient, n [27]. If we take the limit $n_i \rightarrow \infty$ in Eq. (2), we generate a piecewise linear equation that can be explicitly integrated [33, 34, 37, 38, 47]. This observation forms the basis for the generalization of this example to a much more general class of equations.

We repeat these simple examples from earlier work since they form a major justification for the elaborations we report below, and if they are understood well, then much of what follows should be clear. The dynamics in the differential equations are in qualitative agreement with the dynamics in the synthetic genetic networks in *E. coli.* One important aspect of biology is to understand the ways in which the organization and structure of the control networks can be used to predict the dynamics. Thus, we would like to develop methods that could be used to predict the dynamic behaviors just demonstrated without integrating or carrying out the stability analysis of the differential equations.

III. A DIFFERENTIAL EQUATION

A Boolean switching network with N elements is represented by

$$X_i(t+1) = F_i(X_{i_1}(t), X_{i_2}(t), \ldots, X_{i_K}(t)), \qquad i = 1, \ldots, N \tag{3}$$

where $F_i(X_{i_1}(t), X_{i_2}(t), \ldots, X_{i_K}(t)) \in \{0, 1\}$ and K is the number of inputs to each element. This is a discrete time and discrete state space system which, therefore, must eventually reach a fixed point or cycle under iteration.

Since biological systems are not believed to have clocking devices that simultaneously update the network, a differential equation would be a more suitable class of mathematical model. The logical structure of Eq. (3) can be captured by a differential equation [33, 34, 38]. To a continuous variable $x_i(t)$, we associate a discrete variable $X_i(t)$,

$$X_i(t) = 0 \quad \text{if } x_i(t) < 0; \qquad \text{otherwise } X_i(t) = 1 \tag{4}$$

For any logical network, we define an analogous differential equation,

$$\frac{dx_i}{dt} = -\gamma_i x_i + f_i(X_{i_1}(t), X_{i_2}(t), \ldots, X_{i_K}(t)), \qquad i = 1, \ldots, N \tag{5}$$

where $f_i(X_{i_1}(t), X_{i_2}(t), \ldots, X_{i_K}(t))$ is a scalar whose sign is negative (positive) if the corresponding logical variable $F_i(X_{i_1}(t), X_{i_2}(t), \ldots, X_{i_K}(t))$ is 0 (1). Notice that f_i makes explicit the control function h_i in Eq. (1). In the biochemical context, all variables will be positive. By a change of variables, we can make all variables positive. This would also lead to a positive value for the thresholds that are used to define the Boolean state and that are taken to be equal to 0 in Eq. (4).

For each variable, the temporal evolution is governed by a first-order piecewise linear differential equation. Let $\{t_1, t_2, \ldots, t_k\}$ denote the *switch times* when any variable of the network crosses 0. The solution of Eq. (5) for each variable x_i for $t_j < t < t_{j+1}$, is

$$x_i(t) = x_i(t_j)e^{-(t-t_j)} + f_i(X_{i_1}(t_{j^*}), X_{i_2}(t_{j^*}), \ldots, X_{i_K}(t_{j^*}))(1 - e^{-(t-t_j)}) \quad (6)$$

where t_{j^*} is any time in (t_j, t_{j+1}). This equation has the following property. All trajectories in a given orthant in state space are directed toward a focal point. If the focal point lies in a different orthant from the initial condition, then, in general, a threshold hyperplane will eventually be crossed. When the threshold hyperplane is crossed, a new focal point will be selected based on the underlying equations of motion. This equation is also a generalization of equations, often called Hopfield networks, that model neural systems [39, 40, 48].

Even though Eq. (5) is more realistic than Eq. (3) as a model for biological systems, this equation still is highly oversimplified. Yet this equation has remarkable mathematical properties that facilitate theoretical analysis. Moreover, there is an expectation, demonstrated in some simple examples like those discussed above, that the qualitative dynamics in the model system will be preserved in more realistic versions, for example when the discontinuous step functions are replaced by continuous sigmoidal functions [33, 34, 37]. As mentioned above, synthetic gene networks have been created that show some of the simple types of dynamical behavior found in our class of networks—in particular, bistability (two fixed points) [26] and oscillation in an inhibitory loop [27].

IV. THE *N*-CUBE

These differential equations admit striking properties that lie at the heart of our analysis [34, 38]. In general, only one variable will cross its threshold at a given time. As a consequence, the flows in the differential equation can be mapped onto a Boolean hypercube, where each vertex of the hypercube represents an orthant of phase space, and each edge of the hypercube represents the open boundary between two neighboring orthants. Each edge is directed depending on the orientation of the flow across that boundary. Thus a coarse-grained picture of the allowed dynamics can be appreciated from the digraph on the N-dimensional hypercube (the N-cube).

As an example, consider the truth tables that are appropriate for the toggle switch with two variables, and the repressilator, shown in Table I. The associated N-cubes are shown in Fig. 3. The orientations on edges are determined using the following algorithm. First define the Hamming distance between two Boolean

TABLE I
Boolean Switching network for Networks that Correspond to the Structure of (a) Fig. 3a, and (b) Fig. 3b.[a]

X_1	X_2 (t)	X_1	$X_2(t+1)$
0	0	1	1
0	1	0	1
1	0	1	0
1	1	0	0

(a)

X_1	X_2	$X_3(t)$	X_1	X_2	$X_3(t+1)$
0	0	0	1	1	1
0	0	1	0	1	1
0	1	0	1	1	0
0	1	1	0	1	0
1	0	0	1	0	1
1	0	1	0	0	1
1	1	0	1	0	0
1	1	1	0	0	0

(b)

[a] Differential equations associated with these truth tables are defined by Eqs. (4) and (5) with $\gamma_i = 1$, where we assume that $f_i = 1$ is associated with $F_i = 1$ and $f_i = -1$, with $F_i = 0$.

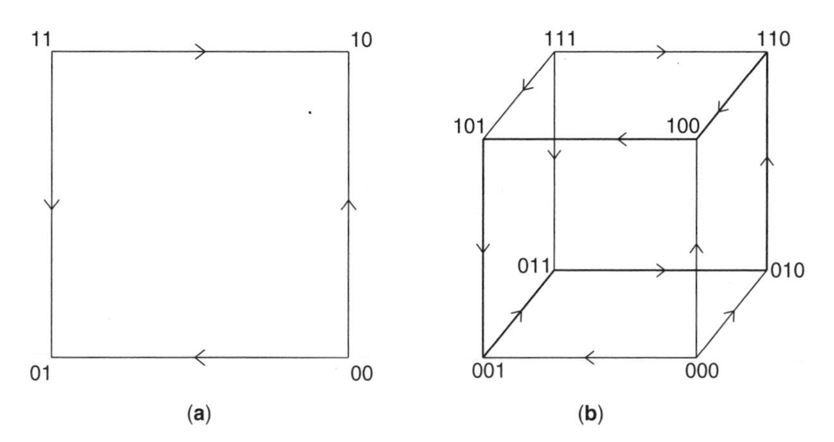

Figure 3. Hypercube structures for the bistable toggle switch of Table Ia and, Ib, respectively.

states to be the number of digits in which they differ. From each Boolean state at time t in the left-hand column of the truth table, determine the Hamming distance to the next state at time $t+1$ in the right-hand column of the truth table. Now draw directed edges from each vertex to all adjacent vertices that lie on a shortest (undirected) path from the vertex corresponding to the state at time t to the vertex corresponding to the state at time $t+1$. This will generate a number of directed edges corresponding to the Hamming distance between the two states.

This construction has special properties for the situation in which the differential equations have no self-input; that is, X_i is NOT one of the inputs of F_i in Eq. (5) [36]. In this case, the Hamming distance from each column of the truth table at time t to the same column at time $t+1$ is 2^{N-1}, so that the total Hamming distance for the whole truth table is $N \times 2^{N-1}$. This number is identical to the number of edges in the N-cube. Furthermore, from the condition of no self-input, no edge can be directed in two different orientations. Consequently, on the N-cube, each edge is directed in a unique orientation and there is a 1:1 correspondence between the N-cube and the truth tables of Boolean networks with no self-input. Furthermore, the N-cube representation that shows the allowed flows between neighboring orthants for any differential equation of the form in Eq. (5) can be constructed. Conversely, if thresholds are known, then by observing dynamic behavior, it should be possible to determine the underlying logical structure of the network.

An underlying motivation for our mathematical analysis is to answer the following questions: *Once the logical structure in Eq. (3) of a network is set, then what are the possible dynamics in the associated differential equation, Eq. (5), for any choice of parameters? What are the dynamics for each particular set of parameters?* We are only able to answer this question in some limited cases.

A. Numerical Integration, Types of Attractors, and Decoding the N-Cube

Because of their simple mathematical structure, these differential equations admit a simple method for integration. The method consists of setting an initial condition and determining the set of times, τ_i, when $x_i = 0$ from Eq. (6). However, if the system is in an orthant such that none of the variables will ever cross 0 (i.e., the focal point coordinates f_i have the sign of x_i for each i), the system will approach a steady state in that orthant. Otherwise, the system will cross to a new orthant at the time $t_{\min} = \min\{\tau_i\}$. Once again, using Eq. (6), the equations can be analytically integrated, and then the process is iterated.

These equations can display a whole range of qualititatively different types of dynamics, including fixed points (nodes or stable foci), limit cycles, chaos, and quasi-periodicity. In this section we briefly describe these different types of behavior, and in the next section we develop analytic methods to analyze particular networks.

Nodes occur when a focal point lies in its own orthant, as described above. Clearly they are asymptotically stable, because all trajectories in that orthant must converge to the focal point. The signature for a stable node in the N-cube graph is a vertex with N-edges directed toward it (e.g., the states 01 and 10 in Fig. 3a).

Stable foci occur when two or more variables switch in some sequence but approach zero as the number of switchings approaches infinity, while all other variables (if any) converge to some (possibly nonzero) value. Thus, stable foci typically have an associated cyclic sequence of orthants through which an approaching trajectory passes. A necessary condition for a stable focus is a cycle on the N-cube digraph. We say that a cycle has dimension k if k is the dimension of the smallest hypercube on which the given cycle can be drawn.

Limit cycles are closed trajectories toward which at least some nearby trajectories converge and they of course have an associated cyclic sequence of orthants, through which the limiting trajectory passes. When the focal points have coordinates that are all ± 1 [i.e., all $f_i = \pm 1$ in Eq. (5)], it is possible for a limit cycle to involve simultaneously switching variables, but this is a nongeneric situation that we do not consider further here [44]. A necessary condition for a limit cycle is once again a cycle in the N-cube. Sharper results are available for a class of cycles called cyclic attractors. In a cyclic attractor, each vertex on the cycle is adjacent to $N - 2$ vertices not on the cycle, and all the edges from these vertices are directed toward the cycle. The cyclic attractor in two dimensions is associated with a stable focus, whereas all cyclic attractors in higher dimensions are associated with stable limit cycle oscillations (see Ref. 38 for a proof). For example, there is only one three-dimensional cyclic attractor, and this is the cycle found for the repressilator (Fig. 3b). However, this class of stable limit cycles represents only a very small fraction of the actual number.

Quasi-periodicity arises if there are two or more disjoint stable limit cycle attractors in which none of the variables of one of the cycles receive inputs from the variables involved in the other cycles, and the periods of all cycles are noncommensurate.

Chaotic dynamics are by definition aperiodic dynamics in deterministic systems with sensitive dependence on initial conditions. Although many refer to chaotic dynamics in Eq. (3), this does not conform to standard nonlinear dynamics terminology since such systems have a finite state space and must necessarily cycle. However, in Eq. (5), chaotic dynamics are possible and have been demonstrated numerically and analytically in some example networks [42, 46]. We have no way to predict whether any particular logical structure is capable of generating chaotic dynamics for some set of parameters. A necessary condition for chaotic dynamics is that there is a vertex that lies on at least two different cycles. In the next section, we develop mathematical techniques to analyze the dynamics.

V. ANALYSIS

A. Linear Fractional Maps

In many cases the dynamics in Eq. (5) are amenable to theoretical analysis. The main theoretical insight is that if all the decay constants γ_i are equal, the maps that take the flows from one orthant boundary to the next have a simple form called a *linear fractional map*:

$$M(\mathbf{y}) = \frac{A\mathbf{y}}{1 + \langle \phi, \mathbf{y} \rangle} \tag{7}$$

where $\mathbf{y} \in \mathbf{R}^N$ is a point on the initial orthant boundary (in an N-dimensional network), A is an $N \times N$ matrix, $\phi \in \mathbf{R}^N$, and $\langle \phi, \mathbf{y} \rangle$ represents a vector dot product between ϕ and $\mathbf{y}$. A and ϕ depend on the focal point coordinates f_i of the flow for the orthant being traversed. The composition of two linear fractional maps of the same dimension is once again a linear fractional map. As a consequence of this property, if there is a cycle, then we can analytically (usually with the assistance of a computer) compute the return map for a given cycle of orthants starting on a particular orthant boundary crossing (Poincaré section). Since we start and end on the same orthant boundary, we can suppress that coordinate in A, ϕ, and $\mathbf{y}$, so that the dimension of the return map for a cycle will be one less than the dimension of the network. The remainder of this section deals only with the case in which all the decay constants are equal so that we can exploit the properties of the return map.

B. Limit Cycles and Stable Foci

Using the return map, it is possible to prove the existence and stability of limit cycles analytically [38, 41, 44]. The linear fractional return map is defined on a region (called a *returning cone*) that may be only part of the orthant boundary, from which trajectories follow the cycle of orthants under consideration. The returning cone is defined in terms of a set of linear inequalities, also calculated from the focal point coordinates. Calling $\{\lambda_i\}$ the eigenvalues of the $(N-1) \times (N-1)$ matrix A and letting λ_1 be the dominant eigenvalue and $\mathbf{v}_1$ its corresponding eigenvector, stable periodic orbits through a given cycle of orthants correspond exactly to the following conditions being met:

1. $\lambda_1 > 1$ (and real).
2. $\lambda_1 \geq |\lambda_j|$, for $j > 1$.
3. $\mathbf{v}_1$ lies in the returning cone (or on its boundary).

If $\lambda_1 = |\lambda_2|$, then stability is only neutral. If $\mathbf{v}_1$ lies on the boundary of the returning cone, then the cycle is degenerate in the sense that somewhere around the cycle there occurs a simultaneous switching of two or more variables in the limit $t \to \infty$.

Thus, if numerical integration suggests convergence to a limit cycle, this can be confirmed or disproved by selecting any orthant boundary crossing along that cycle and calculating the return map and the returning cone, as well as the eigenvalues and eigenvectors of the map. If the above conditions are met, then it is indeed a stable limit cycle. If not, then integration should be continued as the trajectory must eventually deviate from the purported cycle. Of course, these calculations to verify the existence of a stable limit cycle must also generally be done by computer and so accuracy must be considered. In the class of networks with all focal point coordinates at ± 1 and no self-input, the matrix A and the vector ϕ in the return map have integer components and $\det(A) = 1$.

If a repeating sequence of orthants is checked by the above procedure and the dominant eigenvalue, λ_1, is exactly 1, then the cycle corresponds to a stable focus [44]. There must still be an eigenvector in the returning cone, but $\lambda_1 = 1$ and $\det(A) = 1$ imply $|\lambda_j| = 1$ for all j, so it will often happen that λ_1 has an eigenspace of dimension greater than 1, which makes it more difficult in practice to check whether the eigenspace intersects the returning cone.

Limit cycles in networks of dimension as small as four can be surprisingly long and complicated, even when the focal point coordinates are all ± 1. For example, the network with logical structure depicted in Fig. 4 has an asymptotically stable limit cycle in which 174 switches take place before the cycle repeats. Projections of the phase portrait of this limit cycle attractor are shown in Fig. 5. The existence and stability of this limit cycle were confirmed analytically by the methods outlined above. We have found four-dimensional networks with confirmed limit cycles involving as many as 252 switches before repeating. Gedeon [48] has constructed four-dimensional networks with focal point coordinates tuned so that limit cycles of arbitrary length can be obtained.

C. Chaotic Dynamics

Numerical simulations in some networks, even some with all $f_i = \pm 1$, seem to show irregular behavior that persists indefinitely. However, we also know that there are very long and complicated limit cycles such as the one in the previous subsection. So it is not clear *a priori* whether the apparently irregular behavior in such networks truly reflects chaotic attractors or whether we are seeing long transients or very long stable periodic orbits. There are, however, analytic methods to demonstrate in particular examples that trajectories do not approach such long stable periodic orbits. The results for the existence and stability of limit cycles from the previous sections are useful in this regard.

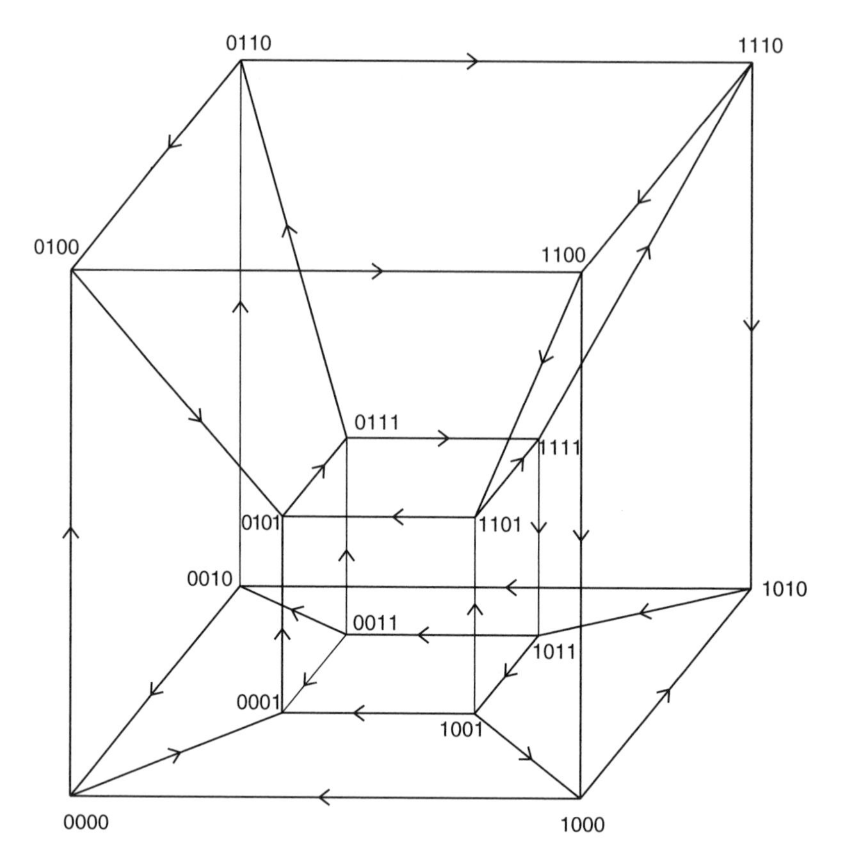

Figure 4. Hypercube structure of a four-dimensional network with a periodic orbit following a cycle on which 174 switches take place before repeating. The edges traversed by the cycle are marked by bold lines.

If we look at a particular orthant boundary on such a trajectory, consider it as a Poincaré section, and determine the cycles followed by the trajectory that pass through this boundary, there will in general be several of them that we can label *A, B, C*, and so on. The trajectory can then be represented as a sequence of these cycles, a "word" made up of the symbols or letters *A, B*, ..., at least as long as the trajectory always returns to the starting boundary. We can calculate the returning cones for each of these cycles.

It is often possible to show that either the union of these cones or some subset of them form a "trapping region." That is, there are cones T_A for the *A* cycle, T_B for the *B* cycle, and so on, such that the image of $T = \cup_k T_k$ (where k ranges over the letters of the possible cycles) is in T. We define a (convex) "cone"

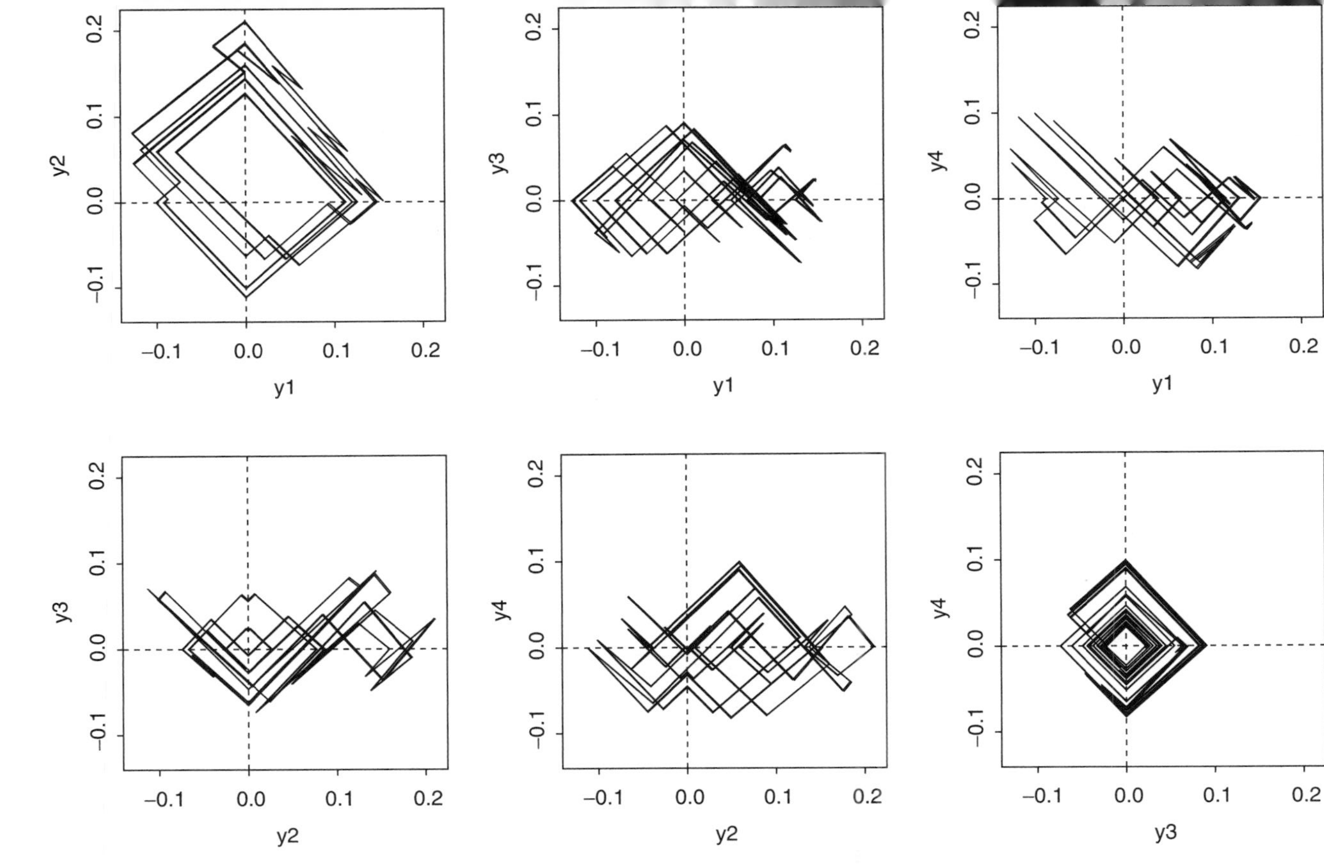

Figure 5. Phase space projections (in each pair of variables) of a periodic orbit with switching sequence of length 174 on the four-dimensional network of Fig. 4.

(following Berman and Plemmons [49, p.2]) as a set in $\mathbf{R}^N$ closed under nonnegative linear combinations. Recall that there is a linear fractional map describing the returning point for every point in T_A, a different linear fractional map for points in T_B, and so on. The image of T_A may crosscut the regions T_k, but we are supposing that it lies completely in the region T, and similarly for the image of each T_k. Such a region was found, for example, in the network described by Mestl et al. [42], where only two cycles were involved. The number of cycles may of course depend on our choice of Poincaré section. In practice, we look for the boundary giving the fewest cycles.

Given that we have found such a trapping region involving two or more cycles, the question of the existence of a stable limit cycle comes down to whether or not there is a symbolic word whose corresponding mapping has a stable fixed point. Now the crucial property of the linear fractional maps associated with cycles, namely that their compositions are also linear fractional maps of the same form, comes into play. The matrix in the numerator of the map, Eq. (7), for this composite cycle is the product (in reverse order) of the matrices for the individual component cycles, which we can also represent by the letters *A, B, C*, and so on. Thus, the cycle with symbolic word C, B, A, B has a Poincaré map with matrix *BABC* in the numerator. The scaling represented by the denominator guarantees that trajectories starting on the same ray converge under iteration (while staying on the same ray), so the dynamics are essentially those of the matrix multiplications, modulo the radial direction. It is necessary in the end to show that the radial convergence is not to the origin if we are to have a true chaotic attractor. An immediate consequence of the observation that all points on a ray must converge under iteration is that chaotic dynamics is not possible for 3-cubes since in three dimensions, the projection of a ray onto a surrounding 2-sphere must approach a fixed point or periodic path [41].

This identification of the dynamics on the Poincaré section with matrix products, in the light of the result of the previous subsection, suggests an approach to ruling out the existence of stable periodic orbits. In order for a stable periodic orbit to exist in a trapping region, *T*, in an orthant boundary, there must be a word *W* whose corresponding matrix has its dominant eigenvector in *T*. The strategy then, is to show that dominant eigenvectors of all words in *A, B, C,* etc., are confined to some region of $\mathbf{R}^{(N-1)}$ **not** including *T*. This is made feasible by generalizations of the Perron–Frobenius Theorem, which guarantee that the dominant eigenvector of a matrix lies in a proper (i.e., closed, with $K \cup (-K) = \{0\}$ and nonempty interior [49]) cone, *K*, if that cone is invariant under the matrix multiplication—that is, if $AK \subset K$. Then a proper cone *K* that is invariant for **all** our cycle matrices, *A, B, C*, ... simultaneously, is invariant for any word *W* composed of these matrices, and therefore the dominant eigenvector of any such word lies in *K*. To ensure that the dominant eigenvector is unique (so that we can't have another dominant eigenvector outside of *K* and therefore possibly in T), we need words W to map K into the **interior** of K. If some of the individual cycle matrices map some boundary points of K to boundary points of K, then we require at least that words beyond some finite length map K to its interior and then we can check short words individually for dominant eigenvectors in T.

The method of proof of nonexistence of limit cycles in a trapping region T is thus to look for a proper cone, K, that is invariant for all the cycles possible from T and ensure that it does not intersect T (except at the origin). This sounds like a difficult task, and indeed it is not trivial, but for 3×3 matrices and few cycles (2 is easiest), it is often possible to construct such a common invariant cone from eigenspaces of the matrices involved, their intersections, and their images. An initial attempt to tackle this problem of common invariant cones in general can be found in Ref. 50. Some specific examples of $N = 4$ networks are shown to have trapping regions with no stable limit cycles by the above approach in Ref. 46.

For example, consider the four-dimensional network

$$
\begin{aligned}
\dot{x}_1 &= -x_1 + 2(\bar{X}_3 X_4 + X_2 X_3) - 1 \\
\dot{x}_2 &= -x_2 + 2(X_1 \bar{X}_3 X_4 + \bar{X}_1 X_3 X_4 + \bar{X}_1 \bar{X}_3 \bar{X}_4) - 1 \\
\dot{x}_3 &= -x_3 + 2(\bar{X}_1 X_2 + X_1 X_4) - 1 \\
\dot{x}_4 &= -x_4 + 2(X_2 \bar{X}_3 + \bar{X}_1 X_3) - 1
\end{aligned}
\tag{8}
$$

where $\bar{X}_i = 1 - X_i$. Note that the focal point coordinates are all ± 1. Figure 6 shows the hypercube for this network. An example trajectory is depicted in Fig. 7. If we consider a Poincaré section on the boundary $(+, +, +, 0)$, the trajectory in Fig. 7 follows only two cycles of orthants, which we call A and B:

$$
\begin{aligned}
A &: 1110 \to 1010 \to 0010 \to 0000 \to 0100 \to \mathbf{0110} \to 0111 \to 1111 \\
B &: 1110 \to 1010 \to 0010 \to \mathbf{0011} \to \mathbf{0001} \to 0000 \to 0100 \to \mathbf{0101} \\
&\quad \to 0111 \to 1111
\end{aligned}
$$

Cycle A is marked by bold lines in Fig. 6. This pair of cycles has a trapping region in the $(+, +, +, 0)$ boundary [46]. Furthermore, there is a proper cone, K, invariant for both matrices A and B (for the cycles A and B above) which therefore contains dominant eigenvectors of every word in A and B. This cone is depicted in Fig. 8. The points marked u_A, v_A, and w_A in the figure are the eigenvectors of matrix A, corresponding to eigenvalues in decreasing order of magnitude, projected onto the unit sphere. The points marked u_B, v_B, and w_B are the eigenvectors of the matrix B in a similar order. The points marked z_1 and z_2 are intersections of planes, each spanned by two eigenvectors of one of the

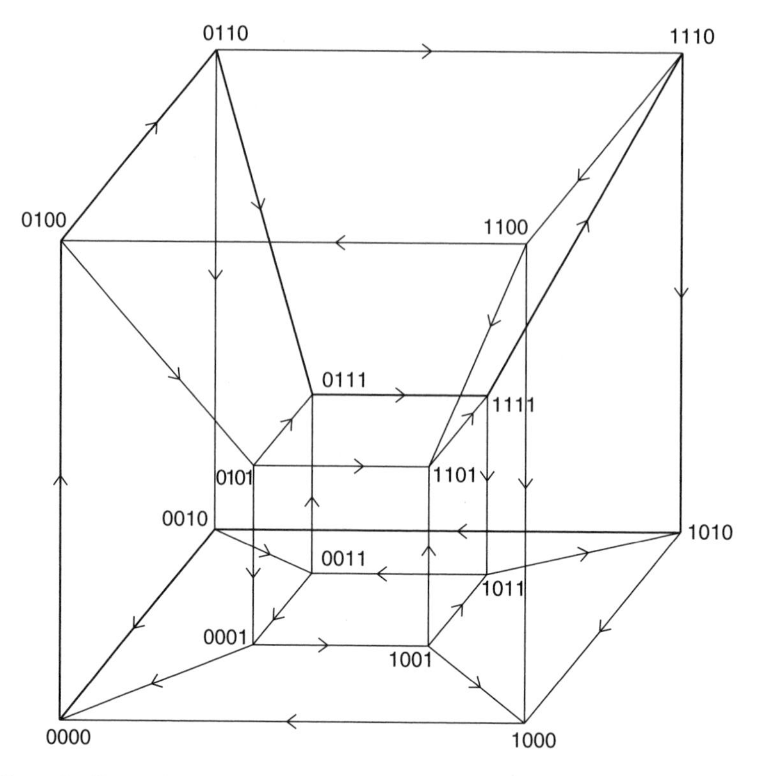

Figure 6. Hypercube structure of a four-dimensional network with a chaotic attractor. One of the two cycles followed by the chaotic trajectory is marked by bold lines.

matrices. The common invariant cone, K, is the one generated by the vectors $u_B, z_2, -Az_2$, and z_1 (i.e., the cone of all non-negative multiples of these vectors). Most importantly, K does not intersect the trapping region, which therefore contains no dominant eigenvectors of words, and so no cycle composed of cycles A and B can possibly have a corresponding stable limit cycle. Trajectories that get into the trapping region stay there but never reach a limit cycle or a fixed point [46].

VI. COMBINATORICS AND DISCRETE MATHEMATICAL APPROACHES

Our approach blends discrete methods and nonlinear dynamics. The previous sections showed how methods of nonlinear dynamics can be used to analyze dynamics in particular networks. However, the discrete nature of the underlying

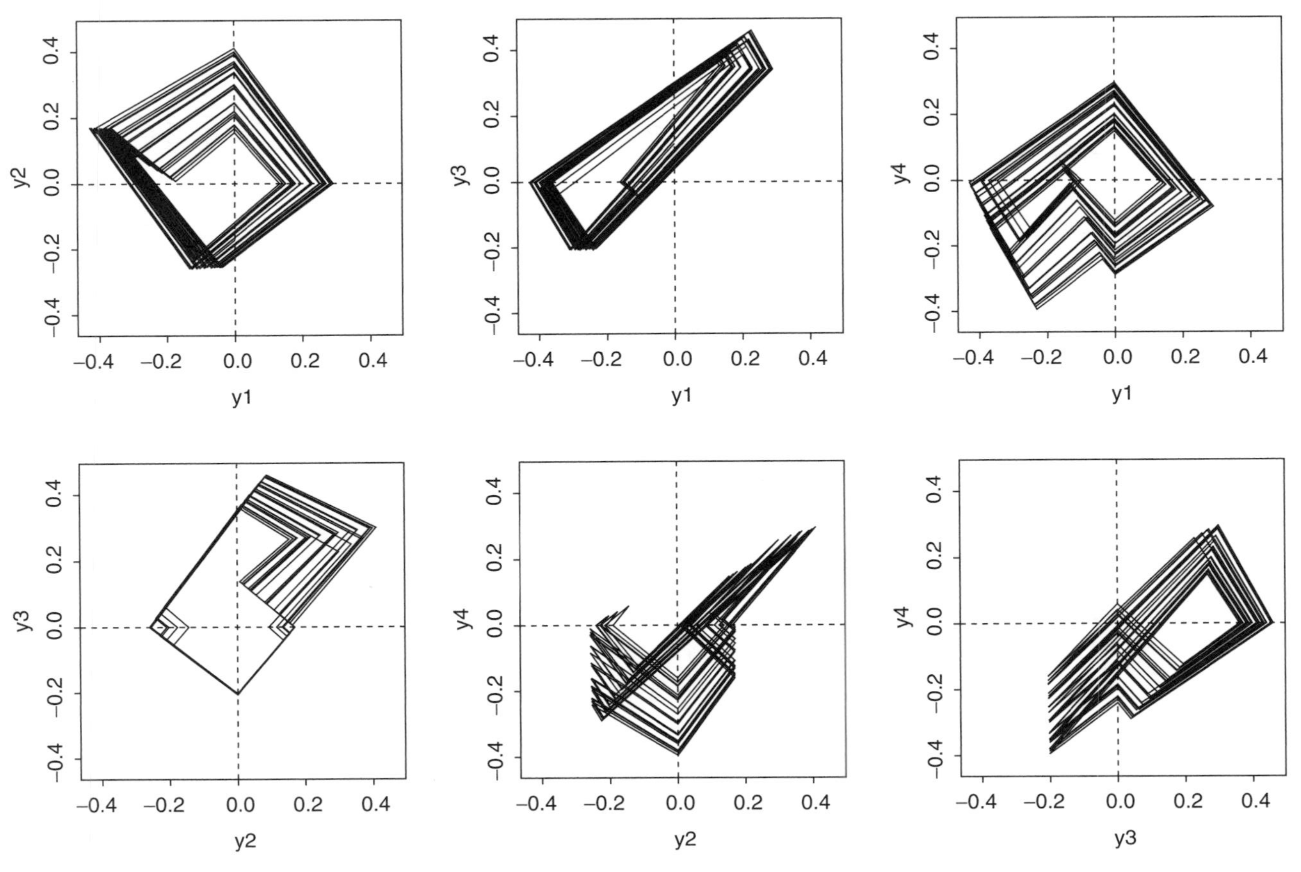

Figure 7. Phase space projections (in each pair of variables) of a chaotic attractor on the four-dimensional network of Fig. 6.

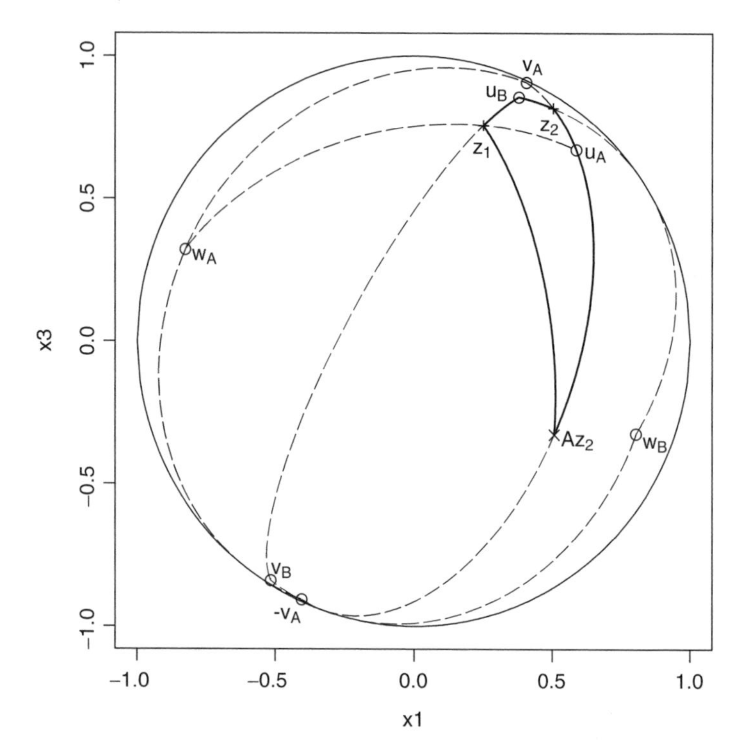

Figure 8. Invariant cone for the matrices A and B corresponding to Poincaré maps for the network of Eq. (8), projected onto the unit sphere, indicated by solid lines. Projections of the invariant subspaces through eigenvectors are indicated by dashed lines.

logical structure of Eq. (3), along with the coarse-graining of the state space, suggests a variety of discrete problems dealing with the classification of the networks and the dynamics in the networks.

A. Structural Equivalence Classes

The classification of logical network structures imposed by the hypercube description depends on the **signs** of the focal point coordinates f_i associated with each orthant of phase space, which leads to the hypercube representation of the allowed flows. We consider that two different networks are in the same dynamical equivalence class if their directed N-cube representations can be superimiposed under a symmetry of the N-cube. For example, in three dimensions there is only one cyclic attractor (see Fig. 3b), but this can appear in eight different orientations on the 3-cube. From a dynamical perspective, exactly the same qualitative dynamics can be found in any of these networks provided the focal points are chosen in an identical fashion. However, from a biological perspective, the underlying biochemistry for the networks can be different. Thus, the cyclic loop of three inhibitory elements in the repressilator leads to a cyclic attractor on the 3-cube (Fig. 3b). But so does a feedback inhibition loop of three elements in which there are two sequential activating interactions and a single inhibitory step.

Using the Polya enumeration theorem, it is possible to compute the number of dynamical equivalence classes as a function of N [34, 45]. A lower limit on this number can be obtained by dividing the number of different networks by the order of the symmetry group of the N-cube. Thus, the number of dynamical equivalence classes is greater than or equal to

$$\frac{2^{N \times 2^{N-1}}}{N!2^N}$$

The results using this computation are shown in Table II. Furthermore, the lower limit is accurate to at least seven significant figures for dimensions 5 and higher.

To get an idea of the relative numbers of different types of dynamic behaviors, we have carried out simulation of randomly selected networks with $f_i = \pm 1$. Out of the 11223994 structural equivalence classes in four dimensions, about 8 out of 1000 networks have stable limit cycles, while about 1 out of 1500 appear to be chaotic. The others go to stable foci or stable nodes. Amongst the periodic networks, a few have very long periods and complex switching sequences as discussed above in Section V. B. Thus, surprisingly complex but nevertheless periodic switching behavior is possible.

With networks of five units, the number of structural equivalence classes is already astronomical. We investigated dynamical behaviors of higher-dimensional networks by a numerical survey of networks with $N = 3$ to 9 units and $K = 1$ to $N - 1$ inputs per unit [51]. This showed that, at least for N in this range, the effect of increasing the network connectivity is to decrease the probability of converging to a stable node but increasing the probability of converging to a stable focus, while the probability of finding a limit cycle or a chaotic attractor seems to peak at about $K = 2$ or 3. We also surveyed networks

TABLE II
Numbers of Distinct Digraphs on the N-Cube, Considering Symmetries

N	Actual Number	Lower Bound
1	1	1.000
2	4	2.000
3	112	85.333
4	11 223 994	11 184 810.666
5	314 824 455 746 718 261 696	314 824 432 191 309 680 913.066

with much larger N in the case where $K = 2$. As the number of network units increases, the number of stable foci and stable nodes both decrease, while the number of limit cycles and chaotic attractors increase, though the probability of chaos remained below 10% for N up to 200. Finally, in randomly constructed networks where N is large, say greater than 50, and $K > 8$, the usual circumstance is that either the dynamics are chaotic or there are exceedingly long transient irregular dynamics [52].

B. Symbolic Dynamics of Bifurcations and Chaotic Dynamics

The above analysis of chaotic dynamics (Section V.C) in terms of symbol sequences, where symbols represent hypercube cycles (or equivalently the matrices in the cycle maps), shows that the chaotic attractors can be understood in terms of properties of the symbol sequences. The set of possible symbol sequences for one of these chaotic networks can be thought of as a "formal language" in the sense used in computer science [47]. For example, we investigated a chaotic network in which only two cycles were involved, A and B, but for which the sequence BB was not possible. The symbolic dynamics is that of the "golden mean map" of dynamical systems theory [44]. A bifurcation parameter allowed us to shift the dynamical regime from chaotic to periodic, but even within the chaotic range, changes in behavior could be identified, giving novel types of bifurcation of chaotic dynamics. For example, we found a value of the bifurcation parameter at which the "language" of the chaotic attractor changed from a two-letter to a four-letter one. Beyond the bifurcation, symbols C and D were required as two other cycles became possible from the Poincaré map. Another bifurcation occurs when the symbol sequence BB becomes possible [47]. These methods provide new ways to study bifurcations of the allowed symbolic representations of chaotic dynamics and provide an area for future mathematical research.

VII. THE ANALYSIS OF REAL BIOCHEMICAL NETWORKS

In the above, we have given a computational scheme that allows us to define the connections and interactions between components in biochemical networks and to determine the dynamics in the resulting networks. For an arbitrary network, it is not possible to give a precise description of the dynamics without carrying out numerical simulations. However, all the networks obey certain dynamic rules that are set by the structure as embodied in the directed N-dimensional hypercube. Moreover, for networks that show certain structural features, such as cyclic attractors, it is possible to make precise statements about the dynamics even without further mathematical analysis or simulation. In other cases, analytical techniques are available to give insight into the dynamics observed—for example, in the cases in which it is possible to prove limit cycles or chaotic dynamics. We expect that further development of the mathematical techniques given here will enable even more insight into the dynamics of these networks.

Independent of the development of a rigorous mathematical framework, a large number of studies have been carried out modeling real biological control networks by differential equations in which the logical structure is explicitly considered. We do not review all such studies, but representative examples include analysis of gene control in: the immune system [11, 21], the repressilator [27], lambda phage [23], and the toggle switch [26]. In all these cases the underlying logical structure of the network is capable of predicting important qualitative features of the associated differential equations that more precisely model the dynamics.

Since there are many important qualitative differences between real biological systems and our model networks, it is possible that the methods described above will be applicable to only a small range of model systems. Yet, there are reasons to believe that the methods will be applicable to a broad class of systems. In the remainder of this section, we discuss several different ways in which our model equation does not capture the properties of real biological systems, and then we give arguments why the methods may still apply.

The current equations are not realistic for many reasons: (i) Control of gene expression is not on or off but is graded; (ii) there may be time delays associated with synthesis or degradation of gene products, or with the transport of mRNA or transcription factors between different cellular compartments; (iii) decay rates of different gene products are different; (iv) although a single gene product might control expression of many different genes, the threshold levels for activation and/or inhibition may be different for different targets.

We wish to comment briefly on each of these factors and describe its influence on the dynamics of complex networks.

A. Sigmoidal Control of Gene Expression

In initial studies of model gene and biochemical networks, sigmoidal functions, such as the Hill function, were employed to model gene control. Simulations by ourselves and others showed that provided the sigmoid was sufficiently steep, the qualitative features of the dynamics were the same as with the step function control [19, 20, 32, 37]. Thus, for networks in which there are stable limit cycle oscillations, although the period of oscillations might change, there is still an oscillation with the same pattern of activity [37]. Furthermore, for chaotic networks, the statistical properties of the chaotic dynamics appears to be the same for very steep continuous sigmoidal functions, or for step functions [39, 40]. As the sigmoidal functions become steep, the dynamics can pass through a sequence of bifurcations of increasingly complex periodic behaviors before the chaos is established [39, 40]. Mestl and co-workers [53] proved that limit cycles that persist in sigmoid function networks with arbitrarily steep sigmoids must also exist in the limiting step function network. It has not been proven that limit cycles in a step function network persist under perturbation to sufficiently steep sigmoids. We are also not aware of any results concerning the stability of the chaotic dynamics. This is an area ripe for further mathematical studies.

Viewed from the biochemical side, it is interesting to consider the extent to which a logical description (usually employing sigmoids) is an appropriate idealization for biological control. There is a large body of experimental data that demonstrates sigmoidal control and attributes this to cooperative interactions between molecules [1]. In addition, macroscopic kinetics of chemical reaction networks can often generate kinetics that approximate logical control (see, for example, Ref. 54). Furthermore, it is possible that the cooperativity could be enhanced as a consequence of cascading biochemical processes, or the cooperative binding to many control sites in promoter regions of genes [55]. Many biologists still refer to genes as being switched "on" or "off," even though the experimental determination of Hill coefficients for gene activation or inhibition leads to a Hill coefficient of about 2, reflecting the dimeric binding of many transcription factors to promoters (see, for example, Chapter 7 in Ref. 1). Finally, a logical model for the control of gene expression in sea urchin was developed based on the analysis of extensive experimental data [56].

B. Time Delays in Biochemical Networks

Our initial studies of dynamics in biochemical networks included spatially localized components [32]. As a consequence, there will be delays involved in the transport between the nuclear and cytoplasmic compartments. Depending on the spatial structure, different dynamical behaviors could be faciliated, but the theoretical methods are useful to help understand the qualitative features. In other (unpublished) work, computations were carried out in feedback loops with cyclic attractors in which a delay was introduced in one of the interactions. Although the delay led to an increase of the period, the patterns of oscillation remained the same. However, delays in differential equations that model neural networks and biological control systems can introduce novel dynamics that are not present without a delay (for example, see Refs. 57 and 58).

In synthetic pathways, compounds are often sequentially transformed through a series of small chemical changes. In the event that a compound near the end of the pathway is controlling other pathways via nonlinear functions, then there will be time delays introduced between the synthesis of the first compound and the last in the pathway. Depending on the time constants for the various steps, these time delays may be very small compared to other significant time constants in the system. Singular perturbation theory may be used to simplify some aspects of the analysis of these networks, but to our knowledge these techniques have not yet been systematically applied to the modeling of biochemical networks.

Cascades of regulatory steps are also a significant feature of biological control circuits and can introduce significant delays. The interplay between the kinetics of each step, the kinetics of the entire cascade, and the introduced time delays is a topic of current interest [55, 59].

C. Decay Rates of Compounds

If the decay constants, γ_i, of the variables in Eq. (5) are different, as would be expected, then although the dynamical equations are still piecewise linear, and thus can be numerically calculated with ease, the trajectories are no longer straight lines in phase space. The mathematical analyses that have been developed largely depend on the ability to analytically compute the linear fractional Poincaré return map. This is no longer possible when the decay constants are not equal. Although we expect that the stability of qualitative features of the dynamics will, in general, not change under small changes of the decay constants, formal mathematical analyses are lacking and this is an area for further research.

D. The Thresholds for Control of Different Genes May Be Different

The hypercube structure arises as a consequence of the binary nature of the state space assumed in Eq. (4). It is possible to modify the number of thresholds and still have a tractable mathematical problem, though many of the simplifications associated with the hypercube would disappear [23, 60, 61]. In principle, many of the same notions apply, although the area is not yet very developed. Finally, software that is capable of incorporating multiple thresholds into the piecewise linear equations with multiple time constants and thresholds has recently been developed by deJong et al. [62].

VIII. CONCLUSIONS

Even with a variety of modifications that would be needed in order to make these equations more realistic, it should be possible to extend the current mathematical approaches to help understand the connections between the structure and the dynamics.

Certainly, in considering the dynamics in biological systems, a compelling observation is the great apparent robustness of the networks even in different cells, in different species, and under a variety of perturbations from environmental factors or from mutations in the genome. Of course, we are all familiar with the sorts of disastrous circumstances that can arise when the "the basin of attraction" of normal cells is transgressed. But how can we define the structure and dynamics of what is normal?

Independent of the particular mathematical formalisms that might ultimately prove to be useful to describe biochemical processes in complex organisms, it is important to recognize that organisms that are alive at present, have emerged following evolutionary processes at the molecular level as well as the systems level. The resulting organisms exhibit remarkable robustness in the face of solving multiple challenges. The underlying organizational principles that confer this robustness are still poorly understood. An interesting and largely unstudied question is how the structure of genetic networks has been shaped by evolution.

In conclusion, the current formalism captures extremely rich dynamics in a vast class of differential equations modeling biochemical systems and relates these dynamics to the underlying structure of the biochemical networks. The methods have the potential of lending transparency to the functioning of what now appear to be arbitrarily complex networks.

Acknowledgments

This research has been supported by grants from NSERC and from the National Research Resource for Complex Physiologic Signals.

References

1. B. Alberts, A. Johnson, J. Lewis, et al. *Molecular Biology of the Cell, 4th ed.*, Garland Science, New York, 2002.
2. P. Jorgensen, J. L. Nishikawa, B.-J. Breitkreutz, and M. Tyers, *Science* **297**, 395 (2002).
3. P. Uetz, L. Giot, Cagney *Nature* **403**, 623 (2000).
4. J. Jeong, S. P. Mason, A.-L. Barabási, and Z. N. Oltvai, *Nature* **411**, 41 (2001).
5. J. Monod and F. Jacob, *Cold Spring Harb Symp. Quant. Biol.* **XXV**, 389 (1961).
6. M. Sugita, *J. Theor. Biol.* **4**, 179 (1963).
7. S. A. Kauffman, *J. Theor. Biol.* **22**, 437 (1969).
8. S. A. Kauffman, *Origins of Order: Self-Organization and Selection in Evolution*, Oxford University Press, Oxford, 1993.
9. R. Thomas, *J. Theor. Biol.* **42**, 563 (1973).
10. R. Thomas, *Adv. Chem. Phys.* **55**, 247 (1983).
11. R. Thomas and R. D'Ari, *Biological Feedback*, CRC Press, Boca Raton, FL, 1990.
12. B. Derrida and Y. Pomeau, *Europhys. Lett.* **1**, 45 (1986).
13. H. Flyvbjerg, *J. Phys. A: Math. Gen.* **21**, L955 (1988).
14. G. Weisbuch, *Complex Systems Dynamics*, Addison-Wesley, Redwood City, CA, 1991.
15. U. Bastolla and G. Parisi, *J. Theor. Biol.* **187**, 117–133 (1997).
16. A. Bhattacharjya and S. Liang, *Physica D* **95**, 29 (1997).
17. B. Luque and R. V. Solé, *Phys. Rev. E* **55**, 257 (1997).
18. S. Bilke and F. Sjunnesson, *Phys. Rev. E* **65**, 016129 (2001).

19. B. C. Goodwin, *Analytical Physiology of Cells and Developing Organisms*, Academic Press, London, 1976.
20. J. Tyson and H. Othmer, *Prog. Theor. Biol.* **5**, 1 (1978).
21. M. Kaufman, J. Urbain, and R. Thomas, *J. Theor. Biol.* **114**, 527 (1985).
22. H. H. McAdams and L. Shapiro, *Science* **269**, 650 (1995).
23. D. Thieffry and R. Thomas, *Bull. Math. Biol.* **57**, 277 (1995).
24. J. Reinitz and D. H. Sharp, *Mech. Dev.* **49**, 133 (1995).
25. G. von Dassow, E. Meir, E. M. Munro, and G. Odell, *Nature* **406**, 188 (2000).
26. T. S. Gardner, C. R. Cantor, and J. J. Collins, *Nature* **403**, 339 (2000).
27. M. B. Elowitz and S. Leibler, *Nature* **403**, 335 (2000).
28. H. H. McAdams and A. Arkin, *Annu. Rev. Biophys. Biomol. Struct.* **27**, 199 (1998).
29. J. Hasty, D. McMillen, F. Isaacs, and J. J. Collins, *Nature Rev. Genet.* **2**, 268 (2001).
30. H. Bolouri and E. H. Davidson, *BioEssays* **24**, 1118 (2002).
31. H. deJong, *J. Comput. Biol.* **9**, 69 (2002).
32. L. Glass and S. A. Kauffman, *J. Theor. Biol.* **34**, 219 (1972).
33. L. Glass and S. A. Kauffman, *J. Theor. Biol.* **39**, 103 (1973).
34. L. Glass, *J. Chem. Phys.* **63**, 1325 (1975).
35. L. Glass, *J. Theor. Biol.* **54**, 85 (1975).
36. L. Glass, in *Statistical Mechanics and Statistical Methods in Theory and Application*, U. Landman, ed., Plenum, New York, 1977, pp. 585–611.
37. L. Glass and J. S. Pasternack, *Bull. Math. Biol.* **40**, 27 (1978).
38. L. Glass and J. S. Pasternack, *J. Math. Biol.* **6**, 207 (1978).
39. J. E. Lewis and L. Glass, *Int. J. Bif. Chaos* **1**, 477 (1991).
40. J. E. Lewis and L. Glass, *Neural Comput.* **4**, 621 (1992).
41. T. Mestl, E. Plahte, and S. W. Omholt, *Dyn. Stab. Syst.* **10**, 179 (1995).
42. T. Mestl, C. Lemay, and L. Glass, *Physica D* **98**, 33 (1996).
43. R. J. Bagley and L. Glass, *J. Theor. Biol.* **183**, 269 (1996).
44. R. Edwards, *Physica D* **146**, 165 (2000).
45. R. Edwards and L. Glass, *Chaos* **10**, 691 (2000).
46. R. Edwards, *Diff. Eq. Dyn. Sys.* **9**, 187 (2001).
47. R. Edwards, H. T. Siegelmann, K. Aziza, and L. Glass, *Chaos* **11**, 160 (2001).
48. T. Gedeon, *Comm. Pure. Appl. Anal.* **2**, 187 (2003).
49. A. Berman and R. J. Plemmons, *Nonnegative Matrices in the Mathematical Sciences*, Academic Press, New York, 1994.
50. R. Edwards, J. J. Macdonald, and M. J. Tsatsomeros, *Linear Algebra Appl.* **398**, 37 (2005).
51. K. Kappler, R. Edwards, and L. Glass, *Signal Processing* **83**, 789–798 (2003).
52. T. Mestl, R. J. Bagley, and L. Glass, *Phys. Rev. Lett.* **79**, 653 (1997).
53. E. Plahte, T. Mestl, and S. W. Omholt, *Dyn. Stab. Syst.* **9**, 275 (1994).
54. A. Arkin and J. Ross, *Biophys. J.* **67**, 560 (1994).
55. H. Bolouri and E. H. Davidson, *Proc. Natl. Acad. Sci. (USA)* **100**, 9371 (2003).
56. C.-H Yuh, H. Bolouri, and E. H. Davidson, *Science* **279**, 1896 (1998).
57. J. Bélair and S. A. Campbell, *SIAM J. Appl. Math.* **54**, 1402 (1994).

58. J. C. Bastos de Figueiredo, L. Diambra, L. Glass, and C. P. Malta, *Phys. Rev. E* **65** 051905 (2002).

59. N. Rosenfeld and U. Alon, *J. Mol. Biol* **329**, 645 (2003).

60. T. Mestl, E. Plahte, and S. W. Omholt, *J. Theor. Biol.* **176**, 291 (1995).

61. E. Plahte, T. Mestl, and S. W. Omholt, *J. Math. Biol.* **36**, 321 (1998).

62. H. deJong, J. Geiselmann, C. Hernandez, and M. Page, *Bioinformatics* **19**, 336 (2003).

Since this chapter was written, a number of new contributions to the field have been published or come to our attention. Mason et al. demonstrated that an electronic analog of a piecewise-linear network can evolve towards a desired periodic behavior, showing that these behaviors are robust enough to be present in real physical systems, not only idealized models [J. Mason, P. S. Linsay, J. J. Collins, and L. Glass, *Chaos* **14**, 707–715 (2004)]. Killough and Edwards have analyzed and classified bifurcations in piecewise-linear networks [D. B. Killough and R. Edwards, *Int. J. Bifurcat. Chaos* **15**, 395–423 (2205)]. Gouzé and Sari have dealt in detail with singular solutions that can occur in subspaces where more than one variable is at its threshold value, using Filippov solutions [J.-L. Gouzé and T. Sari, *Dynamical Systems* **17**, 299–316 (2002)]. Plahte and Kjøglum approach the same problem by a different method, using a clever blow-up of the threshold intersections to allow calculation of singular solutions as limits of solutions of corresponding smooth systems with sigmoid interactions [E. Plahte and Kjøglum, *Physica D* **201**, 150–176 (2005)]. The last three of these papers deal with simultaneous switching and the last two with the multiple-threshold and non-uniform decay rate case.

ANALYSIS AND CONTROL OF ULTRAFAST DYNAMICS IN CLUSTERS: THEORY AND EXPERIMENT†

VLASTA BONAČIĆ-KOUTECKÝ and ROLAND MITRIĆ

Department of Chemistry, Humboldt-Universität zu Berlin, D–12489 Berlin, Germany

THORSTEN M. BERNHARDT and LUDGER WÖSTE

Institut für Experimentalphysik, Freie Universität Berlin, D–14195 Berlin, Germany

JOSHUA JORTNER

School of Chemistry, Tel Aviv University, 699878 Tel Aviv, Israel

CONTENTS

†This chapter is dedicated to Stuart Rice, who pioneered the field of optical control of chemical dynamics.

Adventures in Chemical Physics: A Special Volume in Advances in Chemical Physics, Volume 132, edited by R. Stephen Berry and Joshua Jortner. Series editor Stuart A. Rice

I. INTRODUCTION

Theoretical and experimental investigations of ultrafast processes in elemental clusters and their control by tailored laser pulses allow us to determine how the interplay of size, structures, and light fields can be used to manipulate optical properties and chemical reactivity of these systems. Moreover, laser-selective femtochemistry [1–11] can be combined with the functionalism of nanostructures, providing new perspectives for the basic research as well as for technological applications. The size regime of clusters in which each atom counts [12–15] is particularly important. In this case the number of atoms and the corresponding structures determine size-selective optical and reactivity properties of clusters. The investigation of the dynamics of these systems with finite density of states is especially attractive since the separation of time scales of different processes is possible [14]. Joint theoretical and experimental time-resolved ultrafast studies carried out on clusters provided findings on the nature and the time scales of processes, such as a geometrical relaxation, internal vibrational energy distribution (IVR), charge separation, and Coulomb explosion [14, 16–21]. Furthermore, due to advances in laser technology, tailored laser fields can be produced by pulse shapers that can control molecular dynamics, guiding it to a chosen target, such as a given fragmentation channel, a particular isomer, or a desired reaction product [11, 19, 22–43].

The role of theory has been essential from conceptual as well as from a predictive point of view. Time-resolved observations are strongly dependent on the experimental conditions, such as laser wavelengths, duration of pulses and their shapes, competition between one- and many-photon processes, strength of the electric field, and so on. Here theory has the task not only to provide insight into the nature of time-dependent processes, but also to identify the conditions under which they can be experimentally observed [19–21, 44–55].

Consequently, theory can be directly involved in conceptual planning of time-resolved experiments.

Other prominent examples are theoretical proposals for different optical control schemes using laser field parameters for the manipulation of ultrafast process pioneered by Rice and Tannor, Shapiro and Brumer, and Peirce, Dahleh, and Rabitz [2, 56–61]. They stimulated control experiments that were carried out first on simple systems such as metallic dimers and trimers [62–84], and later on more complex systems [23–25, 43, 85–89], confirming theoretically proposed concepts. Since tailored laser pulses have the ability to select pathways that optimally lead to the chosen target, their analysis should allow one to determine the mechanism of the processes and to provide the information about the selected pathways (inversion problem). Therefore, theoretical approaches are needed, which are capable of designing interpretable optimal laser pulses for complex systems (e.g., clusters or biomolecules) by establishing the connection between the underlying dynamical processes and their shapes. In this case, the optimal control can be used as a tool for the analysis.

In this chapter we present an overview of ultrafast time-resolved pump-probe spectroscopy and optimal control in moderately complex system. The chapter is structured as following: In Section II, dynamics and ultrafast observables will be addressed in the framework of negative ion-to-neutral-to-positive ion (NeNePo) pump-probe spectroscopy advanced by some of the present authors and their colleagues [90, 91]. This technique includes application of vertical one-photon detachment to prepare a nonequilibrium state of clusters and subsequent investigation of its dynamics by two-photon ionization. Since the dynamics of the nonequilibrium states pertain to transition state spectroscopy [14, 17, 92–103] this approach bridges the cluster dynamics with the real-time exploration of chemical (i.e., thermal) reaction pathways. The interplay between theory and experiment allows us to identify the time scales of different processes, to establish the corresponding mechanisms, and to determine the scope of the technique [19–21, 49, 104]. This will be illustrated for pure and mixed noble metal clusters. Moreover, the example of noble metal oxides will be addressed as a precursor for the future reactivity study. To present the scope of the theoretical approach based on the combination of *ab initio* molecular dynamics (MD) "on the fly" and the Wigner distribution approach, in addition to adiabatic multistate dynamics for ground states needed for NeNePo spectroscopy, the nonadiabatic dynamics involving electronic ground and excited states will also be addressed in Section II. This approach will be used to study the photoisomerization process through the conical intersection for the example of the Na_3F_2 cluster.

In Section III, a new strategy for optimal control in complex systems will be outlined. Tailored pump-dump pulses will be used to drive the photoisomerization process in the Na_3F_2 cluster, avoiding the pathway with high excess energy involving the conical intersection and populating only one selected isomer [19, 51]. This is achieved by introducing a new strategy for optimal control based on an intermediate target in the excited state corresponding to a localized ensemble that provides a connective pathway between the initial step and the target in the ground state. The connection between the shapes of the optimized pulses and the underlying processes will be explored. Summary and outlook for the future joint experimental and theoretical work on controlling the reactivity of noble metal oxide clusters in the framework of resonant two-photon detachment R2PD–NeNePo techniques, which are based on the concept of the intermediate target, will be given in Section IV.

II. DYNAMICS OF ULTRAFAST PROCESSES

Femtosecond spectroscopy enables the real-time interrogation of intra- and intercluster and molecule nuclear dynamics during the geometric transformation along the reaction coordinate. It involves the preparation of the transition state of a chemical reaction by optical excitation of a stable species in a nonequilibrium nuclear configuration in the pump step, along with the probing of its time evolution by laser-induced techniques such as fluorescence, resonant multiphoton ionization, or photoelectron spectroscopy. This approach was pioneered by Zewail and co-workers for bimolecular reactions [92–94]. For an elementary reaction involving the breaking of one bond and the creation of another one, changes in intermolecular separation of $\sim$10 Å are observable on a time scale of 1–10 ps. For this purpose, the duration of the probe step must be 10–100 fs if a resolution of $\sim$0.1 Å has to be achieved. In another approach advanced by Neumark et al. and Lineberger et al. a nonequilibrium or transition states can also be produced by vertical photodetachment of stable negative ions. The transition state of the neutral species can be close to the stable geometry of anions as shown by Neumark and co-workers [17, 95–103], or it can provide the starting point for the isomerization process in the neutral ground state as illustrated by Lineberger and co-workers [105–107]. The vertical one-photon detachment spectroscopy was advanced by introducing the NeNePo pump-probe technique [90, 91], which allows to probe structural processes and isomerization relaxation in neutral clusters as a function of the cluster size. An extension of the NeNePo technique by two-color excitations has also been proposed [108]. Complementary, time-resolved photoelectron spectroscopy [17] became a powerful technique that can be also applied to clusters [109]. Recent developments of time-resolved techniques such as ultrafast electron diffraction [110] and time-domain X-ray absorption [111] allow us to reveal transient molecular structures in chemical reactions of complex systems and in excited states of molecules.

The conceptual framework of ultrafast spectroscopy is provided by theory and simulations allowing the determination of the time scales and the nature of configurational changes as well as internal energy redistribution (IVR) in vertically excited or ionized states of clusters [20, 21, 44–47, 49, 104]. The separation of the time scales of different processes is essential for identifying them in measured spectral features. Moreover, the distinction between coherent and dissipative IVR in finite systems can be addressed as a function of the cluster size. For the investigation of intra- and intercluster dynamics in fs spectroscopy, the generation of the initial conditions and multistate dynamics for the time evolution of the system itself and for the probe or the dump step are needed. For this purpose, two basic requirements have to be fulfilled. The first is the use of accurately determined electronic structure in the ground and excited states as a function of all nuclear coordinates. In the case that the electronic states involved are well separated, the Born–Oppenheimer approximation is valid and the adiabatic dynamics is appropriate. In contrast, if avoided crossings and conical intersections between electronic states are present during the geometric and chemical transformation, breakdown of the Born–Oppenheimer approximation occurs and nonadiabatic effects have to be taken into account. This situation represents an additional theoretical and computational challenge. The second basic requirement is the accurate simulation of ultrafast observables such as pump-probe signals. This involves appropriate treatment of optical transitions such as ultrafast creation and detection of the evolving wave packet or classical ensemble. In the latter case, the dynamics is described by classical mechanics, and the average over sufficiently large number of trajectories has to be made in order to simulate the spectroscopic observables.

Accurately precalculated global *ab initio* energy surfaces of ground and excited states have been limited to systems with few atoms for which quantum dynamics of nuclei is feasible. Therefore, such *ab initio* energy surfaces were used as input data for the investigations of ultrafast dynamics of metallic dimers [62, 63, 66, 112–121]. and trimers [122–129]. In contrast, for larger systems in general, either a few degrees of freedom can be selected for explicit treatment or model potentials have to be used. Both situations are generally not applicable to elemental clusters, in particular with metallic atoms [12], since usually they do not contain a "chromophore type" subunit or do not obey regular growth patterns [130]. An addition of a single atom can produce drastic changes in the properties of their ground and excited states. Therefore, in the majority of cases, all degrees of freedom have to be considered, and semiempirical analytic potentials are usually not suitable since they do not properly describe structural and electronic properties with changing cluster size. Consequently, first principle (*ab initio*) molecular dynamics "on the fly" (AIMD), without precalculation of the energy surfaces, represents an appropriate choice to study ultrafast processes in elemental clusters with heavy atoms for which in the first approximation, the classical description of nuclear motion is acceptable. This method, pioneered by Car and Parrinello [131], is based on the density functional method and the plane wave basis sets and was originally introduced for the dynamics in the electronic ground state. The basic idea is to compute forces acting on nuclei from the electronic structure calculations that are carried out "on the fly" [132]. Related AIMD methods with plane wave basis sets have also contributed significantly to the success in applications to clusters [133].

The important starting step in introducing time into quantum chemistry was the implementation of the analytic energy gradients for the optimization of the geometries in the ground states pioneered by Pulay [134]. These analytic gradients that are now available in the framework of different quantum chemistry methods with Gaussian basis sets [134–137] can be used for fast calculations of forces and are implemented in different MD schemes from which classical trajectories can be computed. Advances in these techniques over the last years provided an excellent basis for applications of *ab initio* ground-state classical MD on small and large systems with controllable accuracy, depending on the method used for calculation of the electronic structure (e.g., different versions of density functional methods with atomic basis sets [138, 139] or other approaches accounting for the electronic correlation effects [140, 141]).

The situation is still very different for *ab initio* adiabatic and nonadiabatic MD "on the fly" involving excited electronic states. In spite of recent efforts and successes [45, 46, 142–146], further development of such theoretical methods that combine accurate quantum chemistry methods for electronic structure with MD adiabatic and nonadiabatic simulations "on the fly" has promise to open many new possibilities for the successful investigation of fs processes. This research area will essentially remove borders between quantum chemistry and molecular dynamics communities, in spite of the fact that each of them has numerous challenging tasks to be accomplished in order to provide a conceptual frame for fs chemistry and fs physics of molecules and clusters. In this context, very intense research is presently going on along two main directions. One is to achieve fast calculations of forces in excited states, as well as of nonadiabatic couplings, at the level of theory accounting for electron correlation effects with controllable accuracy which are suitable for implementation in different adiabatic and nonadiabatic MD schemes "on the fly" [146]. The second is to introduce quantum effects for the motion of nuclei, particularly in the case of nonadiabatic dynamics [147–157], in systems with a considerable number of degrees of freedom, allowing for their identification in spectroscopic observables such as fs signals [146, 148, 149, 152, 158, 159]. In this contribution we deal with systems containing heavy atoms, and therefore the quantum dynamical effects do not play an important role, as will be shown on prototype examples.

A. Multistate Adiabatic Nuclear Dynamics and Simulation of NeNePo Signals

Fs–NeNePo spectroscopy is also attractive for theoretical investigation for the following reasons: (1) It stimulates further development of accurate and efficient methods for adiabatic *ab initio* MD "on the fly" in the ground states and their application for simulation of femtosecond signals; (2) it provides the opportunity to determine conditions under which different processes and their time scales can be observed; and (3) it contributes to establish the scope of this experimental technique [20, 49, 104, 160]. Concerning point 1, the accuracy of electronic structure calculations—using, for example, *ab initio* gradient-corrected density functional approach with Gaussian atomic basis sets for MD "on the fly" (AIMD-GDFT) [161–165]—and the adequacy of the semiclassical Wigner distribution approach for simulation of fs pump-probe signals can be tested by comparing the obtained results with experimental findings and with the full quantum mechanical treatment of the nuclei. The latter is feasible for trimers [44]. Point 2 involves the introduction of experimental conditions for the simulation of fs signals, and therefore influence on revealed processes can be examined. Point 3 is addressed by varying the size and composition of clusters. This makes it possible to identify different processes, such as geometric relaxation, intracluster collisions, IVR, structural information based on vibronic patterns, and isomerization processes, as well as fragmentation dynamics in the pump-probe signals, providing the conceptual frame for the NeNePo spectroscopy. All of these aspects will be illustrated with examples of silver, gold, mixed silver–gold clusters, and silver oxide clusters.

To accomplish this goals, the electronic and structural properties of noble metal clusters will be addressed first. Then the attention will be paid to MD "on the fly" and to the simulation of signals. Furthermore, the analysis of the signals and the comparison with the experimental findings will be presented allowing for the identification of processes and conditions under which they can be observed. Finally, cluster reactivity aspects and the scope of NeNePo spectroscopy will be addressed.

1. *Electronic Structure*

Structural, reactivity, and optical properties of noble metal clusters attracted theoretical [47, 49, 104, 166–179] and experimental studies [178–192] over the years because of their relatively simple electronic nature in comparison with transition metals and their similarity to *s*-shell alkali metals. This is particularly the case for the Ag atom with a large *s–d* gap in contrast to the Au atom. In the latter case, the *s–d* gap is considerably smaller, because the relativistic effects play an essential role—for example, strongly influencing the energy of an *s*-orbital. These differences in electronic structure are also reflected in the structural properties of small silver and gold clusters. Recent theoretical and experimental investigations showed that gold clusters remain planar for larger sizes than do the silver clusters [170, 171, 178, 179]. Increasing interest in gold and silver clusters is due to their newly discovered size-selective reactivity properties toward molecular oxygen and carbon monoxide [172–175, 193–196]. In this context, the mixed silver–gold clusters have also attracted the attention of researchers due to the electronegativity difference between Ag and Au atoms [171, 192] giving rise to charge transfer from Ag to Au. All together, the noble metal clusters represent an attractive research direction for fs chemistry.

Relativistic effective core potentials (RECP) are mandatory for a description of these species. In general, effective core potential (ECP) methods allow one to eliminate core electrons (close to nuclei) from explicit electron correlation treatment which then involves only electrons directly participating in bonding. Usually they were developed in the literature for the Hartree–Fock wavefunctions, and therefore they had to be revisited and carefully tested in connection with the gradient corrected density functional method (GDFT) [173]. GDFT is presently the method of choice for the ground state properties of metallic clusters provided that the use of correlation and exchange functionals allow for the accurate determination of binding energies and structural properties, which is not always the case [173]. This is particularly important for a reliable calculation of the energy ordering of different isomers which can assume related or very different structures with close-lying energies. The presence of a number of isomers for a given cluster size is an important but not always pleasant characteristic of metal clusters (alkali- as well as noble-metal clusters). Therefore the determination of the lowest energy structure is not always an easy task. Consequently, the temperature is a crucial parameter that has to be considered in experiment and theory. Note that only at low temperatures can the mixture of different isomers be avoided, in contrast to the cases at higher temperatures.

In the early work on structural and optical properties of neutral and charged silver clusters, one- and eleven-electron relativistic effective core potentials (1e-RECP and 11e-RECP) with corresponding AO basis sets were developed [166–168]. The first one, which was later revisited in connection with the DFT method [49] employing Becke and Lee, Yang, Parr (BLYP) functionals [197, 198] for exchange and correlation, respectively, is suitable for the description of the ground-state properties. The second one is inevitable for the determination of the excited states of pure silver clusters. Since the *d* electrons are localized at the nuclei of the silver atoms, they almost do not participate in bonding. Their role is only important for the quantitative determination of the energies of the excited states in silver clusters. Recent DFT calculations on structural properties using 19e-RECP and ion mobility experiments carried out on Ag_n^+ [179] clusters have confirmed the early findings [166, 167]. The 1e-RECP for the Au atom is less reliable for studying structural properties of Au clusters because the *d* electrons participate directly in bonding. The use of 1e-RECP for gold clusters might be useful only if the results agree with those obtained from 19e-RECP, due to the fact that the former one is computationally considerably less demanding (for details see Ref. 171). Moreover, for reactivity studies involving oxidized clusters, 19e-RECP is mandatory also for silver clusters due to the activation of *d* electrons by *p* electrons of the oxygen atom [173–175].

2. *Semiclassical Dynamics and Signals*

Semiclassical methods for dynamics combine classical mechanics and quantum mechanics. They are particularly attractive for obtaining insight into complex systems with heavy atoms for which full quantum mechanical treatment of dynamics is not mandatory and is usually prohibitive due to the size of the atoms. Semiclassical methods for dynamics, which make use of classical trajectories with quantized initial conditions, are suitable for applications to such systems. However, the approaches that are able to include approximately the quantum coherence [147] and tunneling effects [148, 199, 200] in the classical MD are very valuable and inevitable for the description of some processes (e.g., motion of light atoms such as proton transfer). Moreover, all the processes that involve transitions between different electronic states require quantum mechanics for the adiabatic or the nonadiabatic dynamics of electrons and should in some manner be incorporated consistently with the dynamics of the nuclei. Semiclassical methods, which contain the superposition of probability amplitudes, are therefore capable of providing an approximate description of quantum effects (e.g., interference, tunneling, etc.) in molecular dynamics [152]. Classical MD in different forms, including *ab initio* MD "on the fly," are now applicable to relatively large systems, and classical trajectories can be used as input in semiclassical approaches for the simulations of the observables. Moreover, in principle, it is also possible to add quantum effects to classical MD simulations "on the fly" [199, 200]. Therefore, we focus on the approaches that are able to make use of classical adiabatic and nonadiabatic AIMD and time-dependent quantum chemistry.

The time evolution of the density operator $\hat{\varrho}(t)$ is given by the quantum mechanical Liouville equation

$$i\hbar \frac{\partial \hat{\varrho}}{\partial t} = \left[\hat{H}, \hat{\varrho}\right] \tag{1}$$

where $\hat{H}$ is the Hamiltonian of the system. This offers an appropriate starting point for establishing semiclassical approaches. Equation (1) has the well-known classical limit in the case of the nuclear dynamics on a single electronic surface, corresponding to the classical Liouville equation of nonequilibrium statistical mechanics:

$$\frac{\partial \varrho}{\partial t} = \{H, \varrho\} \tag{2}$$

Here $\varrho = \varrho(\mathbf{q}, \mathbf{p}, t)$ and $H = H(\mathbf{q}, \mathbf{p}, t)$ are functions of classical phase space variables $(\mathbf{q}, \mathbf{p})$, and

$$\{H, \varrho\} = \frac{\partial H}{\partial \mathbf{q}} \frac{\partial \varrho}{\partial \mathbf{p}} - \frac{\partial \varrho}{\partial \mathbf{q}} \frac{\partial H}{\partial \mathbf{p}} \tag{3}$$

is the Poisson bracket. The classical limit can be derived from Eq. (1) by means of a Wigner–Moyal expansion [201–203] of the quantum mechanical Liouville equation in terms of $\hbar$, which emerges from the replacement of the commutator by the Poisson bracket if the expansion is terminated to the lowest order of $\hbar$:

$$[\hat{A}, \hat{B}] \rightarrow i\hbar\{A, B\} + O(\hbar^3) \tag{4}$$

Higher-order terms in $\hbar$ serve for the introduction of quantum effects in the dynamics.

The semiclassical limit of the Liouville formulation of quantum mechanics, based on Wigner–Moyal representation of the vibronic density matrix, offers a methodological approach suited for the accurate treatment of ultrafast multistate molecular dynamics and the pump-probe spectroscopy using classical trajectory simulations [20, 21, 45, 46, 204]. This approach is characterized by the conceptual simplicity of classical mechanics and by the ability to approximately describe quantum phenomena such as optical transitions by means of the averaged ensemble over the classical trajectories. Moreover, the introduction of quantum corrections can be made in a systematic manner. The method requires drastically less computational effort than full quantum mechanical calculations and provides physical insight into ultrafast processes, being applicable to complex systems. In addition, it can be combined directly with quantum chemistry methods for the electronic structure to carry out the multistate dynamics at different levels of accuracy including precalculated energy surfaces as well as the *ab initio* MD "on the fly." The approach is related to the Liouville space theory of nonlinear spectroscopy in the density matrix representation developed by Mukamel and his colleagues (cf. Ref. 205). Following the proposal by Martens and co-workers [204] and our own formulation [20, 21, 45, 46], the method is briefly outlined in connection with its application to simulations of the time-resolved pump-probe or pump-dump signals, involving first adiabatic and then nonadiabatic dynamics "on the fly."

For the simulation of NeNePo signals, a combination of the *ab initio* molecular dynamics "on the fly" with the vibronic density matrix approach in classical Wigner–Moyal representation offers an adequate approach. This involves (i) the densities of the anionic state forming the initial ensemble, (ii) the densities of the neutral state reached after photodetachment by the pump, (iii) the densities of the cationic state after photoionization of the probe, and (iv) the laser-induced transition probabilities between the latter two states. Densities and transition probabilities can be calculated in the framework of the classical approximation to the Wigner–Moyal transformed Liouville equation for the vibronic matrix by restricting the expansion to the lowest order in $\hbar$, as outlined below. In addition, only the first-order optical transition processes can be taken into account, which is justified for the low laser intensities, as is the case in the NeNePo technique. Assuming zero kinetic energy conditions (ZEKE) for the photodetached electron and for the cation, as well as short laser pulses, which can be well-described with Gaussian pulse envelopes, the analytic expression for time-resolved NeNePo signals can be formulated in a straightforward manner (cf. Ref. 20):

$$
\begin{aligned}
S[t_d] &= \lim_{t\to\infty} P_{22}^{(2)}(t) \\
&\approx \int d\mathbf{q}_0\, d\mathbf{p}_0 \int_0^\infty d\tau_1 \exp\left\{-\frac{(\tau_1 - t_d)^2}{\sigma_{pu}^2 + \sigma_{pr}^2}\right\} \\
&\quad \times \exp\left\{-\frac{\sigma_{\mathrm{pr}}^2}{\hbar^2}[\hbar\omega_{\mathrm{pr}} - V_{IP}(\mathbf{q}_1(\tau_1;\mathbf{q}_0,\mathbf{p}_0))]^2\right\} \\
&\quad \times \exp\left\{-\frac{\sigma_{\mathrm{pu}}^2}{\hbar^2}[\hbar\omega_{\mathrm{pu}} - V_{\mathrm{VDE}}(\mathbf{q}_0)]^2\right\} P_{00}(\mathbf{q}_0,\mathbf{p}_0)
\end{aligned} \tag{5}
$$

Here $\sigma_{\mathrm{pu}}(\sigma_{\mathrm{pr}})$ and $E_{\mathrm{pu}} = \hbar\omega_{\mathrm{pu}}(E_{\mathrm{pr}} = \hbar\omega_{\mathrm{pr}})$ are the pulse durations and excitation energies for the pump and the probe step with the time delay t_d. The quantity $V_{\mathrm{IP}}(\mathbf{q_1}(\tau_1;\mathbf{q_0},\mathbf{p_0}))$ labels the time-dependent energy gaps between neutral and cationic ground states calculated at coordinates $\mathbf{q_1}(\tau_1)$ on the neutral ground state with initial coordinates and momenta $\mathbf{q_0}$ and $\mathbf{p_0}$ given by the anionic thermal Wigner distribution $P_{00}(\mathbf{q_0},\mathbf{p_0})$, and $V_{VDE}(\mathbf{q_0})$ are the vertical detachment energies of the initial anionic ensemble. Therefore, the first step for the simulation of signals involves the generation of $P_{00}(\mathbf{q_0},\mathbf{p_0})$, which can be calculated either for individual vibronic states or for the thermal ensemble assuming the harmonic approximation in the case of low or moderate initial temperatures for which the anharmonicities of nuclei are negligible. Then, the Wigner distribution for each normal mode is given by

$$
P(q,p) = \frac{\alpha}{\pi\hbar}\exp\left[-\frac{2\alpha}{\hbar\omega}(p^2 + \omega^2 p^2)\right] \tag{6}
$$

with $\alpha = \tanh(\hbar\omega/2k_bT)$ and the normal-mode frequency ω, corresponding to the full quantum mechanical density distributions. The ensemble of initial conditions needed for the MD on the neutral ground state energies emerges from sampling of the phase space distribution given by expression (6). For the case of high temperatures for which anharmonicities are important but quantum effects of the initial distribution are not, the phase space distribution can be obtained from a sufficiently long classical trajectory. The analytical expression for NeNePo signals given by Eq. (5) is easy to understand. The last exponential in (5) gives the Franck–Condon transition probability after the initial ensemble is photodetached. Then, the propagation occurs on the neutral state in terms of MD "on the fly," giving rise to the time-dependent ionization energies V_{IP}. The transition to the cationic ground state involves the probe step with a window function given by the second exponential of (5). The signal is obtained by the summation over the entire ensemble, and its time resolution is determined by the pump-probe correlation function with the probe window located around the time delay between two pulses given by the first exponential of Eq. (5). The spectral resolution depends on the duration of both pulses.

However, expression (5) has to be modified if the emitted electrons carry away some amount of kinetic energy. Consequently, for the simulation of the transient photoionization NeNePo signal, the integrations of the populations of the anionic and cationic states over the entire range of possible excess energies E_0 and E_2 have to be carried out in order to provide an approximate treatment of continuum. This leads to the following expression of the NeNePo signals:

$$\begin{aligned} S[t_d] &= \lim_{t\to\infty} P_{22}^{(2)}(t) \\ &\approx \int d\mathbf{q}_0 d\mathbf{p}_0 \int_0^\infty d\tau_1 \exp\left\{-\frac{(\tau_1 - t_d)^2}{\sigma_{pu}^2 + \sigma_{pr}^2}\right\} \int_0^\infty dE_2 \\ &\times \exp\left\{-\frac{\sigma_{pr}^2}{\hbar^2}[\hbar\omega_{pr} - V_{21}(\mathbf{q}_1(\tau_1; \mathbf{q}_0, \mathbf{p}_0))]^2\right\} \int_0^\infty dE_0 \\ &\times \exp\left\{-\frac{\sigma_{pu}^2}{\hbar^2}[\hbar\omega_{pu} - V_{10}(\mathbf{q}_0)]^2\right\} P_{00}(\mathbf{q}_0, \mathbf{p}_0) \end{aligned} \tag{7}$$

The modification of Eq. (7) allows one to use this formulation for the simulation of the fs pump-probe or pump-dump signals involving the ground and excited electronic states for the pump step and the cationic or the ground state for the probing of dynamics in the excited states. Moreover, the expression for the signals can be extended to treat, in addition to adiabatic dynamics, also nonadiabatic dynamics, which will be addressed in Section II.H.3.

To study fs dynamics and to simulate fs signals of metallic clusters as a function of their size using the Wigner distribution approach outlined above, the precalculation of energy surfaces is not practicable (although feasible for trimers). Therefore the *ab initio* MD approach "on the fly" is the method of choice. *Ab initio* molecular dynamics codes, which utilize a Gaussian atomic basis set and gradient-corrected density functional (AIMD-GDF) [161–163, 206, 207], are, in the meantime, implemented in standard programs such as Gaussian [208] or Turbomole [209]. The investigation of the dynamics of atoms is carried out by the integration of the classical equations of motion using the Verlet algorithm [210]. The SCF Kohn–Sham and the accurate calculations of Pulay forces are needed at each time step in order to achieve a satisfactory conservation of the total energy. It is important that all this occurs at low computational demand, because the simulation of pump-probe signal requires, in addition to the calculation of an ensemble of trajectories for the ground state of the neutral species, calculations of the energy gaps between neutral and cationic ground states. The accurate numerical evaluation of the exchange-correlation energy parts of the Kohn–Sham matrix and the exchange-correlation energy derivatives are the most computationally demanding steps if the number of AO basis functions is not extremely large. Therefore, effort has been made to reach a good accuracy at relatively low cost (cf. Refs. 161–163, 206 and 207).

The application of the above-outlined combination of methods will be first illustrated for adiabatic and then for nonadiabatic dynamics. Particular attention will be devoted to demonstrate the importance of the interplay between theory and experiment in the case of NeNePo spectroscopy. In our early theoretical work on NeNePo spectroscopy on the silver trimer [20], it has been shown that the geometric relaxation as well as IVR can be identified in NeNePo signals only under zero-kinetic energy electron conditions (ZEKE). In fact, these conditions have not been achieved in the early experimental work, and therefore the inclusion of a continuum according to Eq. (7) reproduced the experimental signals in which the distinction between geometric relaxation and IVR was smeared out. These findings initiated new NeNePo experiments on mixed Ag_2Au trimers for which ZEKE-like conditions have been achieved and which will be described below.

B. Experimental Setup for NeNePo Spectroscopy

The importance of precise temperature control of the initial cluster ensemble in the NeNePo experiment has been emphasized [104]. Only through the experimental knowledge of the temperature parameter, a detailed comparison with theoretically obtained NeNePo signals becomes possible, and different contributions to the observed nuclear dynamics can be distinguished. Therefore, the original NeNePo experimental setup [90, 211] has been extended to enable the control of cluster temperature in the range between 20 and 300 K [212]. The

(a)

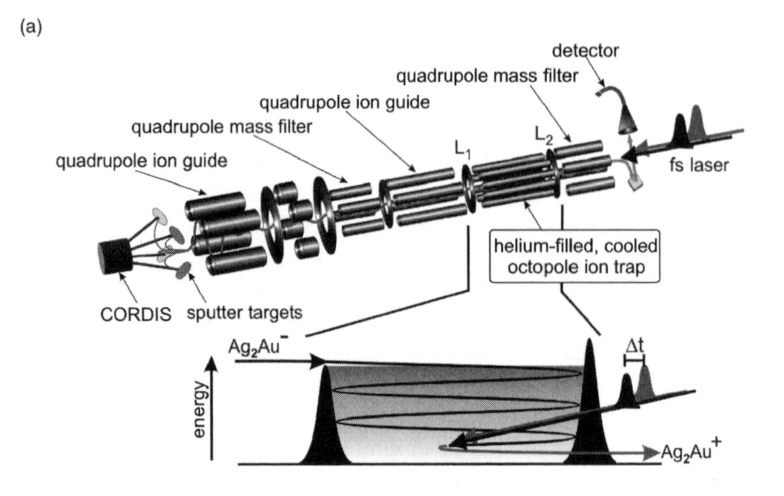
detector
quadrupole mass filter
quadrupole ion guide
quadrupole mass filter
quadrupole ion guide
L1
L2
fs laser
helium-filled, cooled octopole ion trap
CORDIS
sputter targets
Ag2Au−
Δt
energy
Ag2Au+

(b)

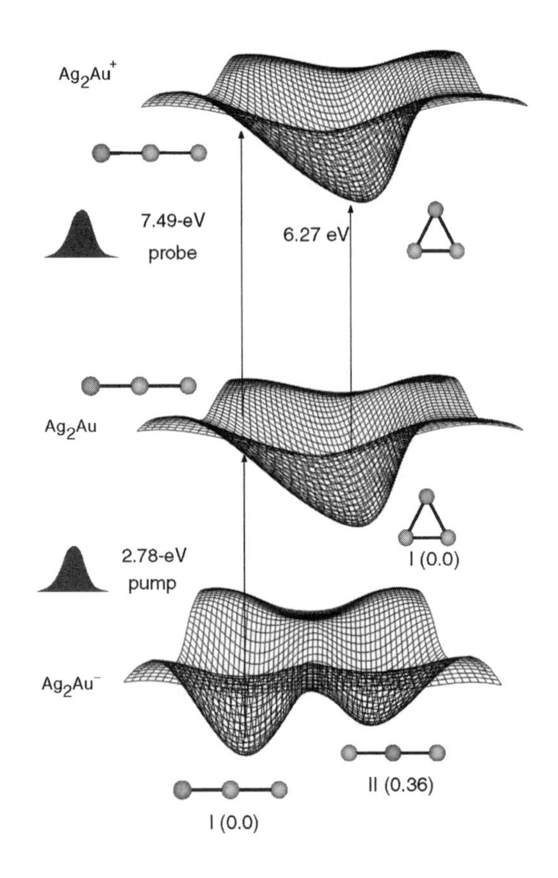
Ag2Au+
7.49-eV probe
6.27 eV
Ag2Au
2.78-eV pump
I (0.0)
Ag2Au−
II (0.36)
I (0.0)

NeNePo experiment was carried out in a helium-filled, temperature-variable radio-frequency (rf)-octopole ion trap. The complete experimental setup is depicted in Fig. 1a [104]. The metal cluster ions are generated by sputtering of metal targets with accelerated xenon ion beams (CORDIS source [213]). The clusters are subsequently mass-filtered and guided into the octopole ion trap (cf. upper part of Fig. 1a). Inside the ion trap the cluster ions rapidly lose energy by collisions with the helium buffer gas (cf. lower part of Fig. 1a), and perfect thermalization is accomplished within a few milliseconds. The ions are spatially confined by the rf field and the electrostatic potential of octopole entrance and exit lenses. The average residence time of the cluster anions in the octopole ion trap before interaction with a laser pulse is on the order of a few hundred milliseconds. The femtosecond laser beams enter the rf-ion trap collinearly with the axis of the apparatus from the opposite side as the cluster ions. The first ultrafast laser pulse (pump) then detaches the excess electron of the anion, resulting in a neutral cluster in the geometry of the anion. This leads to nuclear relaxation dynamics that can be probed in real time by femtosecond time-delayed ionization of the cluster to the cationic state (probe). As soon as cations are prepared inside the ion trap, they will be extracted by the electrostatic field of the octopole entrance and exit lenses and can be mass-analyzed with the final quadrupole mass filter. The recorded ion current at the detector as a function of the pump-probe delay time Δt gives rise to the transient NeNePo signal that reflects the time-dependent ionization probability of the neutral clusters due to the nuclear dynamics initiated by the initial photodetachment. However, only by comparison of transient NeNePo signal with theoretically simulated signals can the conditions be identified, under which a ZEKE-like situation can be achieved as illustrated below.

C. The Trimer: Ag_2Au

We have chosen the example of a mixed silver–gold trimer $Ag_2Au^-/Ag_2Au/Ag_2Au^+$ to demonstrate the ability of our *ab initio* Wigner distribution approach to accurately predict the NeNePo signals, to interpret them, and to identify conditions under which the separation of time scales of processes such as geometric relaxation and IVR can be achieved. This has been realized using the experimental setup at low temperature and close to the zero-kinetic energy

---

Figure 1. (*a*) Schematic representation of the setup for the temperature-controlled NeNePo pump-probe experiment. The upper part illustrates the arrangement of the cluster source, the quadrupole mass filter, and ion guides and of the octupole ion trap with entrance lens L_1 and L_2. The process of cluster cooling and trapping inside the octupole ion trap, as well as the laser-induced NeNePo charge reversal process, is schematically depicted in the lower part of the figure [104]. (*b*) Scheme of the multistate, fs dynamics for NeNePo pump-probe spectroscopy of $Ag_2Au^-/Ag_2Au/Ag_2Au^+$ with structures and energy intervals for pump and probe steps [49, 104].

electron (ZEKE) conditions described above. Furthermore, the aim was to study the influence of the heavy atom on the time scale of fs processes since a comparison with the "light" Ag_3 trimer [20] can be made. The simulations of the NeNePo pump-probe spectra have been performed by using our *ab initio* Wigner distribution approach combined with the *ab initio* MD "on the fly" in the framework of the density functional theory [49, 104] as described above. The mixed Ag_2Au trimer has the following structural properties: The anionic Ag_2Au^- trimer assumes a linear structure with one Au–Ag heterobond. The symmetric isomer with two hetero Ag–Au bonds lies 0.36 eV higher in energy. In the neutral state of Ag_2Au, both linear structures are transition states (with two imaginary frequencies along the degenerate bending mode) between the two equivalent triangular geometries which correspond to the most stable structure. In the cationic state the obtuse triangle is the minimum. It is important to notice that the structural properties of mixed trimers are sensitive to the details of the methodological treatment (choice of RECP and of functionals in DFT procedure). Therefore the inclusion of *d* electrons in RECP is necessary for quantitative considerations. The energetic scheme relevant for NeNePo together with the structural properties of the neutral and the charged Ag_2Au is shown in Fig. 1b.

Since for the simulations the initial temperature of 20 K has been chosen in correspondence with the experimental conditions (Section II.D.), it can be assumed that only the most stable structure is populated in the anionic state. Under these conditions, the harmonic approximation is valid, and therefore the initial conditions for the MD simulations have been sampled from the canonical Wigner distribution for each normal mode independently using Eq. (6). Due to the low temperature, the histograms of the vertical detachment energies (VDE) (or Franck–Condon transition probabilities) assume an almost Gaussian shape centered around 2.78 eV. The experimentally determined adiabatic detachment energy of Ag_2Au^- amounts also to 2.78 eV [192]. The first excited neutral state is separated from the anion by about 4 eV [192]. Pump photon energies of 2.78 eV and 4.00 eV should therefore be suitable to prepare the neutral Ag_2Au in the electronic ground state. The transient NeNePo signals do indeed show the same temporal evolution independent of the pump wavelength in this range.

As already described in the Section II.C, in order to simulate the NeNePo signals, an ensemble of trajectories (e.g., $\sim$500) has to be propagated in the neutral state, and the time-dependent energy gaps to the cationic state have to be calculated along the trajectories to simulate the signals according to the scheme given on the left-hand side of Fig. 2. The energy gaps are presented on the right-hand side of Fig. 2 since they provide visual information about the time evolution of individual processes such as the onset of geometrical changes and of IVR. Within the first 2 ps after the photodetachment, the swarm of energy gaps decreases from 7.5 eV to 6.5 eV, and subsequently all energy gaps exhibit

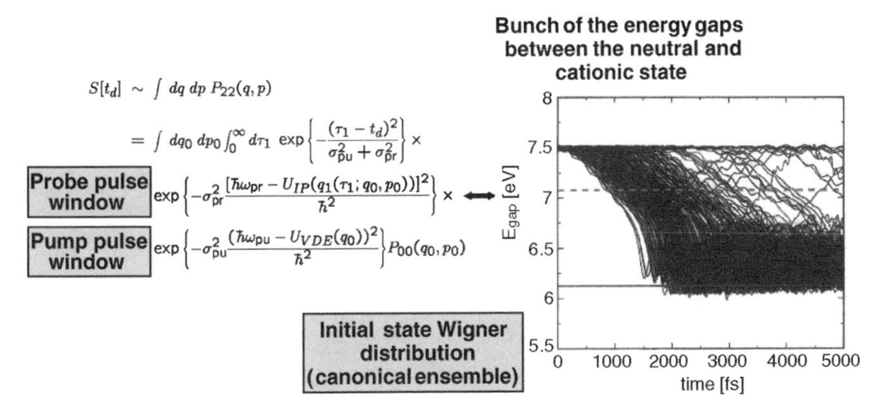

Figure 2. Bunch of the cation-neutral energy gaps of Ag_2Au (*right-hand side*). Energies of 7.09 eV (dashed line) and of 6.1 eV (full line) indicate the proximity of Franck–Condon region and of the minimum of neutral species, respectively, and are used for simulations of signals (compare probe window on the *left-hand side*).

oscillations in the energy interval between 6.1 and 6.5 eV. This allows us to distinguish two different types of processes: (i) the geometric relaxation from the linear toward the triangular structure, taking place within the first 2 ps and (ii) subsequent IVR process within the triangular structure. The minimum energy gap value of ~6.1 eV corresponds to the structure with the closest approach of the terminal silver and gold atoms, which is referred to as an internal collision within the cluster. Therefore the adjustment of the pump-probe energies experimentally allows to probe that processes. Since the highest value of the IP is 7.5 eV, choosing higher probe pulse energy will lead to the signal that is rising very rapidly and that subsequently remains constant due to the contribution of the continuum. Pulse energies between 6.5 eV and 7.5 eV probe the onset of the geometric relaxation processes. Therefore, it is expected that the signals exhibit maxima at delay times when the probe energy is resonant with energy gaps and decrease to zero at later times. Pulse energies below 6.5 eV probe the arrival and the dynamics at the triangular structure. It is to be expected that at ~6.1-eV pulse energies, the signal will rise after ~2 ps and remain constant at later times. This is illustrated also in Fig. 3a, in which the theoretical NeNePo and NeNePo-ZEKE signals are compared with the experimental results at low temperature (T ~20 K) for three energies: $E_{\rm pr} = 7.7$ eV, probing above the ionization threshold of the linear structure, $E_{\rm pr} = 7.09$ eV, probing the Franck–Condon region; and $E_{\rm pr} = 6.10$ eV, probing the triangular geometry region corresponding to the minimum of the neutral Ag_2Au [104].

The lower trace of Fig. 3a displays the NeNePo signal for 6.1-eV two-photon probe energy ($E_{\rm pr2ph}$) measured at 20 K. The measured Ag_2Au^+ ion intensity is lowest around zero time, when pump and probe pulse overlap temporally. The

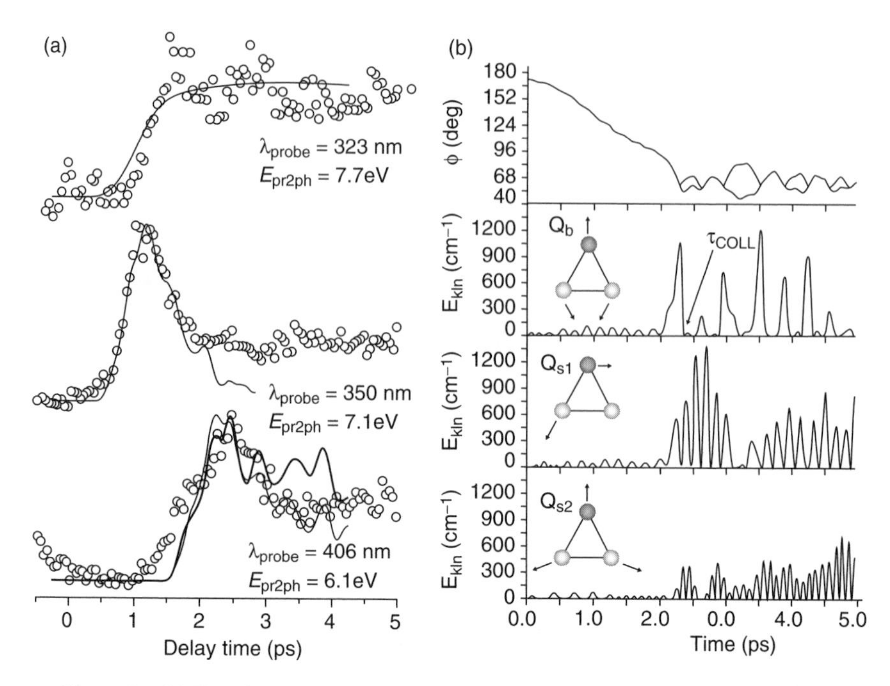

Figure 3. (*a*) Experimental NeNePo signals (open circles) obtained for three different probe pulse wavelengths in comparison with the simulated time-dependent signals (solid lines) for the different probe pulse energies. The ionization probe step is two-photonic as confirmed by power-dependent measurements. The signals are normalized in intensity. *Lower graph*: The experimental data obtained at $\lambda_{pr} = 406$ nm ($E_{pr2ph} = 6.1$ eV) are overlaid by the simulated NeNePo–ZEKE (bold line) and NeNePo (thin line) signals at $E_{pr} = 6.1$ eV. *Middle graph*: Experimental data obtained at $\lambda_{pr} = 350$ nm($E_{pr2ph} = 7.1$ eV) are overlaid by the simulated NeNePo–ZEKE signal at $E_{pr} = 7.1$ eV. *Upper graph*: The experimental data obtained at $\lambda_{pr} = 323$ nm ($E_{pr2ph} = 7.7$ eV) are overlaid by the simulated NeNePo signal at $E_{pr} = 7.41$ eV. A common time zero between experiment and theory has been chosen for all probe energies. The deviation in the time origin corresponds to less than 0.1 eV in the probe energy. (*b*) A single-trajectory example of the evolution of the Ag–Ag–Au bond angle ϕ (*upper panel*) and of the kinetic energy in the three vibrational normal modes Q_b, Q_{s1}, and Q_{s2} (*lower panels*). In the upper panel, two functions are given for ϕ, since for triangular geometries two Ag–Ag–Au bond angles can be defined. The lower curve for ϕ reflects the atom connectivities of the initial linear geometry and therefore monitors the geometrical relaxation until the closest approach of the terminal atoms (internal collision). The upper curve is the larger of the two Ag–Ag–Au bond angles and indicates partial escapes from the potential well of the triangular geometry. In the second panel from the top, the collision time $\tau_{COLL} = 2.36$ ps, determined from the first pronounced rise and the subsequent sharp minimum of the kinetic energy in the bending mode, is marked in the diagram.

signal stays at the same low level for 1.1 ps, then rises gradually until it reaches its maximum value arround 2.5 ps. It subsequently decreases again and remains almost constant at about half of its maximal intensity from 3.5 ps on. The ionization to the cationic state via a two-photon transition is confirmed by the quadratic power dependence of the NeNePo signal intensity [214]. Varying the probe photon energy has a dramatic influence on the temporal evolution of the NeNePo transient signal. The middle trace of Fig. 3a shows the signal obtained with 3.55-eV probe pulse energy, that is, $E_{pr2ph} = 7.1$ eV. Again the Ag_2Au^+ ion intensity is minimal around zero time, but starts to rise already after about 500 fs with a considerably steeper slope than in Fig. 3a to reach a maximum already at 1.1 ps. The signal decreases again comparably fast and stays at a constant level after 2 ps. Finally, the NeNePo signal in the upper trace of Fig. 3a was obtained with a probe energy $E_{pr2ph} = 7.7$ eV. The signal displays no peak, only a fast rise between about 700 fs and 1.5 ps to remain constant afterwards.

The comparison between the theoretically obtained NeNePo signals (solid lines in Fig. 3a) and the measured time-dependent NeNePo ionization efficiencies enables the assignment of the observed pronounced probe energy dependence to the fundamental processes of nuclear dynamics. At $E_{pr2ph} =$ 6.1 eV (Fig. 3a, lower trace) the onset of IVR and the dynamics of Ag_2Au initiated by the collision of the terminal Au and Ag atoms can be probed exclusively. The good agreement between the experimental (open circles) and the simulated NeNePo–ZEKE signals (bold line) in Fig. 3a is apparent, indicating that the experimental signal starts to rise when the system approaches the triangular potential well. The signal maximum can be assigned to the time of intracluster collision at around 2.4 ps followed by IVR in the potential minimum of the neutral triangular geometry. The experimental signal offset at longer delay times is somewhat lower with respect to the maximum than expected from the simulated NeNePo-ZEKE signal. This might be attributed to contributions from the rather similar NeNePo-type signal (thin line in the lower trace of Fig. 3a). This shows explicitly in which regime ZEKE conditions hold in the experiment. The middle trace of Fig. 3a presents the comparison of simulated and experimental transient ion signals at 7.1-eV probe energy. Because the initial peak of the experimental transient is perfectly matched by the simulated NeNePo–ZEKE signal (solid line) at the corresponding wavelength, the experimental conditions in this case allow for direct exclusive probing of the geometrical relaxation of Ag_2Au. The trimer passes through bending angles of $\phi = 166°$ at the signal onset around 500 fs to $\phi = 138°$ at the signal maximum and finally up to $\phi = 96°$ at 2 ps, where the terminal atoms already interact and the intracluster collision is closely ahead (cf. upper trace of Fig. 3b). The experimental signal offset at times later than 2 ps can again be attributed to the imperfect NeNePo–ZEKE conditions. The possible reason for the good agreement of the experimental transient signal with the simulated NeNePo–ZEKE transient signal is most likely due to a particular favorable Franck–Condon overlap in the case of 7.1-eV two-photon probe energy. Finally, at high ionization energy $E_{pr2ph} = 7.7$ eV, a comparably weak experimental transient signal is detected. This signal is in agreement with the simulated NeNePo transient (solid line) at a probe energy of 7.41 eV just below the highest theoretically predicted ionization energy that corresponds to the linear transition state structure (see top trace of Fig. 3a). Thus, the experiment at $E_{pr2ph} = 7.7$ eV apparently monitors the system when it leaves this transition state region. Still there is a considerable signal onset time of about 700 fs, which reflects the very shallow slope of the PES around the linear transition state geometry.

In summary, experiment and theory are in excellent agreement. The simulated signal at $E_{pr} = 7.70$ eV rises after 1 ps and remains constant subsequently without allowing to identify the dynamical processes which take place due to the contribution of the continuum. The signals at $E_{pr} = 7.1$ eV reflect geometric relaxation from linear to triangular geometry of the neutral Ag_2Au. The signals at $E_{pr} = 6.10$ eV are due to IVR.

For an analysis of the vibrational energy redistribution, the kinetic energy was decomposed into normal mode contributions. This was achieved by projecting the atomic velocities at regular time intervals on the non-mass-weighted normal coordinates of the neutral triangular Ag_2Au system. Figure 3b shows a single-trajectory example together with the two distinct Ag–Ag–Au bond angles ϕ. From this representation, valuable insight into the IVR in this model system can be gained. The bond angle ϕ in Fig. 3b decreases from an initial value of about 180° at $t = 0$ to a minimum value of 54° at $t = 2.36$ ps. However, the kinetic energy begins to increase notably only at $t \geq 2.0$ ps, when ϕ falls short of 90°. Accordingly, within the next 360 fs, ϕ decreases much more rapidly and the kinetic energy in the bending mode passes a pronounced maximum. Shortly afterward, the kinetic energy decreases to zero, as the system passes the potential minimum and the terminal atoms subsequently further approach each other ("internal collision"), until the kinetic energy is consumed by running against the repulsive part of the potential. In parallel to the increase of the kinetic energy in the bending mode Q_b, intense oscillations are triggered in the first, antisymmetric stretching mode Q_{s1} and to a smaller extent only in the symmetric stretching mode Q_{s2}. Apparently, the simultaneous gain of kinetic energy of the Q_b and of the Q_{s1} mode is a consequence of the fact that both normal coordinates together make up the major components of the linear-to-triangular geometric relaxation coordinate.

Intense kinetic energy oscillations of the stretching modes appear between two pronounced kinetic energy maxima of the bending mode, when the system is in the deep region of the potential, so that enough energy is available for the stretching modes. This relation is particularly apparent for the antisymmetric stretching mode Q_{s1}, manifesting an extensive energy exchange and a close coupling of these modes. Shortly after the bending mode has passed its maximum kinetic energy, its kinetic energy drops to zero (manifesting internal collision), which for the single trajectory of Fig. 3b occurs at 2.36 ps. Since the other mode energies increase at the same time, IVR is manifested; the drop of the kinetic energy in the bending mode cannot be solely explained by a conversion of kinetic to potential energy in the bending mode. This behavior is also found for other trajectories, whereupon one can generally state that notable IVR sets in at the instant of internal collision.

These results imply that the nature of IVR in Ag_2Au is related to the one found for Ag_3 (cf. Refs. 20 and 21). However, two important aspects should be emphasized. First, time scales are much longer than in the case of Ag_3, due to heavy atom effect. Second, importantly, in contrast to the Ag_3, the experimental results for Ag_2Au reveal for the first time geometric relaxation separated from an IVR process, indicating that the experimental signals are close to the ZEKE-like conditions, which has been proposed by theory as a necessary condition for the separation of time scales of these processes [104].

D. Tetramers: Ag_4 and Au_4

In order to further illustrate the scope of the NeNePo technique and the ability of our theoretical approach to treat more complex systems, two examples, Ag_4 and Au_4, have been chosen for the presentation, because they exhibit qualitatively different structural properties in the anionic state and have common properties in the neutral state. In the case of the silver tetramer, the global minima of the anion and of the neutral cluster assume related rhombic structures. Therefore, after photodetachment at low temperatures ($T \approx 50$ K), which ensures that only the rhombic isomer is populated, the pump step reaches the nonequilibrium rhombic configuration close to the global minimum of the neutral species, as shown on the left-hand side of Fig. 4. Notice that the well-defined initial structure is a necessary condition to observe the time scales of the processes involved in the geometric relaxation of the neutral state, and therefore the experiments should be performed at low temperatures.

For example, in order to monitor the isomerization into the T-form in the neutral ground state of Ag_4, an initial temperature of more than 700 K would be needed. Consequently, for the low-temperature $T = 50$ K initial conditions, the probe in the Franck–Condon region with $E_{pr} = 6.41$ eV reveals the vibrational structure of the rhombic configuration after photodetachment. For the probe with, for example, $E_{pr} = 6.46$ eV, the dynamics in the vicinity of the neutral rhombic structure can be monitored. The simulated NeNePo–ZEKE signal at 6.41 eV for a probe duration of 50 fs, shown in Fig. 4 exhibits oscillations with a vibrational period of ~175 fs, which is close to the frequency of the short diagonal stretching mode, indicating the occurrence of the geometric relaxation along this mode toward the global minimum. The analysis of the signal also reveals contributions from two other modes shown in Fig. 4. In summary, this example illustrates that an identification of the structure of a gas-phase neutral cluster in experimental NeNePo signals is possible due to its vibronic resolution

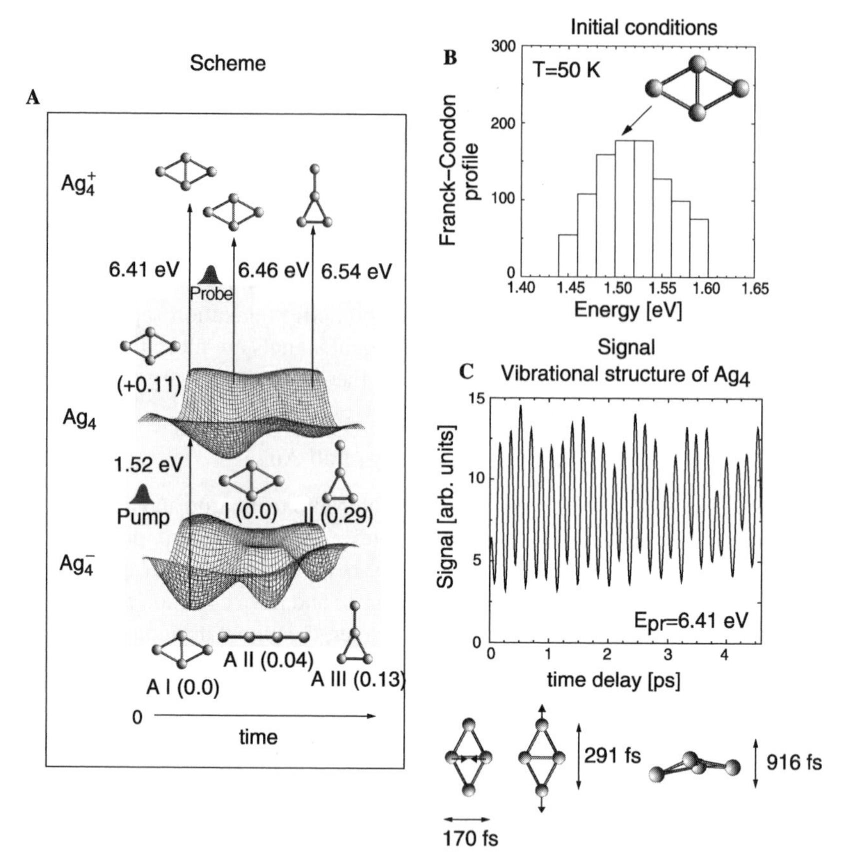

Figure 4. Scheme of multistate fs dynamics for NeNePo pump-probe spectroscopy of Ag_4^-/Ag_4/Ag_4^+ with structures and energy intervals for the pump and probe steps (*A*). Simulated NeNePo–ZEKE signals for the 50 K initial condition ensemble (*B*) at the probe energy of 6.41 eV and a pulse duration of 50 fs (*C*). Normal modes responsible for relaxation leading to oscillatory behavior of the signal are also shown [49].

[49]. The theoretically predicted main features of the pump-probe signals for Ag_4 have been also found experimentally.

One-color NeNePo spectra of Ag_4^- measured at 385 nm and an anion temperature of 20 K are shown in Fig. 5 [212]. Trace A (top right) shows the (uncorrected) mass selected Ag_4^+ yield as function of the delay time between the pump and probe pulses from −4.9 ps to +4.9 ps in steps of 20 fs. Trace A shows a pronounced oscillatory structure, characterized by a period of about 740 fs. The intensity of the maxima decreases for larger delay times and additional, weaker structures are observed at delay times >2.8 ps overlapping the 740-fs

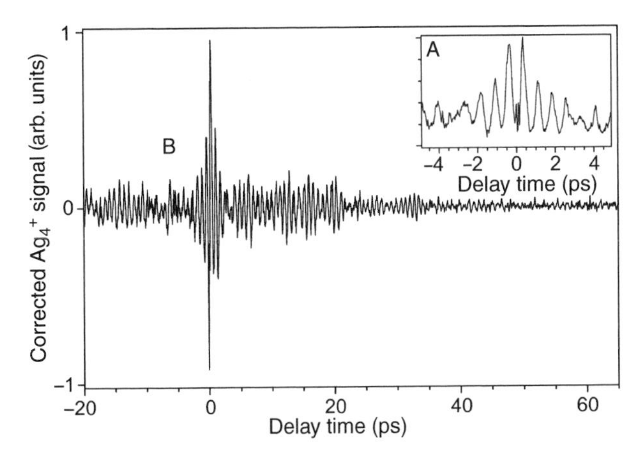

Figure 5. NeNePo spectrum of Ag_4 recorded in a one-color experiment (385 nm) at 20 K ion temperature. Trace A (*insert*) shows the uncorrected, mass-selected Ag_4^+ yield. Trace B is a composite of two measurements, and the signal has been corrected for the FFT analysis. (Figure taken from Ref. 212).

beat structure. Trace B in Fig. 5 is a composite of two measurements covering delay times from −20 ps to +20 ps and +20 ps to +65 ps. The spectra have been baseline corrected and transformed for the following fast Fourier transformation (FFT). Trace B shows that the oscillations in the Ag_4^+ signal intensity extend up to 60 ps, with decreasing intensity. Pronounced partial recurrences are observed. The FFT analysis of this time-resolved signal reveals as dominant feature several peaks centered around $45 cm^{-1}$ [212]. Comparison to photoelectron data of Ag_4^- [106] leads to the conclusion that the oscillations observed in the NeNePo spectra of Ag_4 are due to vibrational wavepacket dynamics in the $2a_g$ mode of either the $3B_{1g}$ or $1B_{1g}$ "dark" electronically excited state of rhombic Ag_4, which is probed by a two-photon ionization step to Ag_4^+. *Ab initio* calculations of the harmonic frequencies of the low-lying electronic states of rhombic Ag_4 support this assignment and confirm the observed pronounced anharmonicity of this vibrational mode of $2\nu_0\chi_0 = 2.65 \pm 0.05\, cm^{-1}$. The $2a_g$ mode was not resolved in the previous anion photoelectron spectroscopy studies, due to its low frequency of $45\, cm^{-1}$ which lies below the resolution of conventional anion photoelectron spectrometers. The results on Ag_4 demonstrate the successful application of femtosecond NeNePo spectroscopy to study the wavepacket dynamics in real time, manifested by a beat structure in the cation yield, in a "purely" bound potential, in contrast to the transition state experiments on the noble metal trimers which connected linear with triangular structures. The spectra of Ag_4 enable the precise characterization of a selected vibrational mode with a resolution, which is superior to that of conventional frequency-domain techniques.

In contrast to Ag_4, the investigation of nuclear dynamics and simulation of NeNePo–ZEKE signals of Au_4 allows one to follow large-amplitude motions induced by the photodetachment, which lead to isomerization because the stable structures of Au_4^- and Au_4 assume very different linear (or closely related zigzag) geometry and rhombic forms, respectively [49]. Again, as in the case of Ag_4, low-temperature initial conditions ($T \approx 50\,K$) ensure that only the linear anionic structure contributes to the initial ensemble, which is photodetached by the probe pulse, as shown in the scheme presented in Fig. 6. The two-photon ionization or probe laser spanning the energy range between ~8.4 and 8.1 eV monitors the initiated relaxation dynamics on the neutral state involving linear, T-form, and rhombic isomers (cf. Fig. 6).

The relaxation dynamics is influenced by the linear local minimum of the neutral species which is energetically reached after the photodetachment. The signal at $E_{pr} = 8.86$ eV, shown in Fig. 6, reflects on dynamics within the local linear isomer reached after photodetachment. It is characterized by oscillations corresponding to one of the symmetric stretching modes that is responsible for a nondephased relaxation process of the initial nonequilibrium ensemble. The intensity of the signal decreases after 1 ps, indicating occurrence of the relaxation process from linear to rhombic structure. Consequently, both signal intensities at $E_{pr} = 8.09$ eV and $E_{pr} = 8.27$ eV increase, reflecting the appearance of other isomers. A temporary identification of the rhombic structure at $E_{pr} = 8.09$ eV is possible only through a small maximum. Otherwise a structureless line shape indicates the presence of both rhombic and T-form isomers due to a large internal vibrational energy. It is interesting to point out that the absence of dephasing during the relaxation dynamics in the vicinity of the nonequilibrium state reached after photodetachment is the signature of the local minimum, whose influence on the dynamics is different from that of the transition state—for example, in the case of Ag_3 and Ag_2Au clusters.

E. Fragmentation of Ag_2O_2 Interrogated by NeNePo Spectroscopy: Reactivity Aspects

Atomic metal clusters, especially clusters of the noble metals, exhibit fascinating reactive properties [215]. This chemical reactivity toward small molecules such as O_2 often strongly depends on the charge state of the clusters. One particular appealing example in this respect is the reactive behavior of the silver dimer toward dioxygen in the gas phase, which has been investigated in detail by different groups [175, 186, 216–221]. Under the conditions of our rf-ion trap experiment, the anionic dimer adsorbs one O_2 molecule in a straightforward association reaction mechanism [175]. Photoelectron spectroscopic studies

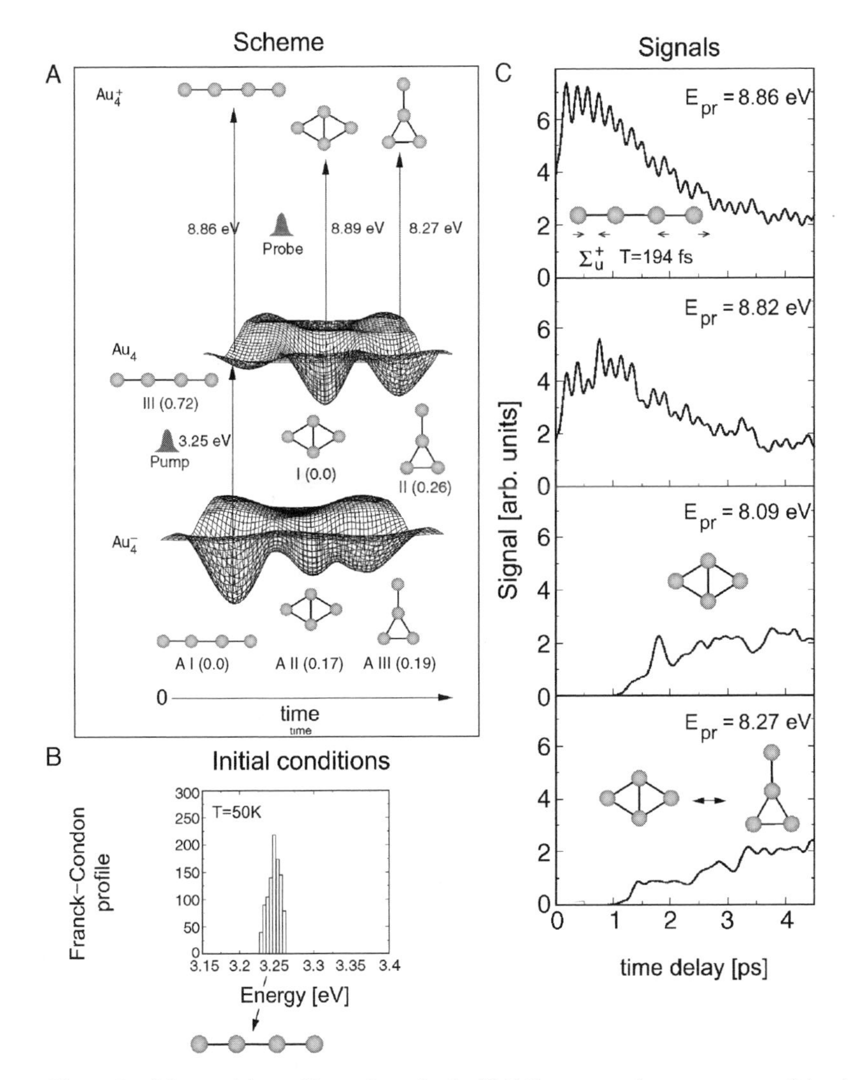

Figure 6. Scheme of the multistate dynamics for NeNePo pump-probe spectroscopy of Au_4^-/Au_4/Au_4^+ with structures and energy intervals for the pump and probe steps (*A*). Simulated NeNePo–ZEKE signals for the 50 K initial condition ensemble (*B*) at different probe energies [49] (*C*).

confirm that the oxygen is molecularly bound to Ag_2^- [221]. In contrast, the positively charged silver dimer shows a strongly temperature-dependent O_2 adsorption behavior: O_2 is first adsorbed molecularly on Ag_2^+, but in an activated reaction step, the O–O bond can dissociate, leading to the adsorption of atomic oxygen at temperatures above 90 K [217]. An NeNePo experiment starting from

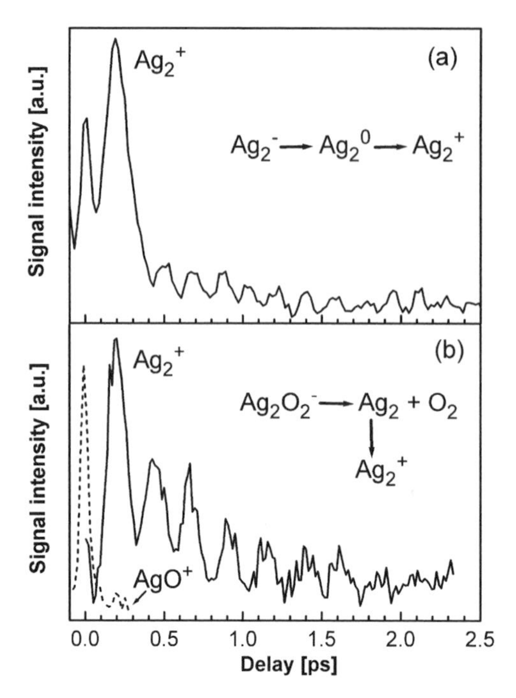

Figure 7. (*a*) NeNePo spectrum of Ag_2 recorded in a one-color experiment (406 nm) at 100 K ion temperature. (*b*) Pump-probe spectra of the NeNePo fragment signals Ag_2^+ (solid line) and AgO^+ (dashed line, magnified by a factor of 10) resulting from neutral Ag_2O_2 dissociation after photodetachment of $Ag_2O_2^-$ (406 nm, 100 K).

the stable $Ag_2O_2^-$ complex is thus expected to probe the real-time nuclear dynamics associated with the change in the reactive O_2 adsorption behavior initiated by the pump-photodetachment step. Figure 7a first displays the experimental NeNePo spectrum of the bare silver dimer without adsorbed oxygen obtained in a one-color pump-probe experiment (406 nm) at 100 K anion temperature. The NeNePo trace exhibits two remarkable features: (i) a pronounced maximum in the recorded Ag_2^+ signal at 190-fs pump-probe delay time and (ii) a distinct vibrational dynamics at longer delay times ($>$400 fs). The amplitude of the vibrational structure at delay times $>$400 fs is about a factor of 10 smaller than the maximum signal. The vibrational period of the observed signal oscillation was determined by FFT analysis to be 180 fs ($\nu = 185\ \mathrm{cm}^{-1}$). The femtosecond NeNePo dynamics detected in the Ag_2^+ signal in Fig. 7a can be understood on the basis of the known spectroscopic properties of Ag_2^- and Ag_2 [189]. Through photodetachment with 406-nm (3-eV) photons, the electronic ground state of Ag_2 (X-$1^1\Sigma_g$) is populated, but also the lowest excited triplet state $1^3\Sigma_u$. This latter triplet state is, however, if at all, only very weakly bound.

The system is thus populated in the repulsive part of the potential energy curve in the dissociation continuum of this state. During the propagation along the triplet potential curve, the wavepacket might be transferred via resonant two-photon transitions at two close-lying locations to the cationic state for detection (via C-$2^3\Pi_u$and B-$1^1\Pi_u$ states of Ag_2). The existence of these resonances explains the strong enhancement of the Ag_2^+ signal at 190-fs delay time. The lack of a periodical revival with comparable amplitude confirms the assumption that the wavepacket further propagates freely to dissociation on the triplet state potential. The wavepacket at the same time prepared on the X-$1^1\Sigma_g$ ground state of Ag_2 oscillates between the inner and outer turning point of the potential, leading to the observed periodic signal at delay times >400 fs with 180-fs oscillation period. The localization of the wavepacket is highest at the turning points of the potential where it can be efficiently transferred into the ground state of Ag_2^+ by irradiation with the probe pulse. The ionization step requires three 406-nm photons, but a resonant transition via the A-$1^1\Sigma_u$ state is possible [222]. If a small partial pressure of O_2 is added to the helium buffer gas inside the rf-ion trap, the complex $Ag_2O_2^-$ is formed immediately, before the silver dimer can interact with the femtosecond laser pulses [175]. Thus, under these conditions, the NeNePo experiment exclusively probes the dynamics of the $Ag_2O_2^-$ cluster complex. Figure 7b shows the result of the NeNePo experiment starting form $Ag_2O_2^-$ performed under otherwise identical conditions as in the case of Ag_2^- (Fig. 7a). It is first interesting to note that at no delay times a signal of $Ag_2O_2^+$ was detected resulting from the NeNePo process. This indicates that the neutral Ag_2O_2 formed by photodetachment is unstable and rapidly dissociates, being in accordance with gas-phase reactivity measurements which show that the neutral complex Ag_2O_2 is not bound [216]. Surprisingly, two fragmentation paths seems to exist, leading to the formation of the product ions Ag_2^+ and AgO^+ (solid and dashed lines in Fig. 7b, respectively). The Ag_2^+ signal is a factor of 10 larger than the AgO^+ signal. The AgO^+ signal exhibits only a peak at zero delay time with an FWHM corresponding to the cross-correlation of the laser pulses (80 fs). Most likely, AgO^+ arises from fragmentation in the cationic state—that is, by decay of $Ag_2O_2^+$, which is generated by vertical multiphoton transition form $Ag_2O_2^-$. The much more intense Ag_2^+ signal shows pronounced vibrational dynamics, which is significantly different from bare Ag_2 (cf. Fig. 7a). First, the amplitude of the observed vibrations is much larger than that of the long delay time dynamics of pure Ag_2. Second, the vibration is damped, and third, the NeNePo spectra do not show the short time scale dynamics apparent form Fig. 7a for bare Ag_2. FFT analysis of the oscillatory dynamics in Fig. 7b leads to a vibrational period of 240 fs ($\nu = 141\ cm^{-1}$), which is significantly red-shifted in comparison to the 180-fs period of bare Ag_2. The amplitude of the vibration can be fitted in good approximation by an exponentially damped sine function with a lifetime of 650 $\pm$50 fs, which gives an approximative time scale for the dephasing of the wavepacket. The red shift of the vibration points toward the substantial influence of the fragmentation of the oxygen ligand on the dynamics. Qualitatively, the fragmentation of the O_2 molecule apparently leaves the silver dimer fragment with higher vibrational levels populated. Due to the anharmonicity of the potential, the wavepacket thus exhibits a red shift in its vibrational frequency [222]. This observation reflects on the strong influence of the molecular adsorbate on the metal cluster structure. Such adsorbate-induced structural changes, geometric as well as electronic, have recently been identified as the origin for the cooperative adsorption of multiple adsorbate molecules on small noble metal clusters. This cooperative action is regarded essential to the catalytic activity of the gas-phase noble metal clusters in, for example, the CO combustion reaction [175, 220]. In the particular case of the negatively charged silver clusters Ag_n^- with odd n, the joint experimental and theoretical work of some of the present authors showed that a weakly bound first O_2 cooperatively promotes the adsorption of a second O_2 molecule, which is then differently bound with the O_2 bond elongated and thus activated for further oxidation reactions such as CO combustion [175]. The potential prospects of these intriguing catalytic properties of free noble metal clusters for real-time laser spectroscopic investigations and photoinduced control of catalytic reactions will be discussed in the final section of this chapter.

F. The Scope of NeNePo Spectroscopy

Finally, the question can be raised: What general information can be inferred from simulated NeNePo–ZEKE signals on the multistate energy landscapes and dynamics? First, our theoretical simulations allowed us to establish the connections between three objectives: the structural relation of anionic and neutral species, the influence of the nature of the nonequilibrium state reached after photodetachment, and the character of subsequent dynamics in the neutral ground state. Three different situations can be encountered in which (i) transition state, (ii) global minimum, and (iii) local minimum can influence the dynamics after photodetachment. Second, different types of relaxation dynamics can be identified in NeNePo–ZEKE signals. Moreover, (iv) the fragmentation and signature of fragments can be also identified.

(i) In cases where the anionic structure is close to a transition state of the neutral electronic ground state (e.g., trimers), large-amplitude motion toward the stable structure dominates the relaxation dynamics. In other words, the dynamics is incoherent but localized in phase space. IVR can be initiated as a consequence of the localized large-amplitude motion. Large-amplitude structural relaxation after the transition state is responsible for a pronounced single peak in NeNePo–ZEKE signals at a given time delay and probe excitation wavelengths. In addition, subsequent IVR processes can be identified but only under ZEKE-like conditions since the integration over the continuum of electron kinetic energies leads to the loss of the fine features in the signals.

(ii) In cases where the anionic structure is close to the global minimum (i.e., the stable isomer) of the neutral electronic ground state, vibrational relaxation reflecting the structural properties of the neutral stable isomer (e.g., Ag_4) takes place. The dynamics can be dominated by a single (e.g., Ag_4) or only by few modes that are given by the geometric deviations between anionic and neutral species. Other modes and anharmonicities weakly contribute, leading to dephasing on a time scale up to several picoseonds (longer than 2 ps for Ag_4). Vibrational relaxation gives rise to oscillations in NeNePo signals (for different pulse durations) which can be analyzed in terms of normal modes. This allows us to gain indirect information about vibrational spectra of a neutral cluster and use them as a fingerprint for the identification of the structure.

(iii) In cases where the anionic structure (the initial state) is close to a local minimum (energetically high-lying isomer) of the neutral electronic ground state (e.g., Au_4), the local minimum governs the dynamics after the photodetachment. Vibrational relaxation within the local minimum is likely to dominate the ultrashort dynamics (on a time scale of less than 1 ps for Au_4). Nondephased regular vibrational relaxation has been shown in the case of Au_4, where the pronounced activation of only one stretching mode takes place since the normal modes of the anionic and neutral species are almost identical.

Moreover, the local minimum can act as a strong capture area for nuclear motion with time scales up to several picoseconds. As a consequence, isomerization processes toward other local minima and/or toward the global minimum structure are widely spread in time. In other words, structural relaxation dynamics is characterized as being incoherent and delocalized in phase space. Signals exhibit (at different excitation wavelengths of the probe laser) fingerprints of vibrational relaxation within the local minimum, providing structural information. After systems escape from the local minima, the time scales for the beginning of structural relaxation can be identified by the onset of signals at given probe wavelengths ($\approx$ 1ps Au_4), although the relatively structureless signals of low intensity can reflect the delocalized character of the structural relaxation.

(iv) The fragmentation patterns as well as the characteristics of fragments can also be identified in NeNePo signals. From this information we infer that the NeNePo technique in connection with the optimal control schemes provides a promising powerful technique to introduce the control of the chemical reactivity of clusters, such as the oxidation of CO by noble metal oxide clusters, which is of relevance for heterogeneous catalysis.

G. Multistate Nonadiabatic Nuclear Dynamics in Electronically Excited and Ground States

As already pointed out, *ab initio* nonadiabatic MD "on the fly" involving excited states and simulation of observables is demanding from theoretical and computational point of view and still needs further developments. Therefore, in order to meet the requirements on high accuracy and realistic computational demand, the choice of the systems has to be made for which the description of electronic excited states and nonadiabatic coupling is particularly simple. This is the case for nonstoichiometric alkali-halide clusters with one excess electron (e.g., Na_nF_{n-1}). Their structural and optical properties have attracted the attention of many theoretical and experimental studies [54, 55, 223–252] due to the localization of the excess electrons, which are not involved in ionic bonding. The prototypes for a particularly simple situation concerning the description of excited states are nonstoichiometric sodium fluoride clusters with a single excess electron (e.g., Na_nF_{n-1}). In this case, a strong absorption in the visible–infrared energy interval occurs due to the excitations of the one excess electron placed in a large energy gap between occupied (HOMO) and unoccupied (LUMO) one-electron levels that resemble the "valence" and the "conductance" bands in infinite systems. Therefore, these clusters offer the opportunity to explore the optical properties of finite systems with some bulk characteristics such as F-color centers. Moreover, a simple but accurate description of the excited states is possible to achieve in the framework of the one-electron "frozen ionic bonds" approximation. In this method, the optical response of the single excess electron can be explicitly considered in the field of other $(n - 1)$ valence electrons that are involved in strongly polar ionic Na–F bonding [45].

The calculation of excited-state energies and of gradients based on the "frozen ionic bonds" approximation (as outlined in Ref. 45) is, from a computational point of view, considerably less demanding in comparison with other approaches such as RPA, CASSCF or CI, and provides comparable accuracy. Therefore, this approach allows to carry out adiabatic molecular dynamics in the excited state, by calculating the forces "on the fly" (cf. Ref. 45) applicable to relatively large systems. This is particularly convenient for the simulation of time-dependent transitions for which an ensemble of trajectories is needed. Moreover, the fast computation of nonadiabatic couplings "on the fly" allows one also to carry out nonadiabatic MD as outlined in Ref. 46. Of course, the application is limited to systems for which the "frozen ionic bonds" approximation offers an adequate description.

Based on *ab initio* classical trajectories and assuming Gaussian femtosecond envelopes for the laser fields, analytic expressions for the time-resolved pump-probe and pump-dump signals in the framework of the Wigner distribution approach are given by Eq. (6) (cf. Ref. 20). This *ab initio* Wigner distribution approach to adiabatic dynamics has been outlined above to illustrate the scope of the NeNePo approach. It has been extended for nonadiabatic dynamics in Ref. 46, which will be briefly described below. In addition to methodological aspects, the study of the dynamics in the first excited state of Na_3F_2 and the radiationless transition to the ground state allow for the prediction and verification of the consequences of conical intersections in fs pump-probe signals in the gas phase without the necessity to consider the environment. The latter external medium effects complicate the issue, as, for example, in the case of photochemistry in solution or in the case of the cis–trans photoisomerization of the visual pigment due to the influence of the protein cavity [253, 254].

Therefore, the photoisomerization in the Na_3F_2 cluster through a conical intersection will be addressed first and then in Section III the new strategy for optimal control will be applied in order to suppress the passage through the conical intersection and to selectively populate one of the chosen isomers.

H. Photoisomerization Through a Conical Intersection in the Na_3F_2 Cluster

The goal is to show that the breaking of bonds in excited states leading to the conical intersection can be identified in observables such as fs pump-probe signals. Since this will be illustrated by the example of Na_3F_2, first (1) the optical properties, then (2) the characterization of the conical intersection of this cluster will be given, and subsequently (3) the nonadiabatic couplings and the nonadiabaticity will be addressed. After the formulation of an analytic expression for the fs signals in the framework of the *ab initio* Wigner distribution approach (4), the analysis of the nonadiabatic dynamics and of the signals (5) will provide the information about the time scales of the different processes such as bond-breaking and the passage through conical intersection which can be identified in the pump-probe spectra.

1. *Optical Response Properties*

The absorption spectra obtained for both isomers of Na_3F_2 using the "frozen ionic bond" approximation are shown in Fig. 8 and compare well with those calculated by taking into account all valence electrons [245]. The lowest-energy isomer I, with the ionic Na_2F_2 subunit to which the Na atom is bound (forming Na–Na and Na–F bonds), gives rise to the low-energy intense transition in the infrared. This is a common feature found for Na_nF_{n-1} clusters due to the localized excitation of the one-excess electron, as mentioned above. In contrast, the transition to the first excited state of isomer II (C_{2v}) with the Na_3 subunit, which is bridged by two F atoms, has a higher energy close to the energies of transitions usually arising from excitations in metallic subunits. After the vertical transition at the geometry of isomer I, the geometric relaxation in the first excited

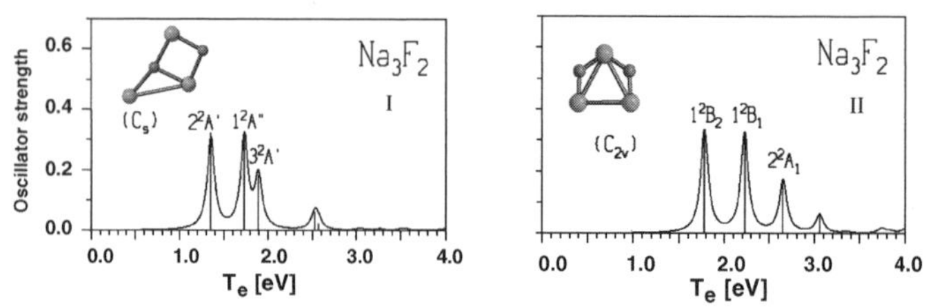

Stationary absorption spectra
Oscillator strength
0.6
0.4
0.2
0.0
Na3F2
I
(Cs)
2²A'
1²A"
3²A'
0.0
1.0
2.0
3.0
4.0
Te [eV]
Na3F2
II
(C2v)
1²B2
1²B1
2²A1
0.0
1.0
2.0
3.0
4.0
Te [eV]

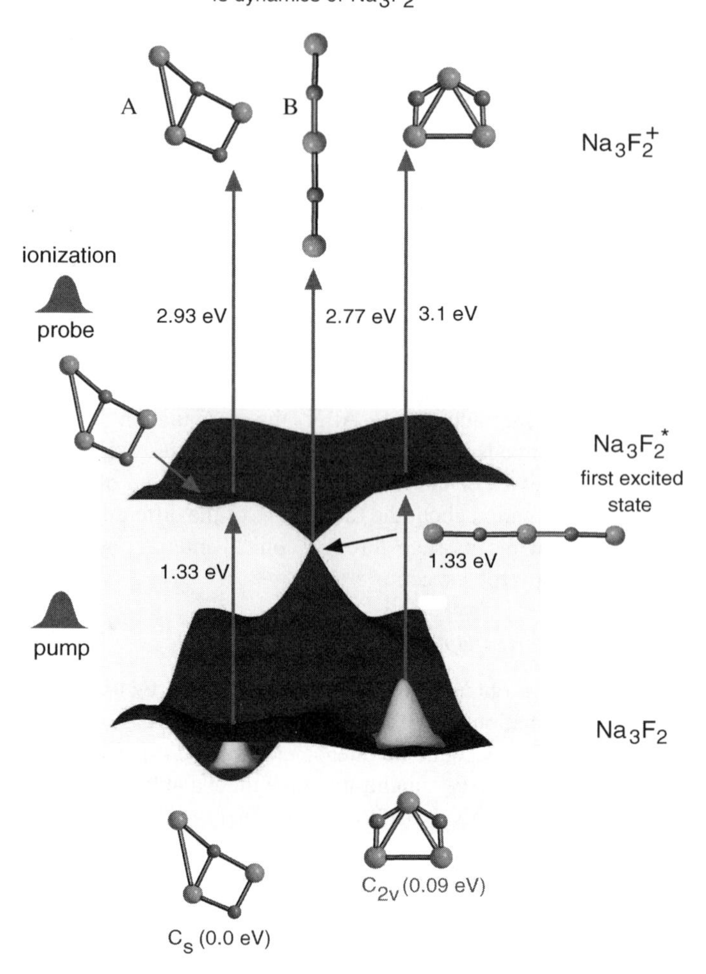

NeExPo pump–probe scheme for multistate
fs dynamics of Na3F2
A
B
Na3F2+
ionization
probe
2.93 eV
2.77 eV
3.1 eV
Na3F2*
first excited
state
1.33 eV
1.33 eV
pump
Na3F2
Cs (0.0 eV)
C2v (0.09 eV)

state takes place, involving a breaking of the Na–Na bond which leads to the first local minimum of the excited state (cf. Fig. 8) with a moderate lowering of the energy. Afterwards, the relaxation process proceeds to the absolute minimum with the linear geometry corresponding to the conical intersection for which a further considerable decrease of energy takes place. The linear geometry of the conical intersection is also reached after vertical transition to the first excited state at the geometry of the second isomer with C_{2v} structure. Accordingly, the investigation of the dynamics in the first excited state involves the breaking of metallic and ionic bonds starting from isomer I, and just metallic bonds starting from isomer II, as well as the passage through the conical intersection.

Consequently, one expects strong thermal motions within the ensemble, leading to phase space spreading and IVR. All processes can be monitored by a second ionizing probe pulse with excitation energies between ~2.9 eV and ~4.8 eV, as shown by the scheme given in Fig. 8. The first value is close to the initial Franck–Condon transition region and probes the relaxation dynamics on the potential surface of the first excited electronic state before the branching process, which is due to the conical intersection, does occur. The ground-state dynamics after passage through the conical intersection allows us to monitor processes involved on the ground-state potential surface.

2. *Conical Intersection*

The algorithm introduced by Robb and co-workers [255] is very useful for the determination of the lowest structure and the energy at the intersection seam as well as for analyzing the topology of the intersection in the space spanned by the internal degrees of freedom. The results obtained for the linear geometry of Na_3F_2, which has $N = 10$ internal degrees of freedom, show that the displacements in eight out of the 10 directions almost do not change the energetic separation of the surfaces, while displacements in the orthogonal plane characterized by two directions, X_1 and X_2, strongly remove the energy degeneracy. X_1 is the gradient difference vector, and X_2 involves the coupling vector between the two states. In other words, the ground-state reaction pathways starting in the plane X_1X_2 connect the excited-state reactants with the two ground-state products. Thus, the intersection of the ground and the first excited state has the shape of a double cone, with respect to X_1 and X_2, where the apex spans an eight-dimensional hyperline along which the energy is degenerate. The intersection seam is therefore (N – 2)-dimensional because it is characteristic for *conical intersections*.

Figure 8. Absorption spectra for two isomers I and II of Na_3F_2 obtained from one electron "frozen ionic bonds" approximation [46] (upper part). Scheme of the multistate fs dynamics for NeExPo pump-probe spectroscopy of Na_3F_2 including conical intersection with structures and energy intervals for the pump and probe steps [46]. See color insert.

The analysis of the wavefunctions of the ground and the first excited state in the close neighborhood of the conical intersection yields positive and negative linear combinations of two "valence bond-like" ("VB") structures $Na^+–F^-–Na^+–F^-–\dot{Na} \pm \dot{Na}–F^-–Na^+–F^-–Na^+$. One of them contributes dominantly to the ground, and the other one contributes to the first excited state, thus giving rise to two states with different symmetries. The location of the excess electron is indicated by the dot above the sodium atom. Of course, at the point of the conical intersection, the arbitrary linear combination of the above "valence bonds" structures is possibly due to degeneracy. The two "VB" structures differ in the translocation of the single excess electron or of the charge from one to the other end of the linear system. In other words, the length of the linear chain is sufficiently long to allow for an energy gap closing, in analogy to the dissociation limit of the H_2^+ molecule for which the degeneracy of the ground and excited state occurs due to equal energies of $\dot{H}–H^+$ and $H^+–\dot{H}$ structures. We conclude that the presence of the conical intersection in Na_3F_2 through which the isomerization process can take place is the consequence of the electronic structure properties. Therefore, due to general characteristics, it can be found for other systems by designing the analogous electronic situation.

In fact, the analogy can be drawn to conical intersections found in organic photochemistry involving biradicaloid species, which are generated by partial breaking of double hetero bonds due to geometric relaxation in the singlet excited states. The condition for the occurrence of conical intersections in so-called "critical biradicals" has been formulated in the framework of the two-orbital two-electron model and can be fulfilled in the case that the electronegativity difference between the two centers is sufficient to minimize the repulsion between the ground and the excited states [256]. In fact, it has been confirmed experimentally that the conical intersection is responsible for the cis–trans isomerization of the retinal chromophore in the vision process [253, 254].

Moreover, investigation of the nonadiabatic dynamics through the conical intersection of the Na_3F_2 cluster has advantages. The system has 10 degrees of freedom and permits the calculation of an ensemble of trajectories based on the accurate *ab initio* description of the excited and ground electronic states and on corresponding MD. Thus it provides the conceptual framework for fs observables such as fs pump-probe signals, which will be addressed below.

3. *Nonadiabatic Dynamics*

The breakdown of the Born–Oppenheimer approximation, due to avoided crossings or conical intersections between two electronic states, and the consideration of nonadiabatic couplings and nonadiabaticity will be now outlined, and the analytic expressions for the fs signals involving nonadiabatic dynamics in the framework of the semiclassical Wigner distribution approach will be presented.

In order to address nonadiabatic transitions in complex systems involving avoided crossings and conical intersections between electronic states, semiclassical methods based on *ab initio* multistate nonadiabatic dynamics are suitable for the simulation of fs pump-probe signals. For this purpose, in addition to the calculation of forces in the electronic ground and excited states, the computation of coupled electronic states "on the fly" in the adiabatic or diabatic representation is required. Furthermore, the choice of the approach to nonadiabatic dynamics must be made. These ingredients can then be combined with the Wigner–Moyal representation of the vibronic density matrix, which allows one to determine the fs signals. The electronic part, concerning *ab initio* calculations of forces in the excited states and nonadiabatic couplings, in the framework of "frozen ionic bonds" approximation valid for Na_3F_2 are given in Ref. 46. For review of nonadiabatic dynamics "on the fly" (cf. Ref. 146).

Since we consider the systems with all degrees of freedom, the most simple choice of treatment of the nonadiabatic dynamics is limited either to the classical-path methods or to surface hopping methods [257]. They are characterized by problems arising from the approximations that the trajectories propagate in the mean-potential or in the state specific potential, respectively [258–260]. In general, nonadiabaticity involves changes in the population of adiabatic states with changing nuclear configurations. In this way, the electronic distribution influences the trajectories. The simplest way to include such electron–nuclei feedback is to use the mean-field (Ehrenfest) method. It is assumed that the system evolves on an effective potential that can be obtained as an average over adiabatic states weighted by their state populations. The problem with this approach is that the system, which was prepared initially in a pure adiabatic state, will be in a mixed state after leaving the nonadiabatic region. Therefore, the adiabatic nature of the involved states does not prevail even in the asymptotic region [146]. Moreover, the microscopic reversibility is not preserved (cf. Ref. 146). The improvement to the Ehrenfest method is to include decoherence, assuming that the trajectories finish in a pure state after leaving the region of coupled states. This is possible by introducing the continuous surface switching procedure CSS [149].

In contrast, the basic feature of the surface-hopping methods is that the propagation is carried out on one of the pure adiabatic states, which is selected according to its population, and that the average over the ensemble of trajectories is performed. The molecular dynamics with quantum transitions (MDQT) version of the fewest-switches surface hopping method, as introduced by Tully [257], is based on the assumption that the fraction of trajectories on each surface is equivalent to the corresponding average quantum probability determined by coherent propagation of quantum amplitude. Furthermore, a choice between adiabatic and diabatic representation has to be made. In the former case the nonadiabatic couplings have to be calculated, and in the latter case the overlap between the wavefunctions of two states is needed in the framework of the method used for calculations of the electronic structure. For example, the MD as well as nonadiabatic couplings calculated "on the fly" can be directly connected with MDQT and then used to simulate fs signals. In what follows, we briefly outline the concept involving the adiabatic representation.

The time-dependent wavefunction $\Psi(t, \boldsymbol{r}, \boldsymbol{R})$, which describes the electronic state at time t, is expanded in terms of the adiabatic electronic basis functions ψ_j of the Hamiltonian with complex-valued time-dependent coefficients

$$\Psi(t, \boldsymbol{r}, \boldsymbol{R}) = \sum_{j=o}^{M} c_j(t)\psi_j(\boldsymbol{r}; \boldsymbol{R}) \tag{8}$$

The adiabatic states are also time-dependent through the classical trajectory $R(t)$. Substitution of this expansion into the time-dependent Schrödinger equation, multiplication by ψ_k from the left, and integration over $\boldsymbol{r}$ yields a set of linear differential equations of the first order for the expansion coefficients, which are equations of motion for the quantum amplitudes:

$$\mathrm{i}\dot{c}_k(t) = \sum_j \left[\epsilon_j \delta_{kj} - \mathrm{i}\dot{\boldsymbol{R}}(t) \cdot \langle \psi_k | \nabla_{\boldsymbol{R}} | \psi_j \rangle \right] c_j(t) \tag{9}$$

Here ϵ_j are the eigenvalues of the Hamiltonian, and $\langle \psi_k | \nabla_{\boldsymbol{R}|\psi_j\rangle}$ are nonadiabatic couplings.
The system of equations (9) has to be solved simultaneously with the classical equations of motion for the nuclei

$$M\ddot{\boldsymbol{R}} = -\nabla_{\boldsymbol{R}} E_m(R) \tag{10}$$

where the force is the negative gradient of the potential energy of the "current" mth adiabatic state. The hopping probabilities g_{ij} between the states are determined by

$$g_{ij} = 2\frac{\Delta t}{c_i c_i^*}[\mathrm{Im}(c_i^* c_j \varepsilon_i \delta_{ij}) - \mathrm{Re}(c_i^* c_j \dot{\mathbf{R}} \langle \psi_i | \nabla_{\boldsymbol{R}|\psi_j\rangle}] \tag{11}$$

and can occur randomly according to the fewest-switches surface hopping approach introduced by Tully [257]. This approach has been designed to satisfy the statistical distribution of state populations at each time according to the quantum probabilities $| c_i |^2$ using the minimal number of "hops" necessary to achieve this condition.

However, this internal consistency is not always maintained, as analyzed in the literature [259]. One of the often noticed reasons for the internal inconsistency in MDQT is the presence of classically forbidden transitions. The energy conservation is achieved in MDQT during the transition by adjusting the classical velocities in the direction of the nonadiabatic coupling vector [257]. The transition is classically forbidden, if there is not enough velocity in this direction. In this case, two alternatives are commonly used. Either this velocity component is inversed or it remaines unchanged. The existence of classically forbidden transitions may lead to an inconsistency between the fraction of trajectories in each state and the averaged quantum probability. Another reason for the internal inconsistency in MDQT is the divergence of independent trajectories. For example, in the case that two surfaces substantially differ, the trajectories on the lower state can diverge and follow different pathways after leaving the nonadiabatic coupling region. Since in standard MDQT the quantum amplitudes are propagated coherently for each trajectory, in the case that some trajectories diverge, the coherent propagation can lead also to an inconsistency between the fraction of trajectories in each state and the corresponding average quantum probability. The analysis of the reasons for these inconsistencies and the proposals for improving them can be found in Ref. 259. The conclusion can be drawn that in order to obtain the time evolution of the population, the fraction of trajectories is more reliable than the averaged quantum probabilities. Thus, it is better to use the fraction of trajectories for the simulation of the pump-probe signals. Problems with surface hopping methods are particularly pronounced for systems involving an extended nonadiabatic coupling region or when tunneling processes as well as a large number of recrossings occur in this region.

4. *Pump-Probe Signals*

For the determination of the pump-probe signal accounting for the passage through the conical intersection, the expression for the cationic occupation $P_{22}^{(2)}$ given by Eq. (5) has to be modified. This is due to necessity in considering that the propagation of the ensemble starts in the excited state but can hop to the ground state according to the fewest-switches hopping algorithm. Therefore, not only the common averaging over the whole ensemble of the initial conditions due to the Wigner approach is required, but also, for a given initial condition, an averaging over trajectories obtained from different random numbers according to the hopping algorithm must be carried out [46]. Consequently, the coordinates and momenta of the propagated state can be labeled $\mathbf{q}_x^\nu$ and $\mathbf{p}_x^\nu$, where x is either the excited or the ground state, as determined by the hopping procedure. The quantities ν numerate the set of random numbers used in the hopping algorithm, satisfying the same initial condition. Therefore, the average over the number of hoppings N_{hop} has to be performed, and for the cationic population for nonadiabatic dynamics the following expression yields

$$
\begin{aligned}
P_{22}^{(2)}(t) &= \int d\mathbf{q}d\mathbf{p}\, P_{22}^{(2)}(\mathbf{q},\mathbf{p},t) \\
&\sim \int d\mathbf{q}_0 d\mathbf{p}_0 \int_0^t d\tau_2 \int_0^{t-\tau_2} d\tau_1 \\
&\frac{1}{N_{\text{hop}}}\sum_{\nu} \exp\left[-\sigma_{\text{pr}}^2 \frac{[\hbar\omega_{\text{pr}} - V_{+1,x}(\mathbf{q}_x^{\nu}(\tau_1;\mathbf{q}_0,\mathbf{p}_0))]^2}{\hbar^2}\right] \\
&\times \exp\left[-\sigma_{\text{pu}}^2 \frac{[\hbar\omega_{\text{pu}} - V_{10}(\mathbf{q}_0,\mathbf{p}_0)]^2}{\hbar^2}\right] \\
&I_{\text{pu}}(t-\tau_1-\tau_2) I_{\text{pr}}(t-\tau_2-t_d) P_{00}^{(0)}(\mathbf{q}_0,\mathbf{p}_0)
\end{aligned}
\tag{12}
$$

which is a modification of Eq. (5) valid for the adiabatic dynamics. The quantity $V_{+1,x}$ labels the energy gap between the propagating state and the cationic state at the instant of time. From this expression, the pump-probe signal can be calculated after integration over the pump-probe correlation function $\int_0^{\infty} d\tau_2 I_{\text{pu}}(t-\tau_1-\tau_2) I_{\text{pr}}(t-\tau_2-t_d)$ is performed explicitly:

$$
\begin{aligned}
S[t_d] &= \lim_{t\to\infty} P_{22}^{(2)}(t) \\
&\sim \int d\mathbf{q}_0 d\mathbf{p}_0 \int_0^{\infty} d\tau_1 \exp\left\{-\frac{(\tau_1 - t_d)^2}{\sigma_{\text{pu}}^2 + \sigma_{\text{pr}}^2}\right\} \\
&\times \frac{1}{N_{\text{hop}}}\sum_{\nu} \exp\left\{-\frac{\sigma_{\text{pr}}^2}{\hbar^2}[\hbar\omega_{\text{pr}} - V_{+1,x}(\mathbf{q}_x^{\nu}(\tau_1;\mathbf{q}_0,\mathbf{p}_0))]^2\right\} \\
&\times \exp\left\{-\frac{\sigma_{\text{pu}}^2}{\hbar^2}[\hbar\omega_{\text{pu}} - V_{10}(\mathbf{q}_0,\mathbf{p}_0)]^2\right\} P_{00}^{(0)}(\mathbf{q}_0,\mathbf{p}_0)
\end{aligned}
\tag{13}
$$

According to expression (13), the initial ground-state density $P_{00}^{(0)}$ is promoted to the first excited-state with the Franck–Condon transition probability given by the last exponential of Eq. (13). The propagation, the passing through the conical intersection, and the probe transition to the cationic state are described by the second exponential. This expression can be generalized for more than two states by introducing in Eq. (13) the sum of weighting factors corresponding to transition moments between the electronic states involved, for which also time-dependent energy gaps have to be calculated. The probe pulse window, being located around the time delay t_d between the pump and the probe pulse and the resolution of the signal determined by the square of the pulse durations, are given by the first exponential. Because it is required in the Wigner distribution approach, an ensemble average over the initial conditions has to be performed. The latter can be obtained from a sampling of the initial vibronic Wigner distribution $P_{00}^{(0)}$ of the ground electronic state.

Of course, the basically inherent problems of surface hopping methods can be overcome by using other semiclassical formulations. For example, semiclassical schemes advanced in the framework of stationary phase approximation [261, 262] involve the linearized semiclassical initial value representation [152–154], the semiclassical multi-surface hopping propagator approach [263–265], the multiple spawning method [157–266], the quantum-classical density matrix approach involving a hybrid MD–Monte Carlo algorithm with momentum jumps [267, 268], and the semiclassical multistate Liouville dynamics in diabatic and adiabatic representation [147, 151, 269, 270]. The majority of these methods is computationally more demanding and so far is usually tested and applied to model systems. Regarding the connection with time-dependent quantum chemistry through classical trajectories "on the fly," it is of particular interest to mention the semiclassical multistate Liouville dynamics in diabatic and adiabatic representation [147, 151, 269, 270] and the multiple spawning method [157]. Both approaches allow introduction of the quantum effects in nonadiabatic dynamics.

However, in many systems involving radiationless isomerization processes through conical intersection from the first excited state, the quantum effects are washed out due to a high excess of energy, resulting in the high-temperature situation. In such cases, in spite of this limitation, the approach described above is reliable and practicable. It provides full information about underlying ultrafast processes from the analysis of the simulated signals, as will be shown for the example of the Na_3F_2 cluster. The investigation of the nonadiabatic dynamics at the conical intersection between the first excited state and the ground state separating two isomers of Na_3F_2 (cf. Fig. 8) offers an excellent opportunity to simulate fs pump-dump signals at a high level of accuracy. It also allows one to identify the time scales of different ultrafast processes, including different kinds of bond breaking as well as radiationless transitions. For this purpose it is adequate to use the combination of the Wigner–Moyal representation of the vibronic density matrix and *ab initio* multistate molecular dynamics in the ground state and in the first excited state without precalculation of energy surfaces including the computation of the nonadiabatic couplings "on the fly." Analogous to adiabatic dynamics, an analytic formulation of non adiabatic coupling in the framework of the "frozen ionic bonds" approximation, valid for nonstoichiometric alkali-halide clusters with one excess electron, is used for calculation of nonadiabatic couplings "on the fly" and is outlined in Ref 46. In the framework of the "frozen ionic bond" approximation, all requested ingredients such as gradients of energies as well as nonadiabatic couplings are available. They have been formulated in Ref. 45 and 46 and can be straightforwardly inserted in Eqs. (8)–(11) and used for the nonadiabatic dynamics "on the fly" (e.g., passage through the conical intersection) [46].

In order to obtain the initial conditions, a canonical thermal ensemble of 50 K can be determined by the Wigner distribution function of the electronic ground state including all normal modes $\omega_i, i = 1, \ldots, 10$, of the C_s structure, corresponding to the total minimum of energy according to Eq. (6). The set of, for example, 100 initial conditions can be obtained by sampling the Wigner distribution function with respect to the coordinates $\mathbf{q}_0$ and momenta $\mathbf{P}_0$, which can be used for the classical trajectory simulations on the first excited state of Na_3F_2. The finite temperature of 50 K causes thermal deviations from the energy minimum C_s structure.

5. *Analysis of the Nonadiabatic Dynamics and of the Signals*

Important aspects of the analysis of the nuclear dynamics will be first addressed. The simulation of the classical trajectory ensemble, consisting of a large number of sampled phase space points, can be started on the first excited electronic state using initial conditions described above. The geometric relaxation (over the local minimum) toward the linear structure corresponding to the conical intersection and its passage through the conical intersection as well as the subsequent relaxation dynamics on the electronic ground state can be visualized by considering the phase space density of the cluster ensemble shown in Fig. 9 for different propagation times. Initially at $t = 0$ fs, the phase space density is localized corresponding to the C_s structure (cf. Fig. 9). During the subsequent $\sim$90 fs, the distance between the Na–Na atoms elongates, indicating the bond-breaking between both sodium atoms, and corresponds to a local minimum on the first excited state (cf. Fig. 9). Consecutive ionic bond-breaking between the Na and the F atoms of the Na_2F_2 subunit can be observed after 220 fs (cf. Fig. 9) together with a small delocalization of the phase space density. After $\sim$400 fs, the region of the conical intersection corresponding to the linear structure is reached (cf. Fig. 9). This triggers the branching of the phase space density from the excited electronic state to the ground state. At this stage, the system gains an additional kinetic energy of $\sim$0.67 eV. Due to this large vibrational excess energy, strong anharmonicities between the vibrational modes are present, which are responsible for the phase space spreading. The subsequent relaxation dynamics on the electronic ground state is characterized by an even larger phase space spreading, particularly after 800 fs. This is due to the fact that the vibrational excess energy rose to $\sim$1.3 eV, which corresponds to an equilibrium temperature of $\sim$3400 K (cf. Fig. 9). However, in spite of increasing phase space spreading, structural information of the cluster ensemble can be gained up to a propagation time of $\sim$800 fs by considering the center-of-mass positions of the atomic phase space distributions in Fig. 9. In particular, the

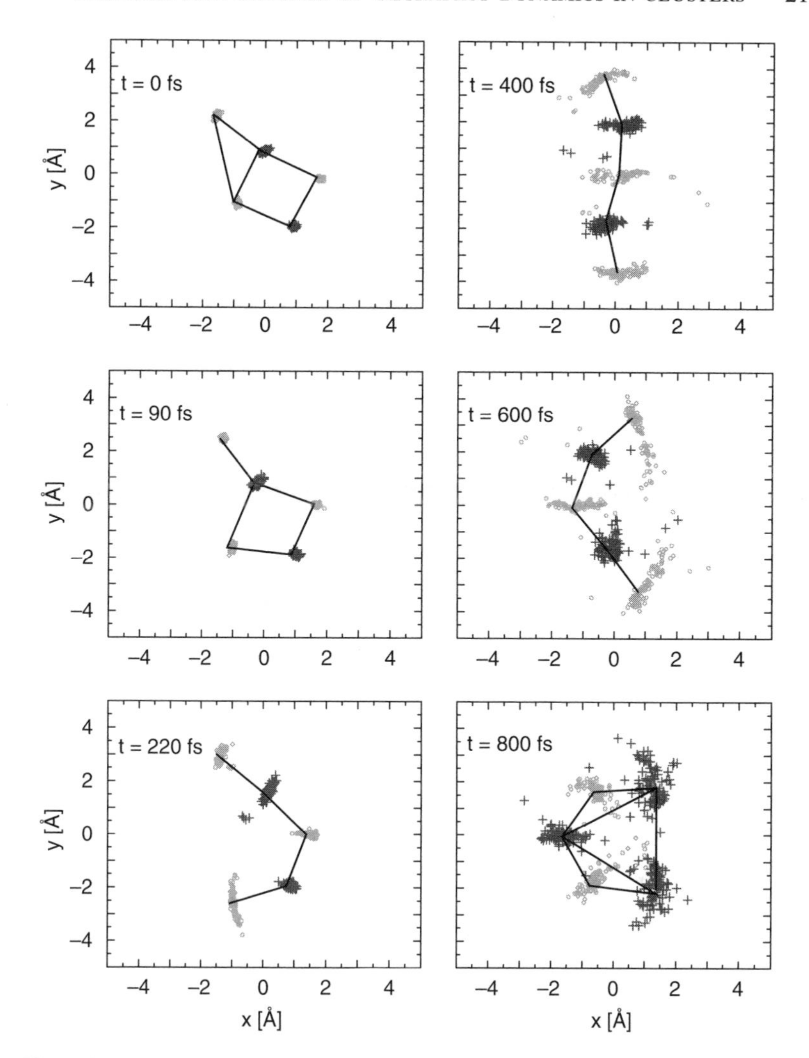

Figure 9. Snapshots of the phase space distribution (PSD) obtained from classical trajectory simulations based on the fewest-switches surface-hopping algorithm of a 50 K initial canonical ensemble [46]. Na atoms are indicated by black circles, and F atoms are indicated by gray crosses. Dynamics on the first excited state starting at the C_s structure ($t = 0$ fs) over the structure with broken Na–Na bond ($t = 90$ fs) and subsequently over broken ionic Na–F bond ($t = 220$ fs) toward the conical intersection region ($t = 400$ fs), Dynamics on the ground state after branching of the PSD from the first excited state leads to strong spatial delocalization ($t = 600$ fs). The C_{2v} isomer can be identified at ~800 fs in the center-of-mass distribution. See color insert.

"center of-mass-geometry" at 800 fs is close to the C_{2v} structure. However, due to the phase space spreading, there are also considerable deviations, and even geometries close to the C_s structure are involved in the phase space distribution of the cluster ensemble. As shown below, one can obtain detailed information about the branching ratio between these structures as well as about energetic distributions in the cluster ensemble from pump-probe signals. For times beyond 1 ps, no structures can be identified in the phase space distribution. The ensemble is geometrically completely delocalized at least up to the propagation time of 2.5 ps, which is understandable due to the large vibrational excess energy.

In summary, the dynamics through the conical intersection represents an elementary physical event for the cluster ensemble in the sense that it initiates the transition from structurally and energetically localized pattern involving consecutive metallic and ionic bond breaking processes to energy delocalized pattern. Thus, the molecular dynamics might be divided into a reversible and an irreversible part separated by the passage through the conical intersection.

Simulations of signals are based on Eq. (13), with energy gaps obtained from the classical trajectory simulations using the fewest switching surface hopping algorithm [Eqs. (8)–(11)] for the ensemble at an initial temperature of 50 K [Eq. (6)]. In order to obtain comprehensive information on the dynamical processes of Na_3F_2, a zero pump pulse duration ($\sigma_{pu} = 0$) is suitable, which involves a complete excitation of the ground-state ensemble prepared at the initial temperature. The ultrafast structural relaxation processes involving the bond-breaking can be resolved using a probe pulse duration of 50 fs. The simulated signals are shown for four different excitation energies (wavelengths) of the probe pulse in Fig. 10, which allow us to analyze the underlying processes:

(i) $E_{pr} = 2.8$ eV and $E_{pr} = 3.0$ eV correspond to transition energy values between the first excited and the cationic state at the time of the Na–Na metallic and the Na–F ionic bond-breaking, respectively (cf. Fig. 8). Thus the signals for those transition energies provide information on the structural relaxation involving the bond-breaking processes in the first excited state of Na_3F_2 before the conical intersection is reached. In fact, they exhibit maxima at ~90 fs and ~220 fs (cf. Fig. 10), in agreement with the time scales for the metallic and ionic bond breaking obtained from the analysis of the phase space distribution shown in Fig. 9. Both signal intensities decrease rapidly after 0.4–0.5 ps, indicating the branching of the phase space density from the first excited electronic state to the ground state due to the conical intersection.

(ii) $E_{pr} = 4.3$ eV and $E_{pr} = 4.8$ eV (cf. Fig. 10) correspond to transition energies between the ground state and the cationic state at the C_s geometry and the C_{2v} geometry, respectively. In such a way, the signals shown in Fig. 10 monitor the ratio of both isomers in the phase space

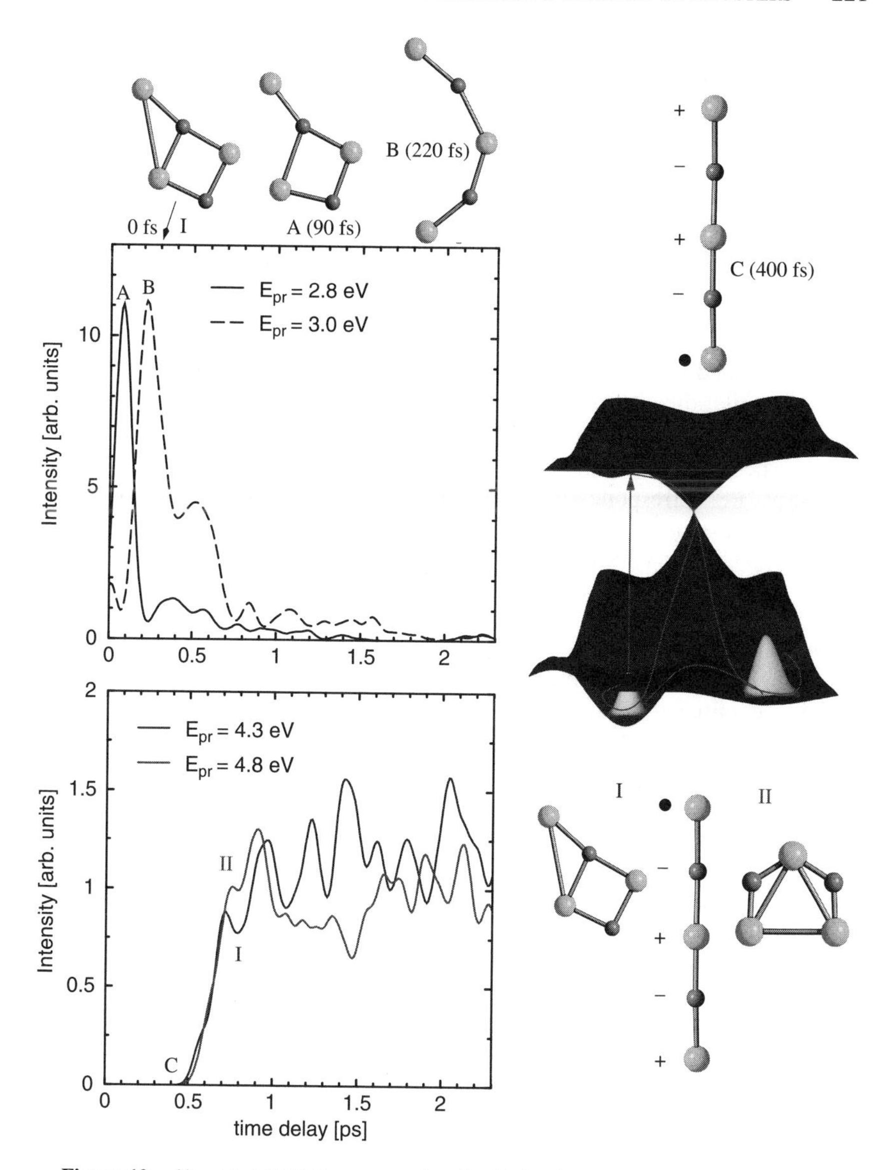

Figure 10. Simulated NeExPo pump-probe signals for the 50 K initial temperature Na_3F_2 ensemble at different excitation energies of the probe laser monitoring the geometric relaxation on the first excited state involving bond-breaking processes and passage through the conical intersection as well as geometric relaxation and IVR processes on the ground state after the passage (left-hand side). The isomerization through the conical intersection is schematically illustrated on the right-hand side [46]. See color insert.

distribution after the passage through the conical intersection up to a time delay between pump and probe of ~1 ps. This time represents the limit up to which structural information can be resolved in the phase space distribution (cf. Fig. 9). For larger time delays, the signals provide only information about the energetic redistribution, thus IVR. In fact, both signals start to increase after an incubation time of ~0.4 ps since the ground state becomes populated, providing the time scale for the passage through the conical intersection (cf. Fig. 10). Furthermore, the signal at $E_{pr} = 4.8$ eV exhibits a maximum at 0.8–0.9 ps, indicating the larger ratio of the C_{2v} structure in correspondence with the results obtained from the phase space distribution (cf. Fig. 9). This signal drops rapidly after 0.9 ps and the signal at $E_{pr} = 4.3$ eV increases, indicating that the population of the C_s structure is larger at 0.9–1.0 ps (cf. Fig. 10). The latter time dependence also exhibits oscillatory features beyond 1 ps, i.e. corresponding to the IVR regime. This leads to the conclusion that a somewhat periodic energy flow is present in the cluster ensemble. However, in view of the high vibrational excess energy, these oscillations cannot be attributed to particular normal modes.

In summary, these results provide information about the dynamics of the Na_3F_2 system in full complexity. They show that distinct ultrafast processes, which are initiated by the Frank–Condon pump pulse transition to the first excited electronic state, are involved in the dynamics of the Na_3F_2 cluster. These include geometric relaxation, consecutive bond-breaking of metallic and ionic bonds, passage through the conical intersection, and IVR processes [46]. Moreover, the time scales of these processes can be identified in the pump-probe signals, and each of them can be selectively monitored by tuning the probe excitation energy. However, in order to populate only one of the isomers, the pathway has to be found which avoids a large excess of energy disposal through the conical intersection. This offers the opportunity to tailor laser pulses that will drive the system into the desired target, and it will be addressed in Section III. Similar situations can be expected in considerably larger systems, provided that the characteristic electronic aspects remain preserved.

III. CONTROL OF ULTRAFAST PROCESSES

The conceptual framework underlying the control of the selectivity of product formation in a chemical reaction using ultrashort pulses rests on the proper choice of the time duration and the delay between the pump and the probe (or dump) step or/and their phase, which is based on the exploitation of the coherence properties of the laser radiation due to quantum mechanical interference effects [56, 57, 59, 60, 271]. During the genesis of this field, single-parameter control was proposed. Within the Brumer–Shapiro phase-control scheme [59, 60, 271], constructive and destructive interference between different light-induced reaction pathways is used in order to favor or to suppress different reaction channels. The other scheme, introduced by Tannor and Rice [56, 57], takes the advantage of differences in potential energy surfaces of different electronic states and therefore uses the time parameter for control. The pump pulse brings the system to nonequilibrium configurations from which transformations such as bond stretchings take place. If the probe or dump laser is timed properly, different pathways to dissociation of one of the stretched bonds can be achieved. Both single-parameter control schemes were experimentally confirmed [62–65, 120, 272–278]. Another single control parameter is a "linear chirp" [279, 280] corresponding to a decrease or increase of the frequency as a function of time under the pulse envelope. This was the first step toward shaping the pulses in the framework of so-called many-parameter optimal control theory (OCT). Tannor and Rice have first variationally optimized electric fields [281]. Then the optimal control theory was applied to molecular problems by Rabitz and co-workers [61, 282–284], and by Rice, Tannor, Kosloff, and co-workers [285, 286]. Technological progress due to fs pulse shapers allowed the manipulation of ultrashort laser pulses [22, 27, 28, 30, 31]. Finally, a closed-loop learning control (CLL) was introduced by Judson and Rabitz [287], opening the possibility to apply optimal control to more complex systems. Since potential energy surfaces of multidimensional systems are complicated and mostly not available, the idea was to combine an fs-laser system with a computer-controlled pulse shaper to produce specific laser fields acting on the system initiating photochemical processes. After detection of the product, the learning algorithm [288, 289, 22] was used to modify the field based on information obtained from the experiment and from the objective (the target). The shaped pulses were tested and improved iteratively until the optimal shape for the chosen target was reached. Such a black-box procedure is extremely efficient, but it does not provide information about the nature of the underlying processes that are responsible for the requested outcome. The success of the above-mentioned schemes has been demonstrated by a multitude of control experiments [23–41, 77–82, 290]. However, any multiparameter optimization scheme has a drawback of having a manifold of local solutions that are reachable depending on initial conditions. Intense research activity is directed toward improvements of these aspects, particularly in the closed-loop learning control [291, 292].

The investigation of simple systems offers a possibility to learn how to use control as a tool for analyzing the underlying processes. Therefore, metallic dimers [66–71, 84, 121, 292, 293] and diatomic molecules [292, 294] have been extensively studied. This is due to the fact that they are suitable model systems for establishing scopes of different control schemes and because they became easily accessible to experimental pulse-shaping techniques [72–83]. In fact, experimental work on Na_2 [113], using one-parameter-control, provided the first confirmation of the simple Tannor–Rice control scheme [56, 57]. By varying the time delay between the first and second pulses, Gerber and his colleagues investigated competition between ionization and dissociative ionization of Na_2 ($Na_2 \rightarrow Na_2^+ + e^-$ versus $Na_2 \rightarrow Na^+ + Na + e^-$). Consequently, the ratio of molecular to atomic ion products $Na_2{}^+/Na^+$ oscillates with the change in the time delay between the pump and probe pulses with the period determined by the motion of the wavepacket on the $2^1\Pi_g$ state. Control over the branching between the Na^+/Na_2^+ channels was also achieved by using the given laser wavelengths. In these experiments, a different sequence of states in Na_2, which involves double minimum potential energy surface ($^1\Sigma_u^+$), was reached [278]. Similarly, the variation of the delay time between the pulses was used by Herek, Materny, and Zewail [64] to switch between different channels for the photofragmentation of NaI, leading to the same product.

Encouraged by the confirmation of the control concept, two-parameter control was considered in order to manipulate different processes in dimers and diatomic molecules. In addition to the pump-probe time delay, the second control parameter involved the pump [72, 73] or probe [66, 67] wavelength, the pump-dump delay [69, 74, 75], the laser power [121], the chirp [68, 76], or the temporal width [70] of the laser pulse. Optimal pump-dump control of K_2 has been carried out theoretically in order to maximize the population of certain vibrational levels of the ground electronic state using one excited state as an intermediate pathway [71, 292–294]. The maximization of the ionization yield in mixed alkali dimers has been performed first experimentally using closed-loop learning control [77, 78, 83] (CLL) and then theoretically in the framework of optimal control theory (OCT) [84].

Since experimentally and theoretically optimized pulses obtained from OCT and CCL are available for NaK [84], first, it was possible to show under which conditions the shaped pulses are reproducible, and second, the connection between the forms of the shaped pulses and different ionization pathways was established. This allowed one determination of the mechanism for the maximization of the ionization yield under the participation of several excited states. The agreement between experimentally and theoretically optimized pulses, which was independent from the initial guess, showed that the shapes of the pulses can be used to deduce the mechanism of the processes underlying the optimal control. In the case of optimization of the ionization process in NaK, this involved a direct two-photon resonant process followed by a sequential one-photon processes at later times. These findings obtained for the simple system are promising for the use of shapes of tailored pulses to reveal the nature of processes involved in the optimal control of more complex systems. This will be addressed below.

Until recently, the limitation in the theory was imposed by difficulties in precalculating multidimensional potential surfaces of large clusters. In order to bypass this obstacle, *ab initio* adiabatic and nonadiabatic MD "on the fly" without precalculation of the ground- and excited-state energy surfaces is particularly suitable provided that an accurate description of the electronic structure is feasible and practicable. Moreover, this approach offers the following advantages. The classical–quantum mechanical correspondence between trajectory and a wavepacket is valid for short pulses and short time propagation. MD "on the fly" can be applied to relatively complex systems; moreover, it can be implemented directly in the procedures for optimal control. This allows us to identify properties that are necessary for assuring the controllability of complex systems and to detect mechanisms responsible for the obtained pulse shapes. In that context, the Liouville space formulation of optimal control theory developed by Yan, Wilson, Mukamel, and their colleagues [294–306]-in particular, its semiclassical limit in the Wigner representation [297, 51]—is very suitable in spite of its intrinsic limitations. For example, quantum effects such as interference phenomena or tunneling and zero-point vibrational energy are not accounted for. The study of clusters with varying size offers an ideal opportunity to test these concepts and methods as well as to investigate conditions under which different processes can be experimentally controlled and observed.

The ultimate goal of optimal control is not only to advance maximum yield of the desired process but also to use the shapes of the tailored pulses to understand the processes which are responsible for driving a complex system to the chosen target.

Optimal control theory has a broad spectrum of applications that will not be addressed in their completeness here. This includes research directions such as laser cooling of internal degrees of freedom of molecules and quantum computing (cf. Refs. 307–311 for examples of metallic dimers) since they usually require inclusion of additional methodological aspects.

A. Optimal Control and Analysis of Dynamic Processes in Complex Systems

It is still an open, central issue if and under which conditions optimal control involving more than one electronic state can be achieved for systems with increasing complexity. For these systems, energy landscapes of the ground and excited states can substantionally differ from each other or they can exhibit very complicated features. In this context there are several open basic questions, which should be addressed. An important question concerns the existence of a connective pathway between the initial state and the region of the energy lanscape (objective) which is reached via a different electronic state. In addition, even if such connective pathway does exist, the optimal path must be found and the method used for nuclear dynamics and for tailoring laser pulses should involve the realistic computational demand. Therefore, the development of new strategies for optimal control is required. An attractive possibility offers the concept of the intermediate target [51, 312] in the excited state. It is defined as a localized ensemble (wavepacket) corresponding to the maximum overlap between the forward propagating ensemble on the electronic excited state (starting from the initial state) and the backwards propagated ensemble from the objective in the ground state at optimal time delay between both pulses.

The classical nuclear dynamics is the only realistic approach in the case that separation of active from passive degrees of freedom cannot be made for complex systems and therefore a large number of them have to be treated explicitly. Furthermore, quantum corrections can be also introduced under the given circumstances. As will be shown below, the classical MD "on the fly" can be extremely useful for realization of new strategies for optimal control such as the construction of the intermediate target. The role of the intermediate target is to guarantee the connective pathway between the initial state and the objective and to select the appropriate parts of both energy surfaces involved. This issue is directly related to the inversion problem [117, 313–316].

In the case of the pump-dump control for two-phase unlocked ultrafast fields in the weak response regime, we have shown that the intermediate target serves first to optimize the pump pulse. This leads to the decoupled optimization of the pump and the dump pulses, which is very advantageous from the computational point of view. An appropriate formalism for the realization of the strategy for optimal control of complex systems based on the concept of the intermediate target is the density matrix formulation of the OCT. It combines the Wigner–Moyal representation of the vibronic density matrix with *ab initio* molecular dynamics (MD) "on the fly" in the electronic excited and the ground states without precalculation of both energy surfaces. This method, called the *ab initio* Wigner distribution approach, was outlined in Section II, first in connection with NeNePo spectroscopy which involves ground-state adiabatic dynamics and then for nonadiabatic dynamics involving the excited and ground electronic states. When adequate quantum chemical procedures can be used for dynamics in excited states, this method is also suitable to treat complex systems. Moreover, due to available analysis based on MD, the shapes of the optimized pulses can be directly interpreted and connected with the underlying ultrashort processes. After the outline of the theoretical basis for this optimal control strategy, our new strategy for optimal control using intermediate target will be applied to optimize the pump and dump pulses for driving the isomerization process in the nonstoichiometric Na_3F_2 cluster, avoiding conical intersection between the ground and the first excited state and maximizing the yield in the second isomer.

B. Intermediate target as a New Strategy for Optimal Control in Complex Systems

The goal of the optimal control strategy described here is to optimize temporal shapes of phase-unlocked pump and dump pulses (i.e., pump and dump pulses) and the time delay between them, which drive the system starting from the lowest energy isomer over the first excited state to the second isomer—that is, the objective.

The analytic form of the pump and dump pulses in the optimal phase-unlocked pump-dump control is $\epsilon_{P(D)}(t) = E_{P(D)}(t)\exp(-i\omega_{eg}t) + E^{*}_{P(D)}(t)$ $\exp(i\omega_{eg}t)$, where $E_{P(D)}$ is a slowly varying envelope of the fields, and ω_{eg} is the energy difference between the minima of the excited and the ground states. The objective in the ground state is represented in the Wigner formulation by an operator $\hat{A} = A(\Gamma)|g\rangle\langle g|$. $A(\Gamma)$ is the Wigner transform of the objective in the phase space $\Gamma = \{q_i, p_i\}$ of coordinates and momenta, and $|g\rangle\langle g|$ is the ground electronic state projection operator. $A(\Gamma)$ can be defined, for example, as

$$A(p,q) = \prod_{i=1}^{N} \frac{1}{\sqrt{2\pi}\Delta q_i} e^{-\frac{(q_i-\bar{q}_i)^2}{2(\Delta q_i)^2}} \Theta\left(E_{\min} - \sum_{i=1}^{N} \frac{p_i^2}{2m_i}\right) \tag{14}$$

where $\bar{q}_i$ represents Cartesian coordinates of the second isomer, and Δq_i represents the corresponding deviations. The role of the step function Θ is to insure that the kinetic energy is below the lowest isomerization barrier E_{min}. This corresponds to the spatial localization of the phase space density and arbitrary distribution of momenta. The optimized pulses can be obtained from the functional

$$J(t_f) = A(t_f) - \lambda_P \int_0^{t_f} |E_P(t)|^2 dt - \lambda_D \int_0^{t_f} |E_D(t)|^2 dt \tag{15}$$

where $A(t_f)$ is the yield at the time t_f, which for weak fields can be calculated in second-order perturbation theory [295, 303, 317]. It involves the propagated excited- and ground-state ensembles induced by the pump and dump pulses, the time-dependent energy gaps between the two states, and the initial distribution of the phase space in the Wigner representation. Optimal field envelopes can be obtained by calculating the extrema from the control functional (15) by using the variation procedure [295–298, 300–304, 317–320]. This leads to the pair of coupled integral equations for the field envelopes:

$$\int_0^{t_f} d\tau' M_P(\tau,\tau';E_D)E_P(\tau') = \lambda_P E_P(\tau) \tag{16}$$

$$\int_0^{t_f} d\tau' M_D(\tau,\tau';E_P)E_D(\tau') = \lambda_D E_D(\tau) \tag{17}$$

The integral kernels corresponding to response functions are given by

$$M_P(\tau, \tau'; E_D)$$
$$= \iint d^2\Gamma_0 \int_0^{t_f} d\tau'' \int_0^{\tau''} d\tau''' A(\Gamma_g(t_f - \tau''; \Gamma_e(\tau''' - \tau; \Gamma_0))) e^{i(\omega_{eg} - U_{eg}(\Gamma_e(\tau''' - \tau; \Gamma_0)))(\tau'' - \tau''')}$$
$$e^{i(\omega_{eg} - U_{eg}(\Gamma_0))(\tau - \tau')} \rho_{gg}(\Gamma_0) E_D(\tau''') E_D^*(\tau''), \qquad \tau \geq \tau' \tag{18}$$
$$M_D(\tau, \tau'; E_P)$$
$$= \iint d^2\Gamma_0 \int_0^{\tau'} d\tau'' \int_0^{\tau''} d\tau''' A(\Gamma_g(t_f - \tau; \Gamma_e(\tau' - \tau''; \Gamma_0))) e^{i(\omega_{eg} - U_{eg}(\Gamma_e(\tau' - \tau''; \Gamma_0)))(\tau - \tau')}$$
$$e^{i(\omega_{eg} - U_{eg}(\Gamma_0))(\tau'' - \tau''')} \rho_{gg}(\Gamma_0) E_P(\tau''') E_P^*(\tau''), \qquad \tau \geq \tau' \tag{19}$$

Γ_e and Γ_g correspond to propagated excited- and ground-state ensembles, and U_{eg} is the time-dependent energy gap between the excited and the ground state. Since both equations depend on the pump and dump pulses, they are coupled and can, in principle, be solved iteratively yielding optimized pump and dump pulses. However, this is computationally unrealistic even for systems of moderate complexity because the coupled classical simulations on the ground and excited states have to be performed. The calculation of objective A in Eqs. (18) and (19) requires the propagation of the ensemble on the ground state Γ_g, starting at different initial conditions. These conditions are obtained from the propagated ensemble Γ_e of the excited state at each time step. Therefore, the strategy involves decoupling of Eqs. (18) and (19), which is possible only in the short pulse regime on the fs time scale, and the necessary steps are outlined below.

(i) In the zero-order approximation of an iterative procedure and in the ultrafast regime, it is justified to calculate the kernel functions M_P and M_D with strongly temporally localized pulse envelopes $E_P \approx \delta(t)$ and $E_D \approx \delta(t - t_d)$. Then the zero-order response functions take the following forms:

$$M_P^{(0)}(\tau, \tau') = \iint d^2\Gamma_0 A(\Gamma_g(t_f - t_d; \Gamma_e(t_d - \tau; \Gamma_0)))$$
$$e^{i(\omega_{eg} - U_{eg}(\Gamma_0))(\tau - \tau')} \rho_{gg}(\Gamma_0), \quad \tau \geq \tau' \tag{20}$$
$$M_D^{(0)}(\tau, \tau') = \iint d^2\Gamma_0 A(\Gamma_g(t_f - \tau; \Gamma_e(\tau'; \Gamma_0)))$$
$$e^{i(\omega_{eg} - U_{eg}(\Gamma_e(\tau'; \Gamma_0)))(\tau - \tau')} \rho_{gg}(\Gamma_0), \quad \tau \geq \tau' \tag{21}$$

The equations for pump and dump pulses now become decoupled. Consequently, the pump pulse optimization involves the propagation on the excited state $\Gamma_e(t_d - \tau; \Gamma_0)$ from $\tau = 0$ until $\tau = t_d$ starting with Γ_0 (initial ensemble)

[Eq. (20)]. For the dump optimization, according to the Eq. (21), the dynamics on the ground state has to be carried out $\Gamma_g(t_f - \tau; \Gamma_e(t_d))$ for $\tau' = t_d$ until t_f with the initial conditions given by the ensemble of the excited state $\Gamma_e(t_d)$ at t_d which corresponds to the intermediate target. $\Gamma_e(t_d)$ at t_d can be determined from the maximal overlap between a forward-propagated ensemble from the first isomer on the excited state and a backwards-propagated ensemble on the ground state from the second isomer.

(ii) Equations (16) and (20) yield an optimal pump pulse that localizes phase space density at the intermediate target.

(iii) The optimized dump pulse projects the intermediate target to the ground state and optimally localizes the phase space density into the objective (second isomer) at a final time t_f. This means that the connective pathway between the initial state and the objective is guaranteed by the intermediate target at a time t_d. For this purpose, the function $A(\Gamma_g(t_f - t_d); \Gamma_e(t_d))$ must have nonvanishing contributions as follows from Eqs. (20) and (21). This procedure can be continued iteratively, but it is most likely that the zeroth- and first-order iterations lead to sufficient accuracy. In summary, the concept of the intermediate target represents a new strategy that ensures the connective pathway between the initial state and the objective, and moreover it allows to reach the objective with maximal yield optimizing pump and dump pulses independently. This allows the application of the optimal pump-dump control to complex systems without restricting the number of degrees of freedom and ensures controllability, provided that the intermediate target can be found, which is illustrated below in the next section.

C. Optimal Control of Photoisomerization in Na_3F_2

The isomerization in Na_3F_2 through a conical intersection between the first excited and the ground state is a nonselective process due to the high internal energy ($\sim$0.65 eV), which populates almost equally both isomers in the ground state and does not allow for selective population of the second isomer [46]. Therefore, the optimal control strategy described above, which is based on the concept of the intermediate target, represents an adequate tool to find the optimal pathway allowing one to populate isomer II with maximal yield and to suppress of the pathway through the conical intersection [51].

For this purpose, several steps are needed. First, the initial ensemble of isomer I has to be generated. Then the intermediate target involving excited- and ground-state dynamics has to be determined. Finally, the pump and dump pulses have to be optimized. For the initial ensemble a 50 K canonical ensemble in the ground state of isomer I in the Wigner representation can be constructed using, for example, a set of $\sim$1000 randomly sampled coordinates and momenta. In the pump step (photon energy of 1.33 eV), the ensemble is first propagated on the excited state (e.g., for 300 fs). In order to determine the intermediate target and the optimal time delay t_d, the ensemble has to be dumped to the ground state (in steps of, e.g., 25 fs) and subsequently propagated (e.g., 1 ps). It can be shown that isomer II is reached by the ensemble at $t_d = 250$ fs, and the residence time of 500 fs, at least, can be achieved. The ensemble-averaged geometry that determines the coordinates of the intermediate target is shown in Figs. 11 and 12. Note that the "geometry" of the

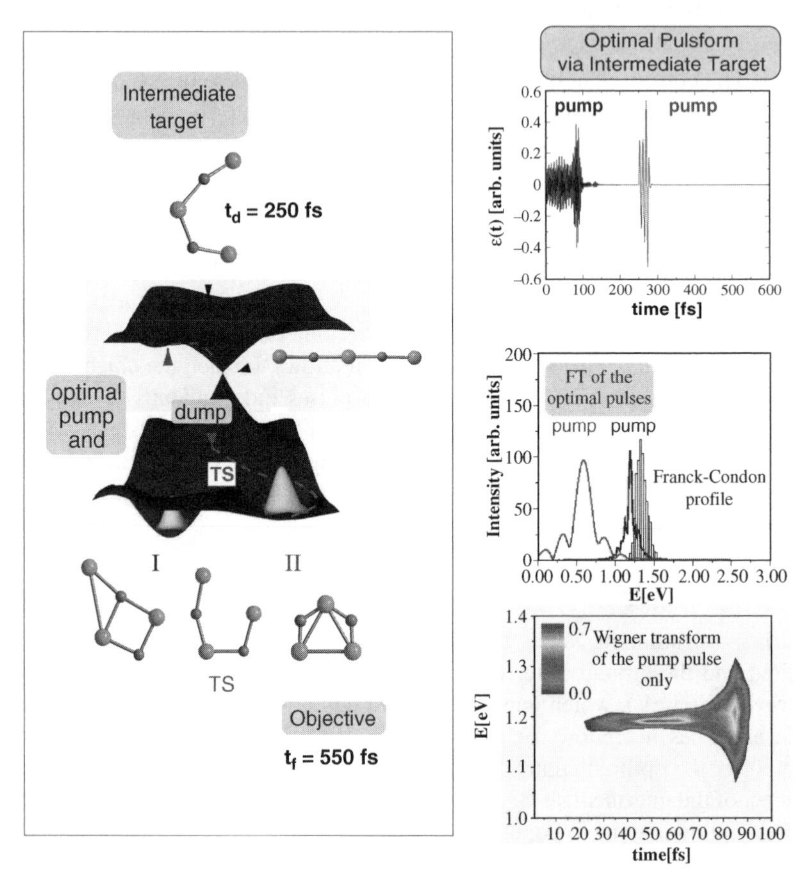

Figure 11. *Left-hand side*: Scheme for pump-dump optimal control in the Na_3F_2 cluster with geometries of the two ground-state isomers and of the transition state separating them, the conical intersection, and the intermediate target. *Upper panel, right-hand side*: The optimal electric field corresponding to the pump and dump pulses [51]. The mean energy of the pump pulse is 1.20 eV and the mean energy of the dump pulse is 0.6 eV. *Middle panel, right-hand side*: Fourier transforms of the optimal pump and dump pulses and the Franck–Condon profile for the first excited state corresponding to the excitation energy $T_e = 1.33$ eV. *Bottom panel, right-hand side:* Wigner transform of the optimal pump pulse. See color insert.

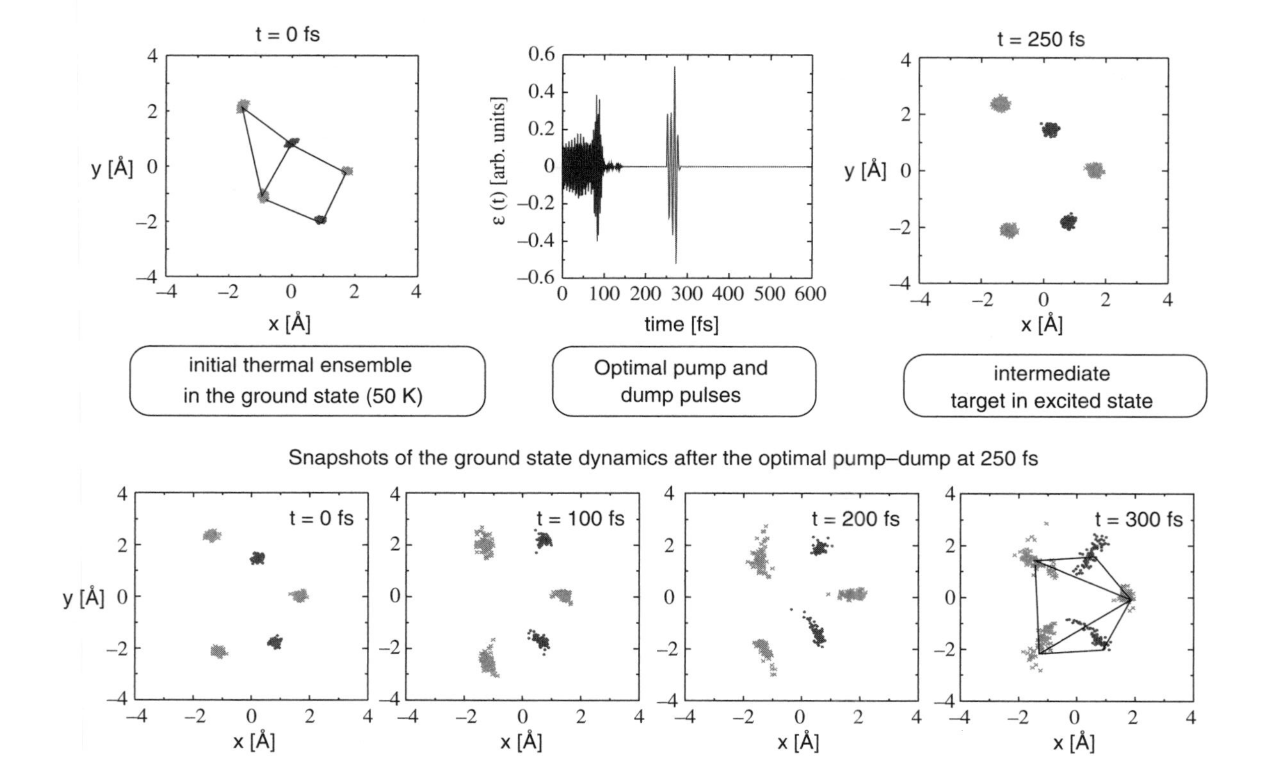

Figure 12. *Upper panels*: Initial thermal ensemble, optimal pump and dump pulses, intermediate target. *Lower panels*: Snapshots of the dynamics obtained by propagating the ensemble corresponding to the intermediate target after the optimized pump-dump at 250 fs on the ground state showing the localization of the phase space density in the basin corresponding to isomer II [51]. See color insert.

intermediate target is closely related to that of the transition state separating the two isomers on the ground-state. The role of the intermediate target to ensure the connective pathway from the initial state to the objective over the excited state is evidenced by its relation to the transition state which separates both ground-state isomers. The average kinetic energy of the intermediate target corresponds to ~75% of the isomerization barrier in the ground state. This guarantees that, after the dump, the ensemble will remain localized in the basin of isomer II.

The optimization of the pump pulse leads to a localization of the phase space density around the intermediate target. The intermediate target operator can be represented in the Wigner representation [Eq. (20)] by a minimum uncertainty wavepacket:

$$A(p_i, q_i) = \prod_{i=1}^{3N=15} \frac{1}{2\pi \Delta p_i \Delta q_i} e^{-\frac{(q_i - \bar{q}_i)^2}{2(\Delta q_i)^2}} e^{-\frac{(p_i - \bar{p}_i)^2}{2(\Delta p_i)^2}} \tag{22}$$

The response function $M(\tau, \tau')$ for the pump pulse [Eq. (20)] can be calculated, for example, on a time grid of 1 fs and can be symmetrized and diagonalized according to Eq. (16). In this case, the largest eigenvalue was obtained to be 0.82, corresponding to the globally optimized pulse which has 82% efficiency to localize the ensemble in the intermediate target.

The optimized pump pulse, shown in Fig. 11, consists of two portions with durations of ~70 fs and ~10 fs, respectively. Fourier and Wigner–Ville transforms of the pump pulse, shown also in Fig. 11, provide physical insight. Comparison of Fourier transform with the Franck–Condon profile of isomer I shows that the excitation of the low-lying vibrational modes at ~1.2 eV of the initial ensemble is dominantly responsible for reaching the intermediate target. This spectral region corresponds to lower-lying vibrational modes that open the C_s structure of isomer I by breaking the Na–Na and one of the Na–F bonds. The Wigner–Ville transform shows that this energetically sharp transition corresponds to the first temporal portion of ~70 fs of the pump pulse. In contrast, a very short second portion after 80–90 fs of ~10 fs is energetically much wider. It is related to tails of the Fourier transform, which are symmetric with respect to the 1.2-eV transition, reflecting equally distributed velocities in the initial ensemble.

The dump pulse optimization leads to a spatial localization of the phase space density in the objective (isomer II). For this purpose, the intermediate target operator [Eq. (22)] can be propagated on the ground state, and the dump pulse is obtained from Eqs. (17) and (21). The largest eigenvalue (e.g., 0.78) can be obtained. This corresponds to 78% efficiency of localization of isomer II. The optimized dump pulse is very short (~20 fs; cf. the part of the signal after $t_d = 250$ fs in Fig. 11). This implies that the time window around t_d for depopulation of the excited state is very short. Otherwise the system would gain a large amount of energy in the excited state (leading to the conical intersection). The Fourier transform of the dump pulse is centered around 0.6 eV, corresponding to the Franck–Condon transition at t_d as shown on the left-hand side of Fig. 11. Finally, in order to illustrate the efficiency of optimized pulses, snapshots of the ground-state ensemble propagated after the dump process are shown in Fig. 12. It can clearly be seen that the phase space density is localized in isomer II (the objective) after $t_d + 200$ fs $= 450$ fs.

Using the strategy for optimal pump-dump control based on the intermediate target, we have shown that the isomerization pathway through the conical intersection can be suppressed and that optimized pulses can drive the isomerization process to the desired objective (isomer II). This means that the complex systems are amenable to control, provided that the intermediate target exists. Furthermore, the analysis of the MD and of the tailored pulses allows for the identification of the mechanism responsible for the selection of appropriate vibronic modes necessary for the optimal control.

In summary, optimal pump-dump control of fs processes, involving two electronic states, requires the identification of the connective pathway between the initial state and the objective. This is possible if the intermediate target in the excited state can be found, which selects the appropriate parts of energy surfaces for the control. This was illustrated for the example of Na_3F_2 for which the optimal pump and dump pulses populate the objective (isomer II) with maximal yield, taking the optimal pathway and avoiding the conical intersection. The identification of the mechanism responsible for the shape of the pulses serves as a guide toward the understanding of theoretically and experimentally obtained tailored fields, which still represents a challenging task for future work. In this way, the control is used not only to achieve desired goal but also to identify and to analyze the underlying ultrafast processes responsible for favoring one pathway and for suppressing the others. Control as a tool for analysis of the dynamics complements the closed-loop learning (CLL) control technique and sheds light on the nature of the "black box."

IV. PERSPECTIVES

Analysis and control of ultrafast processes in atomic clusters in the size regime in which "each atom counts" are of particular importance from a conceptual point of view and for opening new perspectives for many applications in the future. Simultaneously, this research area calls for the challenging development of theoretical and computational methods from different directions, including quantum chemistry, molecular dynamics, and optimal control theory, removing borders between them. Moreover, it provides stimulation for new experiments.

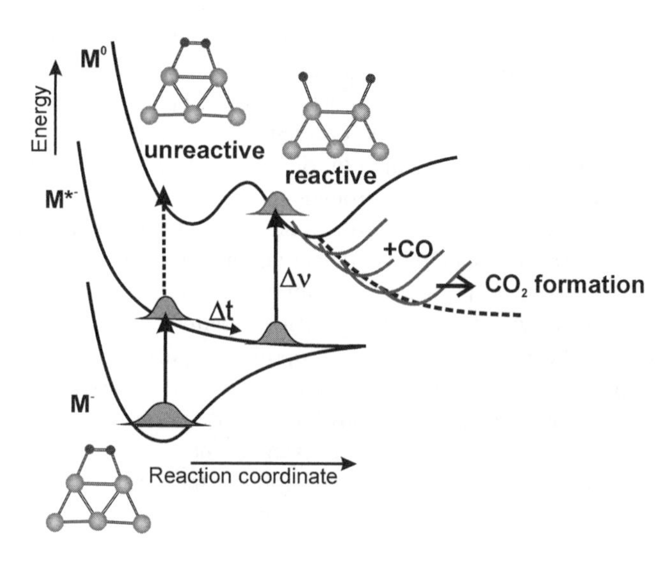

Figure 13. Schematic sketch of a reactive NeNePo control experiment. Control is achieved through two time- and frequency-shifted photodetachment laser pulses employing an anion excited state (M^{*-}) for intermediate wavepacket propagation. The wavepacket is finally prepared on the neutral potential energy surface in a region that corresponds to enhanced reactivity of the system. The aim of the experiment and theory is to find optimal composite pulses, based on the concept of the intermediate target outlined in Section III.A, that accomplish such a reactive activation of M^0. Detection is performed by ionization of the potential reaction products of M0 to the cationic state (not shown in the graphic).

One of these proposals will be now briefly outlined. Aiming to control the dynamics of reactive processes by designed tailored laser pulses, a new resonant two-photon detachment R2PD–NeNePo scheme is proposed which largely relies on the involvement of excited electronic states of the initial anionic complex (cf. Fig. 13). Because bound excited states of anions are very rare, a bound-free transition is expected to be reached by a first near-infrared photon. Consequently, a rapid propagation of the nuclear ensemble on the repulsive excited anion potential energy surface should be initiated. However, before dissociation is completed, a second photon, appropriately shifted in time and frequency, should transfer the wavepacket to the neutral potential surface by photodetachment, possibly reaching the desired reactive nuclear configuration of the neutral cluster complex. Most interestingly, this two-step photodetachment process might be achieved not only through two separate or composite pulses, but via one phase and amplitude modulated broadband ultrafast laser pulse. The idea is based on the new strategy for optimal control using the concept of the intermediate target. This should allow us to find a localized ensemble (wavepacket) in excited anionic state with a maximum overlap with the reactive structures on the neutral ground state for the given time delay. Optimization of the pump pulse on the intermediate target should allow us to control desired reactivity channel.

In conclusion, by changing the size of cluster, and therefore its structural and optical properties, different ultrafast processes can be monitored, and their time scales can be determined. These include bond-breaking, geometric relaxation of different nature, IVR, isomerization, and other reaction channels. These processes can be identified by the analysis of adiabatic or nonadiabatic dynamics and from the simulated fs signals. Therefore, a precise determination can be provided for the conditions for the experimental observation of distinct dynamic processes. This predictive power of theory can be directly used for conceptual planning of experiments, as illustrated by several examples in this review. Moreover, the tailored laser fields obtained in the framework of optimal control theory can drive selected processes, such as direct versus sequential ionization, isomerization toward one of the isomers, or the chosen reaction channel for which particular bond breaking or new bonding rearrangements promote the emanation of the reaction products.

Theoretical methods that combine *ab initio* MD "on the fly" with the Wigner distribution approach, which is based on classical treatment of nuclei and on quantum chemical treatment of electronic structure, represent an important theoretical tool for the analysis and control of ultrashort processes in complex systems. Moreover, the possibility to include, in principle, quantum effects for nuclear motion by introducing appropriate corrections makes this approach attractive for further developments. However, for this purpose, new proposals for improving the efficient inclusion of quantum effects for the motion of nuclei and fast but accurate calculations of MD "on the fly" in the electronic excited states are mandatory. Both aspects represent attractive and important theoretical research areas for the future.

The strategies based on the localization of the wavepacket or its ensemble (e.g., an intermediate target), ensuring the connective pathway between the initial state and the target in complex systems involving at least two different electronic states, are attractive for several reasons. They allow simplification of optimization of pump and dump pulses for complex systems. They also permit selection of important parts of energy surfaces, which makes the inversion problem accessible. Finally, the analysis of the underlying dynamics makes it possible to assign the shapes of optimized pulses to distinct processes, allowing one to unravel the mechanisms responsible for optimal control. This also allows the use of optimal control schemes as tools for analysis of the dynamics of complex systems, which constitutes important conceptual issue with a promising perspective for applications in biomolecules, clusters, or even their complexes.

Due to the structure–reactivity relationships of clusters, the reactive centers can be identified. Furthermore, their size selectivity can be exploited for inducing reactions toward organic and inorganic molecules or for finding the cooperative effects required for promoting these reactions. This research direction opens new roads for using tailored laser fields to drive the laser-induced selective chemical reactions involving clusters. It also takes advantage of the functional characteristics of clusters, inducing a large impact in different application areas.

The exploration of ultrafast molecular and cluster dynamics addressed herein unveiled novel facets of the analysis and control of ultrafast processes in clusters, which prevail on the femtosecond time scale of nuclear motion. Have we reached the temporal boarders of fundamental processes in chemical physics? Ultrafast molecular and cluster dynamics is not limited on the time scale of the motion of nuclei, but is currently extended to the realm of electron dynamics [321]. Characteristic time scales for electron dynamics roughly involve the period of electron motion in atomic or molecular systems, which is characterized by $\tau \simeq 1$ a.u. (of time) = 24 attoseconds. Accordingly, the time scales for molecular and cluster dynamics are reduced (again!) by about three orders of magnitude from femtosecond nuclear dynamics to attosecond electron dynamics. Novel developments in the realm of electron dynamics of molecules in molecular clusters pertain to the coupling of clusters to ultraintense laser fields (peak intensity $I = 10^{16}$–10^{20} W cm^{-2} [322], where intracluster fragmentation and response of a nanoplasma occurs on the time scale of 100 attoseconds to femtoseconds [323].

The exploration of electron dynamics in large finite systems will stem from concurrent progress in theory and experiment, which will focus on analysis and control of various channels of "pure" electron dynamics process, without the involvement of nuclear motion, bypassing the constraints imposed by the Franck–Condon principle [321]. Of considerable interest will be the extension of the conceptual framework and development of experimental tools for the exploration of control of electron dynamics in the attosecond–femtosecond time domain. Current advances in this fascinating research area involve experimental, computational, and theoretical studies of extreme cluster multielectron inner and outer ionization and nanoplasma formation in ultraintense laser fields. They are being interrogated by the utilization of molecular dynamics simulations [322, 323], model calculations [322], real-time pump-probe experiments [324], and the advent of laser pulse shaping [325], opening the new research area of electron dynamics.

Acknowledgments

We thank our co-workers M. Hartmann, A. Heidenreich, J. Hagen, J. Pittner, D. Reichardt, B. Schäfer-Bung, and L. D. Socaciu-Siebert, who have substantially contributed to the theoretical and experimental studies of analysis and control of ultrafast dynamics in clusters, which were included in this chapter. This work has been supported by Deutsche Forschungsgemeinschaft (DFG) SFB 450, "Analysis and control of ultrafast photoinduced reactions."

References

1. A. H. Zewail, *Femtochemistry*; World Scientific, Singapore, 1994.
2. J. Manz and L. Wöste, eds., *Femtosecond Chemistry*, Vol. 1 and 2, VCH Verlagsgesellschaft mbH, Wenheim, Germany, 1995.
3. M. Chergui, ed., *Femtochemistry*, World Scientific, Singapore, 1996.
4. V. Sundström, ed., *Nobel Symposium Book: Femtochemistry and Femtobiology: Ultrafast Reaction Dynamics at Atomic Scale Resolution*, World Scientific, Imperial College Press, London, 1997.
5. A. H. Zewail, J. *Phys. Chem. A* **104**, 5660 (2000).
6. R. N. Zare, *Science* **279**, 1875 (1998).
7. C. V. Shank, *Opt. Lett.* **12**, 483 (1987).
8. G. R. Fleming, *Chemical Applications of Ultrafast Spectroscopy*, Oxford University Press, New York, 1986.
9. G. R. Fleming, T. Joo, M. Cho, A. H. Zewail, V. S. Lehotkov, R. A. Marcus, E. Pollak, D. J. Tannor, and S. Mukamel, *Adv. Chem. Phys.* **101**, 141 (1986).
10. K. Wynne and R. M. Hochstrasser, *Adv. Chem. Phys.* **107**, 263 (1999).
11. T. Brixner and G. Gerber, *Chem Phys Chem* **4**, 418(2003).
12. V. Bonačić-Koutecký, P. Fantucci, and J. Koutecký, *Chem. Rev.* **91**, 1035 (1991).
13. A. W. Castleman, Jr. and K. H. Bowen, Jr., *J. Phys. Chem.* **100**, 12911 (1996).
14. J. Jortner, *Faraday Discuss.* **108**, 1 (1997).
15. U. Landman, *Int. J. Mod. Phys.* **B6**, 3623 (1992).
16. Q. Zhong and Jr. A. W. Castleman, Jr., *Chem. Rev.* **100**, 4039 (2000).
17. A. Stolow, A. E. Bragg, and D. M. Neumark, *Chem. Rev.* **104**, 1719 (2004).
18. T. E. Dermota, Q. Zhong, and A. W. J. Castleman, *Chem. Rev.* **104**, 1861 (2004).
19. V. Bonačić-Koutecký and R. Mitrić, *Chem. Rev.* **105**, 11 (2005).
20. M. Hartmann, J. Pittner, V. Bonačić-Koutecký, A. Heidenreich, and J. Jortner, *J. Chem. Phys.* **108**, 3096 (1998).
21. M. Hartmann, J. Pittner, V. Bonačić-Koutecký, A. Heidenreich, and J. Jortner, *J. Phys. Chem.* **102**, 4069 (1998).
22. T. Baumert, T. Brixner, V. Seyfried, M. Strehle, and G. Gerber, *Appl. Phys. B* **65**, 779 (1997).
23. A. Assion, T. Baumert, M. Bergt, T. Brixner, B. Kiefer, V. Seyfried, M. Strehle, and G. Gerber, *Science* **282**, 919 (1998).
24. T. Hornung, R. Meier, and M. Motzkus, *Chem. Phys. Lett.* **326**, 445 (2000).
25. S. Vajda, P. Rosendo-Francisco, C. Kaposta, M. Krenz, L. Lupulescu, and L. Wöste, *Eur. Phys. J. D* **16**, 161 (2001).
26. R. J. Levies, G. M. Menkir, and H. Rabitz, *Science* **292**, 709 (2001).
27. D. Yelin, D. Meshulach, and Y. Silberberg, *Opt. Lett.* **22**, 1793 (1997).
28. A. Efimov, M. D. Moores, N. M. Beach, J. L. Krause, and D. H. Reitze, *Opt. Lett.* **23**, 1915 (1998).
29. T. Brixner, M. Strehle, and G. Gerber, *Appl. Phys. B* **68**, 281 (1999).
30. E. Zeek, K. Maginnis, S. Backus, U. Russek, M. M. Murnane, G. Mourou, H. C. Kapteyn, and G. Vdovin, *Opt. Lett.* **24**, 493 (1999).
31. E. Zeek, R. Bartels, M. M. Murnane, H. C. Kapteyn, S. Backus, and G. Vdovin, *Opt. Lett.* **25**, 587 (2000).

32. A. Efimov, M. M. D., B. Mei, J. L. Krause, C. W. Siders, and D. H. Reitze, *Appl. Phys. B* **70**, 133 (2000).
33. D. Zeidler, T. Hornung, D. Proch, and M. Motzkus, *Appl. Phys. B* **70**, 125 (2000).
34. D. Meshulach, D. Yelin, and Y. Silberberg,
35. D. Meshulach, and Y. Silberberg, *Nature* **396**, 239 (1998).
36. D. Meshulach, and Y. Silberberg, *Nature* **396**, 239 (1998).
37. T. Hornung, R. Meier, D. Zeidler, K. L. Kompa, D. Proch, and M. Motzkus, *Appl. Phys. B* **71**, 277 (2000).
38. T. C. Weinacht, J. L. White, and P. H. Bucksbaum, *J. Phys. Chem. A* **103**, 10166 (1999).
39. T. C. Weinacht, J. Ahn, and P. H. Bucksbaum, *Nature* **397**, 233 (1999).
40. R. Bartels, S. Backus, E. Zeek, L. Misoguti, G. Vdovin, I. P. Christov, M. M. Murnane, and H. C. Kapteyn, *Nature* **164** (2000).
41. J. Kunde, B. Baumann, S. Arlt, F. Morier-Genoud, U. Siegner, and U. Keller, *Appl. Phys. Lett.* **77**, 924 (2000).
42. M. Dantus and V. V. Lozovoy, *Chem. Rev.* **104**, 1813 (2004).
43. C. Daniel, J. Full, L. Gonzalez, C. Lupulescu, J. Manz, A. Merli, S. Vajda, and L. Wöste, *Science* **299**, 536 (2003).
44. I. Andrianov, V. Bonačić-Koutecký, M. Hartmann, J. Manz, J. Pittner, and K. Sundermann, *Chem. Phys. Lett.* **318**, 256 (2000).
45. M. Hartmann, J. Pittner, and V. Bonačić-Koutecký, *Chem. Phys.* **114**, 2106 (2001).
46. M. Hartmann, J. Pittner, and V. Bonačić-Koutecký, *Chem. Phys.* **114**, 2123 (2001).
47. M. Hartmann, R. Mitrić, B. Stanca, and V. Bonačić-Koutecký, *Eur. Phys. J. D* **16**, 151 (2001).
48. V. Bonačić-Koutecký, M. Hartmann, and J. Pitter, *Eur. Phys. J. D* **16**, 133 (2001).
49. R. Mitrić, M. Hartmann, B. Stanca, V. Bonačić-Koutecký, and P. Fantucci, *J. Phys. Chem. A* **105**, 8892 (2001).
50. S. Vajda, C. Lupulescu, A. Merli, F. Budzyn, L. Wöste, M. Hartmann, J. Pitter, and V. Bonačić-Koutecký, *Phys. Rev. Lett.* **89**, 213404 (2002).
51. R. Mitrić, M. Hartmann, J. Pittner, and V. Bonačić-Koutecký, *J. Phys. Chem. A* **106**, 10477 (2002).
52. V. Bonačić-Koutecký, R. Mitrić, M. Hartmann, and J. Pittner, *Int. J. Quant. Chem.* **99**, 408 (2004).
53. N. E. Henriksen and V. Engel, *Int. Rev. Phys. Chem.* **20**, 93 (2001).
54. M.-C. Heitz, G. Durand, F. Spiegelman, and C. Meier, *J. Chem. Phys.* **118**, 1282 (2003).
55. M.-C. Heitz, G. Durand, F. Spiegelman, C. Meier, R. Mitrić, and V. Bonačić-Koutecký, *J. Chem. Phys.* **121**, 9898 (2004).
56. D. J. Tannor and S. A. Rice, *J. Chem. Phys.* **83**, 5013 (1985).
57. D. J. Tannor and S. A. Rice, *Adv. Chem. Phys.* **70**, 441 (1988).
58. G. K. Paramonov and V. A. Savva, *Phys. Lett.* **97A**, 340 (1983).
59. P. Brumer and M. Shapiro, *Faraday Discuss. Chem. Soc.* **82**, 177 (1986).
60. M. Shapiro and P. Brumer *J. Chem. Phys.* **84**, 4103 (1986).
61. A. P. Peirce, M. A. Dahleh, and H. Rabitz, *Phys. Rev. A* **37**, 4950 (1988).
62. T. Baumert, B. Buhler, M. Grosser, R. Thalweiser, V. Weiss, E. Wiedenmann, and G. Gerber, *J. Phys. Chem.* **95**, 8103 (1991).
63. T. Baumert and G. Gerber, *Isr. J. Chem.* **34**, 103 (1994).
64. J. L. Herek, A. Materny, and A. H. Zewail, *Chem. Phys. Lett.* **228**, 15 (1994).

65. A. Shnitman, I. Sofer, I. Golub, A. Yogev, M. Shapiro, Z. Chen, and P. Brumer, *Phys. Rev. Lett.* **76**, 2886 (1996).

66. H. Schwoerer, R. Pausch, M. Heid, V. Engel, and W. Kiefer, *J. Chem. Phys.* **107**, 9749 (1997).

67. C. Nicole, M. A. Bouchene, C. Meier, S. Magnier, E. Schreiber, and B. Girard, *J. Chem. Phys.* **111**, 7857 (1999).

68. L. Pesce, Z. Amity, R. Uberna, S. R. Leone, and R. Kosloff, *J. Chem. Phys.* **114**, 1259 (2001).

69. Z. W. Shen, T. Chen, M. Heid, W. Kiefer, and V. Engel, *Eur. Phys. J. D* **14**, 167 (2001).

70. G. Grègeire, M. Mons, I. Dimicoli, F. Piuzzi, E. Charron, C. Dedonder-Lardeux, C. Jouvet, S. Matenchard, D. Solgadi, and A. Suzor-Weiner, *Eur. Phys. J. D* **1**, 187 (1998).

71. T. Hornung, M. Motzkus, and J. de Vivie-Riedle, *Chem. Phys.* **115**, 3105 (2001).

72. G. Rodriguez and J. G. Eden, *Chem. Phys. Lett.* **205**, 371 (1993).

73. G. Rodriguez, P. C. John, and J. G. Eden, *J. Chem. Phys.* **103**, 10473 (1995).

74. R. Pausch, M. Heid, T. Chen, W. Kiefer, and H. Schwoerer, *J. Chem. Phys.* **110**, 9560 (1999).

75. R. Pausch, M. Heid, T. Chen, W. Kiefer, and H. Schwoerer, *J. Raman Spectrosc.* **31**, 7 (2000).

76. R. Uberna, Z. Amitay, R. A. Loomis, and S. R. Leone, *Faraday Discuss* **113**, 385 (1999).

77. S. Vajda, A. Bartelt, E.-C. Kaposta, T. Leisner, C. Lupulescu, S. Minemoto, P. Rosenda-Francisco, and L. Wöste, *Chem. Phys.* **267**, 231 (2001).

78. A. Bartelt, S. Minemoto, C. Lupulescu, S. Vajda, and L. Wöste, *Eur. Phys. J. D* **16**, 127 (2001).

79. S. Vajda, C. Lupulescu, A. Bartelt, F. Budzyn, P. Rosendo-Francisco, L. Wöste, in *Femtochemistry and Femtobiology*, A. Douhal and J. Santamaria, eds., World Scientific Publishing, Singapore, 19xx.

80. A. Bartelt, A. Lindinger, C. Lupulescu, S. Vajda, and L. Wöste, *Phys. Chem. Chem. Phys.* **5**, 3610 (2003).

81. C. Lupulescu, A. Lindinger, M. Plewicky, A. Merli, S. M. Weber, and L. Wöste, *Chem. Phys.* **296**, 63 (2004).

82. A. Bartelt, *Steuerung de Wellenpaketdynamik in kleinen Alkaliclustern mit optimierten Femtosekundenpulsen*, Dissertation, Freie Universität Berlin, 2002.

83. J. B. Ballard, H. U. Stauffer, Z. Amitay, and S. R. Leone, *J. Chem. Phys.* **116**, 1350 (2002).

84. B. Schäfer-Bung, R. Mitrić, V. Bonačić-Koutecký, A. Bartelt, C. Lupulescu, A. Lindinger, S. Vajda, S. M. Weber, and L. Wöste, *J. Phys. Chem.* A **108**, 4175 (2004).

85. C. J. Bardeen, V. V. Yakovlev, K. R. Wilson, S. D. Carpenter, P. M. Weber, and W. S. Warren, *Chem. Phys. Lett.* **280**, 151 (1997).

86. T. Brixner, N. H. Damrauer, P. Niklaus, and G. Gerber, *Nature* **414**, 57 (2001).

87. C. Daniel, J. Full, L. González, E.-C. Kaposta, M. Krenz, C. Lupulescu, J. Manz, S. Minemoto, M. Oppel, P. Rosendra-Francisco, S. Vajda, and L. Wöste, *Chem. Phys.* **267**, 247 (2001).

88. N. H. Damrauer, C. Dietl, G. Krampert, S. H. Lee, K. H. Jung, and G. Gerber, *Eur. Phys. J. D.* **20**, 71 (2002).

89. J. L. Herek, W. Wohlleben, R. J. Cogdell, D. Zeidler, and M. Motzkus, *Nature* **417**, 553 (2002).

90. S. Wolf, G. Sommerer, S. Rutz, E. Schreiber, T. Leisner, L. Wöste, and R. S. Berry, *Phys. Rev. Lett.* **74**, 4177 (1995).

91. R. S. Berry, V. Bonačić-Koutecký, J. Gaus, T. Leisner, J. Manz, B. Reischl-Lenz, H. Ruppe, S. Rutz, E. Schreiber, S. Vajda, R. de Vivie-Riedle, S. Wolf, and L. Wöste, *Phys. Rev. Lett.* **101**, 101 (1997).

92. A. H. Zewail, *Faraday Discuss. Chem. Soc.* **91**, 207 (1991).

93. A. Mokhtari, P. Cong, J. L. Herek, A. H. Zewail, *Nature* **348**, 225 (1990).
94. M. Dantus, R. M. Bowman, M. Gruebele, and A. H. Zewail, *J. Chem. Phys.* **91**, 7489 (1989).
95. A. Weaver, R. B. Metz, S. E. Bradforth, and D. M. Neumark, *J. Chem. Phys.* **93**, 5352 (1990).
96. R. B. Metz and D. M. Neumark, *J. Chem. Phys.* **97**, 962 (1992).
97. D. M. Neumark, *Acc. Chem. Res.* **26**, 33 (1993).
98. B. J. Greenblatt, M. T. Zanni, and D. M. Neumark, *J. Chem. Phys.* **111**, 10566 (1999).
99. M. T. Zanni, B. J. Greenblatt, A. V. Davis, and D. M. Neumark, *J. Chem. Phys.* **111**, 2991 (1999).
100. D. M. Neumark, *Annu. Rev. Phys. Chem.* **52**, 255 (2001).
101. B. J. Greenblatt, M. T. Zanni, and D. M. Neumark, *J. Chem. Phys.* **112**, 601 (2000).
102. R. Wester, A. V. Davis, A. E. Bragg, and D. M. Neumark, *Phys. Rev. A* **65**, 051201 (2002).
103. C. Frischkorn, A. E. Bragg, A. V. Davis, R. Wester, and D. M. Neumark, *J. Chem. Phys.* **115**, 11185 (2001).
104. T. M. Bernhardt, J. Hagen, L. D. Socaciu, J. Le Roux, D. Popolan, M. Vaida, L. Wöste, R. Mitrić, V. Bonačić-Koutecký, A. Heidenreich, and J. Jortner, *Chem Phys Chem* **6**, 105 (2005).
105. S. M. Burnett, A. E. Stevens, C. S. Feigerle, and W. C. Lineberger, *Chem. Phys. Lett.* **100**, 124 (1983).
106. K. M. Ervin, J. Ho, and W. C. Lineberger, *J. Chem. Phys.* **91**, 5974 (1989).
107. P. G. Wenthold, D. Hrovat, W. T. Borden, and W. C. Lineberger, *Science* **272**, 1456 (1996).
108. D. W. Boo, Y. Ozaki, L. H. Andersen, and W. C. Lineberger, *J. Phys. Chem. A* . **101**, 6688 (1997).
109. N. Pontius, P. S. Bechthold, M. Neeb, and W. Eberhardt, *Phys. Rev. Lett.* **84**, 1132 (2000).
110. H. Ihee, V. A. Lobatsov, U. M. Gomez, B. M. Goodman, R. Srinivasan, C. Ruan, and A. H. Zewail, *Science* **291**, 458 (2001).
111. L. X. Chen, W. J. H. Jäger, G. Jennings, D. J. Gosztola, A. Munkholm, and J. P. Hessler, *Science* **292**, 262 (2001).
112. T. Baumert, R. Grosser, R. Thalweiser, and G. Gerber, *Phys. Rev. Lett.* **67**, 3753 (1991).
113. T. Baumert, C. Röttgermann, C. Rothenfusser, R. Thalweiser, V. Weiss, and G. Gerber, *Phys. Rev. Lett.* **69**, 1512 (1992).
114. R. de Vivie-Riedle, B. Reischl, S. Rutz, and E. Schreiber, *J. Phys. Chem.* **99**, 16829 (1995).
115. T. S. Rose, M. J. Rosker, and A. J. Zewail, *J. Chem. Phys.* **88**, 6672 (1988).
116. P. Cong, A. Mokhtari, and A. H. Zewail, *Chem. Phys. Lett.* **172**, 109 (1990).
117. M. Gruebele, G. Roberts, M. Dantus, R. M. Bowman, and A. H. Zewail, *Chem. Phys. Lett.* **166**, 459 (1990).
118. L. E. Berg, M. Beutter, and T. Hansson, *Chem. Phys. Lett.* **253**, 327 (1996).
119. J. Heufelder, H. Ruppe, S. Rutz, E. Schreiber, and L. Wöste, *Chem. Phys. Lett.* **269**, 1 (1997).
120. E. D. Potter, J. L. Herek, S. Pedersen, Q. Liu, and A. H. Zewail, *Nature* (*London*) **355**, 66 (1992).
121. R. de Vivie-Riedle, K. Kobe, J. Manz, W. Meyer, B. Reischl, S. Rutz, E. Schreiber, and L. Wöste, *J. Phys. Chem.* **100**, 7789 (1996).
122. T. Baumert, R. Thalweiser, and G. Gerber, *Chem. Phys. Lett.* **209**, 29 (1993).
123. B. Reischl, R. de Vivie-Riedle, S. Rutz, and E. Schreiber, *J. Chem. Phys.* **104**, 8857 (1996).
124. J. Schön, and H. Köppel, *J. Phys. Chem. A* **103**, 8579 (1999).
125. J. Gauss, Dissertation, Freie Universität Berlin, 1992.
126. H. Ruppe, S. Rutz, E. Schreiber, and L. Wöste, *Chem. Phys. Lett.* **257**, 356 (1996).

127. A. Ruff, S. Rutz, E. Schreiber, and L. Wöste, *Z. Phys. D* **37**, 175 (1996).
128. H. Kühling, K. Kobe, S. Rutz, E. Schreiber, and L. Wöste, *J. Phys. Chem.* **98**, 6679 (1994).
129. S. Vajda, S. Rutz, J. Heufelder, P. Rosendo, H. Ruppe, P. Wetzel, and L. Wöste, *J. Phys. Chem. A* **102**, 4066 (1998).
130. D. J. Wales, *Energy Landscapes*, Cambridge University Press, Cambridge, 2003.
131. R. Car and M. Parrinello, *Phys. Rev. Lett.* **55**, 2471 (1985).
132. C. Leforestier, *J. Chem. Phys.* **68**, 4406 (1978).
133. R. N. Barnett and U. Landman, *Phys. Rev. B* **48**, 2081 (1993).
134. P. Pulay, *Mol. Phys.* **17**, 197 (1969).
135. Y. Yamaguchi, Y. Osamura, J. D. Goddard, and H. F. Schaefer III, *A New Dimension to Quantum Chemistry*, Oxford University Press, New York, 1994.
136. J. F. Gaw, Y. Yamaguchi, and H. F. Schaefer III, *J. Chem. Phys.* **81**, 6395 (1984).
137. J. F, Gaw, N. C. Handy, P. Palmieri, and A. D. Esposti, *J. Chem. Phys.* **89**, 959 (1988).
138. D. P. Chong, ed., *Recent Advances in Density Functional Methods*, vol. I, World Scientific Singapore, 1997.
139. V. Barone, A. Bencini, and P. Fantucci, eds., *Recent Advances in Density Functional Methods*, Vol. III, World Scientific, Singapore, 2002.
140. R. J. Bartlett, ed., *Recent Advances in Coupled-Cluster Methods*, Vol. III, World Scientific, Singapore, 1997.
141. K. Hirao, ed., *Recent Advances in Multireference Methods*, Vol. IV, World Scientific, Singapore, 1997.
142. C. Van Caillie and R. D. Amos, *Chem. Phys. Lett.* **308**, 249 (1999).
143. C. Van Caillie and R. D. Amos, *Chem. Phys. Lett.* **317**, 159 (2000).
144. F. Furche and R. Ahlrichs, *J. Chem. Phys.* **117**, 7433 (2002).
145. A. Köhn and C. Hättig, *J. Chem. Phys.* **117**, 7433 (2002).
146. N. L. Doltsinis, and D. Marx, *J. Theor. Comp. Chem.* **1**, 319 (2002).
147. A. Donoso and C. Martens, *J. Phys. Chem.* **102**, 4291 (1998).
148. S. Hammes-Schiffer, *J. Phys. Chem. A* **102**, 10443 (1998).
149. M. D. Hack and D. G. Truhlar, *J. Phys. Chem. A* **104**, 7917 (2000).
150. M. D. Hack, A. M. Wensman, and D. G. Truhlar, *J. Chem. Phys.* **115**, 1172 (2001).
151. A. Donoso and C. C. Martens, *J. Chem. Phys.* **112**, 3980 (2000).
152. W. H. Miller, *J. Phys. Chem.* **105**, 2942 (2001).
153. X. Sun, H. Wang, and W. H. Miller, *J. Chem. Phys.* **109**, 7064 (1998).
154. H. Wang, M. Thoss, and W. H. Miller, *J. Chem. Phys.* **115**, 2979 (2001).
155. G. Stock and M. Thoss, *Phys. Rev. Lett.* **78**, 578 (1997).
156. M. Thoss and G. Stock, *Phys. Rev. A* **59**, 64 (1999).
157. M. Ben-Nun, J. Quenneville, and T. J. Martínez, *J. Phys. Chem. A* **104**, 5161 (2000).
158. J. C. Tully, in *Classical and Quantum Dynamics in Condensed Phase Simulations*, B. J. Berne, G. Ciccoti, and D. F. Coker, eds., World Scientific, Singapore, 1998.
159. J. C. Tully, in *Modern Methods for Multidimensional Dynamics Computations in Chemistry*, D. L. Thompson, ed.,
160. O. Rubner, C. Meier, and V. Engel, *J. Chem. Phys.* **107**, 1066 (1997).

161. P. Fantucci, V. Bonačić-Koutecký, J. Jellinek, M. Wiechert, R. J. Harrison, and M. F. Guest, *Chem. Phys. Lett.* **250**, 47 (1996).

162. D. Reichardt, V. Bonačić-Koutecký, P. Fantucci, and J. Jellinek, *Z. Phys. D* **40**, 486 (1997).

163. D. Reichardt, V. Bonačić-Koutecký, P. Fantucci, and J. Jellinek, *Chem. Phys. Lett.* **279**, 129 (1997).

164. V. Bonačić-Koutecký, J. Pittner, D. Reichardt, P. Fantucci, and J. Koutecký, Metal clusters, in W. Ekardt, ed., John Wiley & Sons, New York, 1999.

165. V. Bonačić-Koutecký, M. Hartmann, D. Reichardt, and P. Fantucci, Recent advances in density functional methods, Part III, in V. Barone, A. Bencini, P. F. eds., World Scientific, Singapore, 2000.

166. V. Bonačić-Koutecký, L. Češpiva, P. Fantucci, and J. Koutecký, *J. Chem. Phys.* **98**, 7981 (1993).

167. V. Bonačić-Koutecký, L. Češpiva, P. Fantucci, and J. Koutecký, *J. Chem. Phys.* **100**, 490 (1994).

168. V. Bonačić-Koutecký, Pittner J. M. Boiron, and P. Fantucci, *J. Chem. Phys.* **110**, 3876 (1999).

169. V. Bonačić-Koutecký, V. Veyret, and R. Mitrić, *J. Chem. Phys.* **115**, 10450 (2001).

170. H. Hakkinen, M. Moseler, and U. Landman, *Phys. Rev. Lett.* **89**, 033401-1 (2002).

171. V. Bonačić-Koutecký, J. Burda, M. Ge, R. Mitrić, G. Zampella, and R. Fantucci, *J. Chem. Phys.* **117**, 3120 (2002).

172. R. Mitrić, M. Hartmann, J. Pittner, and V. Bonačić-Koutecký, *Eur. Phys. J.* **24**, 45 (2003).

173. W. T. Wallace, R. B. Wyrwas, R. L. Whetten, R. Mitrić, and V. Bonačić-Koutecký, *J. Am. Chem. Soc.* **125**, 8408 (2003).

174. M. L. Kimble, A. W. J. Castleman, R. Mitrić, C. Bürgel, and V. Bonačić-Koutecký, *J. Am. Chem. Soc.* **126**, 2526 (2004).

175. J. Hagen, L. D. Socaciu, J. Le Roux, D. Popolan, T. M. Bernhardt, L. Wöste, R. Mitrić, and V. Bonačić-Koutecký *J. Am. Chem. Soc.* **126**, 3442 (2004).

176. C. Sieber, J. Buttet, W. Harbich, C. Félix, R. Mitrić, and V. Bonačić-Koutecký, *Phys. Rev. A* **70**, 041201 (2004).

177. P. Ballone and W. Andreoni, Metal clusters. In W. Ekardt, ed.; John Wiley & Sons, Chichester, 1999.

178. F. Furche, R. Ahlrichs, P. Weis, C. Jacob, S. Gilb, T. Bierweiler, and M. M. Kappes, *J. Chem. Phys.* **117**, 6982 (2002).

179. P. Weis, T. Bierweiler, S. Gilb, and M. M. Kappes, *Chem. Phys. Lett.* **355**, 355 (2002).

180. L. König, I Raibn, W. Schulze, and G. Ertl, *Science* **274**, 1353 (1996).

181. C. Felix, C. Sieber, W. Harbich, J. Buttet, I. Rabin, W. Schulze, and G. Ertl, *Chem. Phys. Lett.* **313**, 105 (1998).

182. I. Rabin, W. Schulze, G. Ertl, C. Felix, C. Sieber, W. Harbich, and J. Buttet, *Chem. Phys. Lett.* **320**, 59 (2000).

183. C. Felix, S. Sieber, W. Harbich, J. Buttet, I. Rabin, W. Schulze, and G. Ertl, *Phys. Rev. Lett.* **86**, 2992 (2001).

184. W. Harbich and C. Felix, *C. R. Phys.* **3**, 289 (2002).

185. M. Haruta, *Catal. Today* **36**, 153 (1997).

186. T. H. Lee, and K. M. Ervin, *J. Phys. Chem.* **98**, 10023 (1994).

187. B. E. Salisbury, W. T. Wallace, and R. L. Whetten, *Chem. Phys.* **262**, 131 (2000).

188. K. J. Taylor, C. L. Pettiette-Hall, O. Cheshnovsky, and R. E. Smalley, *J. Chem. Phys.* **96**, 3319 (1992).

189. J. Ho, K. M. Ervin, and W. C. Lineberger, *J. Chem. Phys.* **93**, 6987 (1990).

190. H. Handschuh, G. Ganteför, P. S. Bechtold, and W. Eberhardt, *J. Chem. Phys.* **100**, 7093 (1994).
191. G. Lüttgens, N. Pontius, P. S. Bechtold, M. Neeb, and W. Eberhardt, *Phys. Rev. Lett.* **88**, 076102 (2002).
192. Y. Negishi, Y. Nakajima, and K. Kaya, *J. Chem. Phys.* **115**, 3657 (2001).
193. A. Sanchez, S. Abbet, U. Heiz, W. D. Schneider, H. Hakkinen, R. N. Barnett, and U. Landman, *J. Phys. Chem. A* **103**, 9573 (1999).
194. L. D. Socaciu, J. Hagen, T. M. Bernhardt, L. Wöste, U. Heiz, H. Hakkinen, and U. Landman, *J. Am. Chem. Soc.* **125**, 10437 (2003).
195. S. A. Varganov, R. M. Olson, M. S. Gordon, and H. Metiu, *J. Chem. Phys.* **119**, 2531 (2003).
196. G. Mills, M. S. Gordon, and H. Metiu, *J. Chem. Phys.* **118**, 4198 (2003).
197. A. D. Becke, *Phys. Rev. A* **98**, 3098 (1998).
198. C. Lee, W. Yang, and R. G. Parr, *Phys. Rev. B* **37**, 785 (1985).
199. A. Donoso, Y. Zheng, and C. Martens, *J. Chem. Phys.* **119**, 5010 (2003).
200. A. Donoso, and C. Martens, *Phys. Rev. Lett.* **87**, 223202 (2001).
201. E. Wigner, *Phys. Rev.* **40**, 749 (1932).
202. M. Hillary, R. F. O'Connel, M. O. Scully, and E. P. Wigner, *Phys. Rep.* **106**, 1984 (1984).
203. J. E. Moyal, *Comb. Philos. Soc.* **45**, 99 (1949).
204. Z. Li., J.-Y. Fang, and C. C. Martens, *J. Chem. Phys.* **104**, 6919 (1996).
205. S. Mukamel, *Principles of Nonliner Optical Spectroscopy*, Oxford University Press, New York, 1995.
206. J. Jellinek, V. Bonačić-Koutecký, Fantucci, M. Wiechert, *J. Chem. Phys.* **101**, 10092 (1994).
207. V. Bonačić-Koutecký, J. Jellinek, M. Wiechert, P., F. *J. Chem. Phys.* , **107**, 6321 (1997).
208. M. J. Frisch, G. W. Trucks, H. B. Schlegel,, et al., *GAUSSIAN 98* Gaussian Inc., Pittsburgh, PA, 1998.
209. R. Ahlrichs, M. Bär, M. Häser, H. Horn, and M. Kölmel, *Chem. Phys. Lett.* **162**, 165 (1989).
210. L. Verlet, *Phys. Rev.* **159**, 98 (1967).
211. T. Leisner, S. Vajda, S. Wolf, L. Wöste, and R. S. Berry, *J. Chem. Phys.* **111**, 1017 (1999).
212. H. Hess, K. R. Asmis, T. Leisner, and L. Wöste, *Eur. Phys. J. D* **16**, 145 (2001).
213. R. Keller, F. Nöhmeier, P. Spädtke, and H. Schönenberg, *Vacuum* **34**, 31 (1984).
214. J. Hagen, Dissertation, Freie Universität Berlin, 2004.
215. T. M. Bernhardt, *Int. J. Mass. Spectrom.* (2005), in press.
216. L. Lian, P. A. Hackett, and D. M. Rayner, *J. Chem. Phys.* **99**, 2583 (1993).
217. L. D. Socaciu, J. Hagen, U. Heiz, T. M. Bernhardt, T. Leisner, and L. Wöste, *Chem. Phys. Lett.* **340**, 282 (2001).
218. M. Schmidt, P. Cahuzac, C. Brechignac, and H. P. Cheng, *J. Chem. Phys.* **118** (2003).
219. M. Schmidt, A. Masson, and C. Brechignac, *Phys. Rev. Lett.* **91**, 243401 (2003).
220. L. D. Socaciu, J. Hagen, J. Le Roux, D. Popolan, T. M. Bernhardt, L. Wöste, and S. Vajda, *Chem. Phys.* **120**, 2078 (2004).
221. Y. D. Kim and G. Ganteför, *Chem. Phys. Lett.* **383**, 80 (2004).
222. L. D. Socaciu-Siebert, J. Hagen, J. Le Roux, D. Popolan, S. Vajda, T. M. Bernhardt, and L. Wöste, *Phys. Chem. Phys.* (2005), submitted.
223. U. Landman, D. Scharf, and J. Jortner, *Phys. Rev. Lett.* **54**, 1860 (1985).
224. D. Scharf, U. Landman, and J. Jortner, *J. Chem. Phys.* **87**, 2716 (1987).

225. G. Rajagopal, R. N. Barnett, and U. Landman, *Phys. Rev. Lett.* **67**, 727 (1991).
226. U. Landman, in *Physics and Chemistry of Finite Systems: From Clusters to Crystals*, Vol. I. P. Jena, S. N. Khanna, and B. K. Rao, eds., Kluwer, Dordrecht, 1992.
227. G. Galli, W. Andreoni, and M. P. Tosi, *Phys. Rev. A* **34**, 3580 (1986).
228. F. Rajagopal, R. N. Barnett, A. Nitzan, U. Landman, E. Honea, P. Labastie, M. L. Homer, and R. L. Whetten, *Phys. Rev. Lett.* **64**, 2933 (1990).
229. R. Pandey, M. Seel, and A. Kunz, *Phys. Rev. B* **41**, 7955 (1990).
230. P. W. Weiss, C. Ochsenfeld, R. Ahlrichs, and M. M. Kappes, *J. Chem. Phys.* **97**, 2553 (1992).
231. V. Bonačić-Kouteck, C. Fuchs, J. Gaus, J. Pittner, and Z. Koutecký *Z. Phys. D* **26**, 192 (1993).
232. H. Hakkinen, R. N. Barnett, and U. Landman, *Chem. Phys. Lett.* **232**, 79 (1995).
233. R. N. Barnett, H. P. Cheng, H. Hakkinen, and U. Landman, *J. Phys. Chem.* **99**, 7731 (1995).
234. C. Ochsenfeld, J. Gauss, and R. Ahlrichs, *J. Chem. Phys.* **103**, 7401 (1994).
235. J. Giraud-Girard and D. Maynau, *Z. Phys. D* **23**, 91 (1992).
236. J. Giraud-Girard and D. Maynau, *Z. Phys. D* **32**, 249 (1994).
237. T. Bergmann, H. Limberger, and T. P. Martin, *Phys. Rev. Lett.* **60**, 1767 (1988).
238. E. C. Honea, M. L. Homer, and R. L. Whetten, *Phys. Rev. B* **63**, 394 (1989).
239. P. Poncharal, J. M. L'Hermite, and P. Labastie, *Chem. Phys. Lett.* **253**, 463 (1996).
240. S. Pollack, C. R. C. Wang, and M. M. Kappes, *Chem. Phys. Lett.* **175**, 209 (1990).
241. Y. A. Yang, C. W. Conover, and L. A. Bloomfield, *Chem. Phys. Lett.* **158**, 279 (1989).
242. P. Xia and L. A. Bloomfield, *Phys. Rev. Lett.* **70**, 1779 (1993).
243. P. Xia, A. J. Cox, and L. A. Bloomfield, *Z. Phys. D* **26**, 1841 (1993).
244. P. Labastie, J. M. L'Hermite, P. Poncharal, and M. Sence, *J. Chem. Phys.* **103**, 6362 (1995).
245. V. Bonačić-Koutecký, J. Pittner, and J. Koutecký, *Chem. Phys.* **210**, 313 (1996).
246. V. Bonačić-Koutecký, J. Pittner, and J. Koutecký, *Z. Phys. D* **40**, 441 (1997).
247. V. Bonačić-Koutecký, and J. Pittner, *Chem. Phys.* **225**, 173 (1997).
248. G. Durand, F. Spiegelman, P. Poncharal, P. Labastie, J. M. L'Hermite, and M. Sence, *J. Chem. Phys.* **110**, 7884 (1999).
249. G. Durand, J. Giraud-Girard, D. Maynau, F. Spiegelman, and F. Calvo, *J. Chem. Phys.* **110**, 7871 (1999).
250. D. Rayane, I. Compagnon, R. Antoine, M. Broyer, P. Dugourd, P. Labastie, J. M. L'Hermite, A. Le Padallec, G. Durand, F. Calvo, F. Spiegelman, and A. R. Allouche, *J. Chem. Phys.* **116**, 10730 (2002).
251. G. Durand, M.-C. Heitz, F. Spiegelman, C. Meier, R. Mitrić, V. Bonačić-Koutecký, and J. Pittner, *J. Chem. Phys.* **121**, 9906 (2004).
252. J. M. L'Hermite, V. Blanchet, A. Le Padellec, B. Lamory, and P. Labastie, *Eur. Phys. J. D* **28**, 361 (2004).
253. R. W. Schoenlein, L. A. Peteanu, R. A. Mathies, and C. V. Shank, *Science* **254**, 412 (1991).
254. M. Ottolenghi and M. Sheves, *Isr. J. Chem.* **35** (1995).
255. M. J. Bearpark, M. A. Robb, and H. B. Schlegel, *Chem. Phys. Lett.* **223**, 269 (1994).
256. J. Michl and V. Bonačić-Koutecký, *Electronic Aspects of Organic Photochemistry*, John Wiley & Sons, New York, 1990.
257. J. C. Tully, *J. Chem. Phys.* **93**, 1061 (1990).
258. D. Kohen, F. H. Stillenger, and J. C. Tully, *J. Chem. Phys.* **109**, 4713 (1998).

259. J.-Y. Fang and S. Hammes-Schiffer, *J. Chem. Phys.* **110**, 11166 (1999).
260. M. D. Hack and D. G. Truhlar, *Phys. Chem. A* **104**, 7917 (2000).
261. F. Webster, P. J. Rossky, and R. A. Friesner, *J. Chem. Phys.* **100**, 4835 (1994).
262. D. F. Coker and L. Xiao, *J. Chem. Phys.* **102**, 496 (1995).
263. M. F. Herman, *J. Chem. Phys.* **103**, 8081 (1995).
264. M. F. Herman, *Int. J. quant. Chem.* **70**, 897 (1998).
265. M. F. Herman, *J. Chem. Phys.* **110**, 4141 (1999).
266. M. D. Hack, A. M. Wensmann, D. G. Truhlar, M. Ben-Nun, and T. J. Martinez, *J. Chem. Phys.* **115**, 1172 (2001).
267. R. Kapra and G. Ciccotti, *J. Chem. Phys.* **110**, 8919 (1999).
268. S. Nielsen, R. Kapral, and G. Ciccotti, *J. Chem. Phys.* **112**, 6543 (2000).
269. C. C. Martens and J.-Y. Fang, *J. Chem. Phys.* **106**, 4918 (1997).
270. A. Donoso, D. Kohen, and C. C. Martens, *J. Chem. Phys.* **112**, 7345 (2000).
271. M. Shapiro and P. Brumer, *Int. Rev. Phys. Chem.* **13**, 187 (1994).
272. R. G. Gordon and S. A. Rice, *Annu Rev. Phys. Chem.* **48**, 601 (1997).
273. L. C. Zhu, V. Kleiman, X. N. Li, S. P. Lu, K. Trentelman, and R. J. Gordon, *Science* **270**, 77 (1995).
274. C. Chen and D. S. Elliott, *Phys. Rev. Lett.* **65**, 1737 (1990).
275. S. M. Park, R. J. Lu, R. J. Gordon, *J. Chem. Phys.* **94**, 8622 (1991).
276. G. Q. Xing, X. B. Wang, X. Huang, and R. J. Bersohn, *Chem. Phys.* **104**, 826 (1996).
277. A. Assion, T. Baumert, V. Seyfried, V. Weiss, E. Wiedenmann, and G. Z. Gerber, *Phys. D* **36**, 265 (1996).
278. T. Baumert, J. Helbing, and G. Gerber, *Adv. Chem. Phys.* **101**, 47 (1997).
279. J. Somlói, V. A. Kazakov, and D. J. Tannor, *Chem. Phys.* **172**, 85 (1993).
280. B. Amstrup, J. D. Doll, R. A. Sauerbrey, X. Szabó, A. Lörincz, *Phys. Rev. A* **48**, 3830 (1993).
281. D. J. Tannor and S. A. Rice, *J. Chem. Phys.* **85**, 5805 (1986).
282. S. Shi and H. Rabitz, *Chem. Phys.* **139**, 185 (1989).
283. H. Rabitz and S. Shi, *Adv. Mol. Vibr. Collision Dyn.* **1A**, 187 (1991).
284. W. S. Warren, H. Rabitx, and M. Dahleh, *Science* **259**, 1581 (1993).
285. R. Kosloff, S. A. Rice, P. Gaspard, S. Tersigni, and D. J. Tannor, *Chem. Phys.* **139**, 201 (1989).
286. S. H. Tersigni, P. Gaspard, and S. A. Rice, *J. Chem. Phys.* **93**, 1670 (1990).
287. R. S. Judson and H. Rabitz, *Phys. Rev. Lett.* **62**, 1500 (1992).
288. D. E. Goldberg, *Genetic Algorithms in Search, Optimization, and Machine Learning*, Addison-Wesley, Reading, MA, 1993.
289. H. P. Schwefel, *Evolution and Optimum Seeking, Wiley, New York, 1995.*
290. A. Assion, T. Baumert, M. Bergt, T.Brixner, B. Kiefer, V. Seyfried, M. Strehle, and G. Gerber, in *Springer Series in Chemical Physics*, Vol. 63. T. Elsässer, J. G. Fujimoto, D. A. Wiersma, and W. Zinth, eds. Springer, Berlin, 1998.
291. H. Rabitz, R. de Vivie-Riedle, M. Motzkus, and K. Kompa, *Science* **288**, 824 (2000).
292. T. Hornung, M. Motzkus, and R. de Vivie-Riedle, *Phys. Rev. A* **65**, 021403 (2002).
293. K. Sundermann and R. de Vivie-Riedle, *J. Chem. Phys.* **110**, 1896 (1999).
294. Z. Shen, V. Engel, R. Xu, J. Cheng, and Y. Yan, *J. Chem. Phys.* **117**, 6142 (2002).

295. Y. J. Yan, R. E. Gillilan, R. M. Whitnell, K. R. Wilson, and S. Mukamel, *J. Phys. Chem.* **97**, 2320 (1993).
296. J. L. Krause, R. M. Whitnell, K. R. Wilson, Y. L. Yan, and S. Mukamel, *J. Chem. Phys.* **99**, 6562 (1993).
297. J. L. Krause, R. M. Whitnell, K. R. Wilson, and Y. L. Yan, in *Femsecond Chemistry*, J. Manz, and L. Wöste, eds., VCH, Weinheim, 1995.
298. B. Kohler, V. V. Yakovlev, J. Chen, J. L. Krause, M. Messina, K. R. Wilson, N. Schwentner, R. M. Whitnell, and Y. L. Yan, J. *Phys. Rev. Lett.* **74**, 3360 (1995).
299. J. Che, M. Messina, K. R. Wilson, V. A. Apkarian, Z. Li, C. C. Martens, R. Zadoyan, and Y. Yan, *Phys. Chem.* **100**, 7873 (1996).
300. C. J. Bardeen, J. Che, K. R. Wilson, V. V. Yakovlev, P. Cong, B. Kohler, J. L. Krause, and M. Messina, *J. Phys. Chem. A* **101**, 3815 (1997).
301. Y. J. Yan, *Annu. Rep. Prog. Chem. Sect. C: Phys. Chem.* **94**,397 (1998).
302. Y. J. Yan, Z. W. Shen, and Y. Zhao, *Chem. Phys.* **233**, 191 (1998).
303. Y. J. Yan, J. Che, and J. L.Krause, *Chem. Phys.* **217**, 297 (1997).
304. Y. J. Yan, J. S. Cao, and Z. W. Shen, *J. Chem. Phys.* **107**, 3471 (1997).
305. R. Xu, J. Cheng, and Y. Yan, *J. Phys. Chem. A* **103**, 10611 (1999).
306. Z. Shen, Y. Yan, J. Cheng, F. Shuang, Y. Zhao, and G. H *J. Chem. Phys.* **110**, 7192 (1999).
307. D. J. Tannor R. Kosloff, and A. Bartana, *Faraday Discuss.* **113**, 365 (1999).
308. D. J. Tannor, A. Bartana, *J. Phys. Chem. A* **103**, 10359 (1999).
309. A. Bartana, R. Kosloff, and D. J. Tannor, *Chem. Phys.* **267**, 195 (2001).
310. J. P. Palao and R. Kosloff, *Phys. Rev. Lett.* **89**, 188301 (2002).
311. C. M. Tesch and R. de Vivie-Riedle, *Phys. Rev. Lett.* **89**, 157901 (2002).
312. R. Mitrić, C. Bürgel, J. Burda, and V. Bonačić-Koutecký, *Eur. Phys. J. D* **24**, 41 (2003).
313. H. Metiu and V. Engel, *J. Chem. Phys.* **93**, 5693 (1990).
314. Z. M. Lu and H. Rabitz, *Phys. Rev. A* **52**, 1961 (1995).
315. Z. M. Lu and H. Rabitz, *J. Phys. Chem.* **99**, 13731 (1995).
316. B. A. Armstrup, G. J. Toth, H. Rabitz, and A. Lörinz, *Chem. Phys.* **201**, 95 (1995).
317. B. Kohler, J. L. Krause, F. Raksi, C. Rose-Petruck, R. M. Whitnell, K. R. Wilson, V. V. Yakovlev, Y. Yan, and S. Mukamel, *J. Phys. Chem.* **97**, 12602 (1993).
318. B. Kohler, J. Krause, F. Raksi, K. R. Wilson, R. M. Whitnell, V. V. Yakovlev, Y. L. Yan, *Acc. Chem. Res.* **28**, 133 (1995).
319. C. J. Bardeen, J. Che, K. R. Wilson, V. V. Yakovlev, V. A. Apkarian, C. C. Martens, R. Zadoyan, B. Kohler, and M. Messina, *J. Chem. Phys.* **106**, 8486 (1997).
320. J. X. Cheng, Z. W. Shen, and Y. J. Yan, *J. Chem. Phys.* **109**, 1654 (1998).
321. J. Jortner, *Philos. Trans. R. Soc. (London)* **356**, 279 (1998).
322. I. Last and J. Jortner, *J. Chem. Phys.* **120**, 1336 (2004).
323. I. Last and J. Jortner, *J. Chem. Phys.* **120**, 1348 (2004).
324. J. Zweiback, T. Ditmire, and M. D. Penny, *Phys. Rev. A* **A59**, R3166 (1989).
325. S. Zamith, T. Marchenko, Y. Ni, S. A. Aseyev, H. G. Muller, and M. J. J. Vrakking, *Phys. Rev. A* **A70**, 011201 (2004).

ULTRACOLD LARGE FINITE SYSTEMS*

JOSHUA JORTNER and MICHAEL ROSENBLIT†

School of Chemistry, Tel Aviv University, Tel Aviv 69978, Israel

CONTENTS

*Dedicated to Stuart Rice, who changed our perception of physical chemistry and other fields of science.

†*Present Address*: Ilse Katz Center for Meso- and Nanoscale Science and Technology, Ben Gurion University of the Negev, Beer Sheva 84105, Israel

Adventures in Chemical Physics: A Special Volume in Advances in Chemical Physics, Volume 132, edited by R. Stephen Berry and Joshua Jortner. Series editor Stuart A. Rice

I. PROLOGUE

Fascinating chemical and physical phenomena are exhibited in the realm of ultracold, large finite systems—that is, molecules [1–5], clusters [6–11], optical molasses [12, 13] and finite Bose–Einstein condensates [14] in the temperature domain of $T < 2.7$ K (Fig. 1). We shall (arbitrarily) take the upper temperature limit for ultracold systems ($T = 2.7$ K) as the current temperature of the expanding universe. The genesis of this research field originated from the exploration of macroscopic ultracold systems (e.g., liquid ^{4}He and ^{3}He), where quantum effects manifest unique features of elementary excitations in boson ^{4}He and fermion ^{3}He systems. These reveal superfluidity and Bose–Einstein condensation with an onset at $T = 2.17$ K for liquid ^{4}He [15–27], and fermion spin pairing with the onset of superfluidity at $T \simeq 3 \cdot 10^{-3}$ K for liquid ^{3}He [20]. Superfluidity and Bose–Einstein condensation (e.g., in liquid ^{4}He) are related, but distinct phenomena, with an onset at the lambda point, $T_\lambda = 2.17$ K, mark the singularity of the specific heat [19–26]. Superfluidity corresponds to the response of the system to a slow movement of its boundaries, with the normal fluid fraction contributing to the moment of inertia of a rotating body within the liquid, while the superfluid fraction corresponds to those atoms not affecting the rotatory motion. Other properties of superfluid ^{4}He are its vanishingly small viscosity, very high heat conductivity (30 times higher than that of a metal, i.e., copper), formation of a He fountain, and film flow and creep [15–26]. Bose–Einstein condensation manifests the critical temperature for the macroscopic occupation of a single quantum ground state of a boson system, with the fraction of the condensed atoms being unity at $T = 0$ [21–26].

Remarkable progress in the research field of macroscopic ultracold systems was made with the advent of the methods of trapping and cooling of atomic clouds in magnetic traps [27], magneto-optical traps [28], optical traps [29,30], and laser cooling of atoms [28, 31–33]. Another significant development pertains to "cold collisions" of atoms in magnetic fields, with Feschbach resonances giving rise to ultracold, highly vibrationally excited diatomic molecules [34–44], while photoassociation methods [45, 46] show promise for

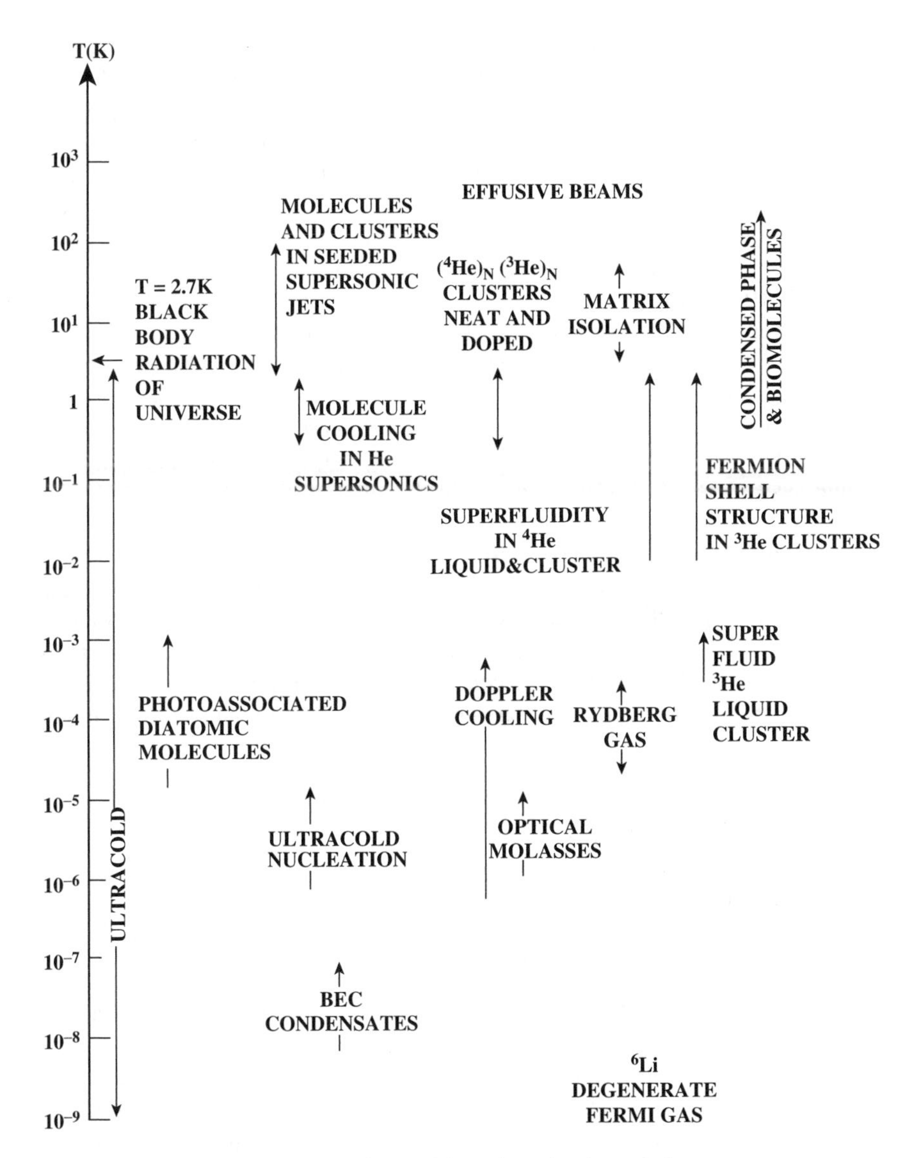

Figure 1. The world of ultracold atomic, molecular, and cluster systems.

the production of ultracold, vibrationally excited [45] and ground vibrational state [46] diatomics. These novel approaches for the production of ultracold atomic or diatomic ensembles gave rise to Bose–Einstein condensation of atoms [47–49] and diatomic molecules [37, 40], which were explored in the temperature domain of 10 nK to 100 μK (10^{-8}–10^{-4} K). Accordingly, the current low–temperature limit for ultracold systems is set at $T \geq 10^{-8}$ K.

Concurrently, the world of ultracold systems has expanded its boundaries during the last decade to encompass ultracold, three–dimensional, large finite systems [e.g., $(^4He)_N$ clusters ($N = 2$–10^7), and $(^3He)_N$ clusters ($N = 25$–10^7)] in the temperature range of $T = 0.1$–2.2 K [6–11, 50–78], finite optical molasses in laser irradiated ultracold atomic gases in the temperature range of 10–100 μK [79], as well as finite Bose–Einstein condensates in the temperature range of 10–100 nK [14, 80].

The genesis of the exploration of ultracold finite systems can be traced to progress in cluster science, which focuses on the energy landscapes, spatial structures and shapes, (rounded off) phase transitions, energetics, nuclear and electronic level structure, spectroscopy, response, dynamics and chemical reactivity of van der Waals elemental, molecular, semiconductor, and metal large finite systems [81–98]. Central issues in this area of cluster chemical physics pertained to the bridging between the properties of molecular systems and infinite condensed phase systems and the utilization of cluster size equations as scaling laws for the nuclear and electronic response of these finite systems [84–87], which serve as precursors of nanostructures. Another bridge for cluster science pertains to the exploration of finite ultracold systems (Fig. 2). The realm of finite ultracold systems in the temperature domain of 10^{-8}–2.7 K, which involves clusters and clouds, falls into the following major classes:

1. Elemental homonuclear clusters (droplets) of $(^3He)_N$ and $(^4He)_N$ at $T = 0.1$–2.2K [6–11, 50–78, 99].
2. Molecular homonuclear clusters (aggregates) of $(H_2)_N$ and $(D_2)_N$ at $T = 0.1$–2.2 K [100–102].
3. Heteroclusters—for example, $(^4He)_{N_1}(^3He)_{N_2}$, $(^4He)_{N_1}(H_2)_{N_2}$, or $(^4He)_{N_1}$ $(H_2)_{N_2}$ at $T = 0.1$–2.2 K [103, 104].
4. Ultracold laser irradiated, finite atomic clouds, referred to as optical molasses—for example, Rb at $T = 10^{-4}$–10^{-6} K [79].
5. Finite ultracold atomic clouds of Bose–Einstein condensates—for example, ^7Li, ^{23}Na, and ^{87}Rb atoms at $T = 10^{-8}$–10^{-7} K [14, 47–49, 80].
6. Finite Bose–Einstein condensates of ultracold clouds of diatomic molecules—for example, $^6Li_2$ or $^{23}Na_2$ at $T = 10^{-8}$–10^{-7} K [45, 46].

Large $(^4He)_N$ ($N = 10^4$–10^7) clusters experimentally studied at $T \simeq 0.4$ K [6–11, 50], small $(^4He)_N$ clusters ($N = 2$–50) [105] studied at $T = 0.1$–1 K, and large $(^3He)_N$ ($N = 10^4$–10^7) clusters studied at $T = 0.15$ K [67, 104] (class 1) are cooled by spontaneous evaporative cooling after the cluster formation in a

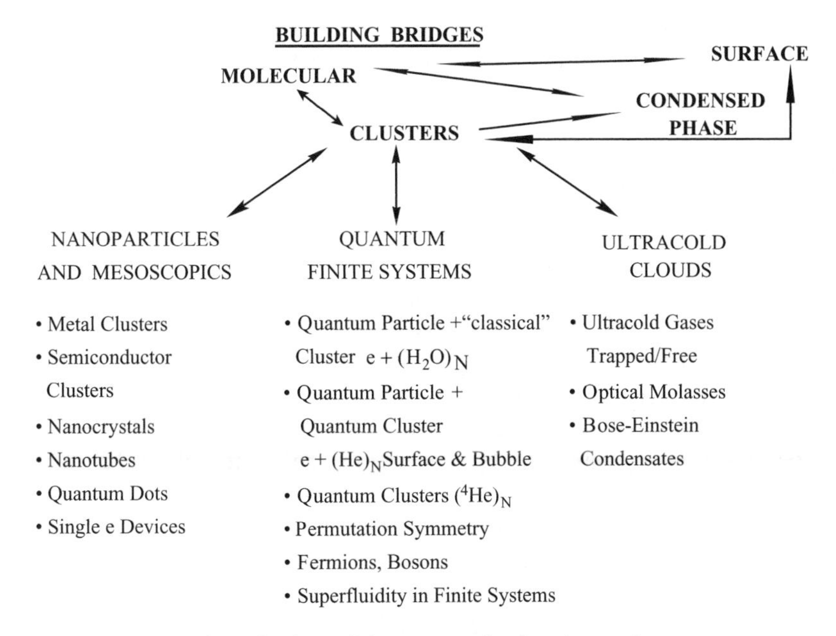

Figure 2. Large finite systems related to cluster science.

free jet expansion. An identical formation mechanism applies for molecular homonuclear clusters (class 2) and heteroclusters (class 3). The ultralow temperature of $T \sim 10$–$100\,\mu$K for optical molasses (class 4) is obtained by evaporative cooling, which is induced and controlled through an external rf confining potential, which lowers the trap depth and lets the hottest atoms escape [106]. Further cooling to extremely low temperatures of $T \sim 10$–$100\,$nK for trapped Bose–Einstein atomic condensates (class 5) is achieved in optical traps. As already pointed out, trapped Bose–Einstein molecular condensates (class 6) can be produced by ultracold collisions in magnetic fields via Feschbach resonances [34–44] and by photoassociation in ultracold atomic gases [45, 46]. Finite ultracold clouds of diatomics (class 6) do not provide the upper limit for the size of molecular constituents; also, future developments in the production of ultracold cluster clouds, produced by ultracold collisions of diatomics, are feasible and should be explored.

In what follows we shall focus on some unique properties and features of finite, large ultracold systems, which can be traced to (a) quantum effects of zero-point energy and kinetic energy for the "light" constituent clusters (classes 1, 2, and 3) and (b) permutation symmetry effects in all systems (classes 1–6) considered herein.

A. Packing and Structure of Boson $(^4He)_N$ Clusters and of Ultracold Atomic Clouds

It will be instructive to compare the properties of boson $(^4He)_N$ clusters (class 1) and of trapped atomic and molecular condensates (classes 4 and 5) [106]. Self-bound $(^4He)_N$ clusters are energetically stable for $N \geq 2$. We shall not dwell here on the interesting properties of small $^4He_2$ and $^4He_3$ quantum molecules—with $^4He_2$ constituting the largest diatomic molecule (bond length $r_e \simeq 52$ Å) [107, 108]—and $(^4He)_3$, which, in addition to a ground state with $r_e \simeq 9.6$ Å [109], should also manifest Efimov three-boson states [110, 111]. Rather, we consider large ($N \sim 10^3$–10^6) clusters, where the ^4He–^4He largest nearest-neighbor distance r_e is close to that for the condensed phase value of $r_e = 3.6$ Å for $(^4He)_N$ and $(^3He)_N$ clusters, which constitute a self-bound fluid, being liquid down to 0 K, and manifest quantum effects (Section I.B). For such large $(^4He)_N$ clusters ($N \simeq 10^3$–10^7) the liquid cluster density in the center of the cluster is $\rho = 2.2 \times 10^{-2}$ Å^{-3}, being close to the density of the macroscopic liquid ^4He [20]. The cluster radius R_0 is approximately given for a step function profile of the density $\rho(r)$ (i.e., $\rho(r) = \rho$ for $r \leq R_0$ and $\rho(r) = 0$ for $r > R_0$) by

$$R_0 = r_0 N^{1/3} \tag{1}$$

where $r_0 = (4\pi\rho/3)^{-1/3}$ is the ^4He constituent mean radius, so that $r_0 = 2.2$ Å and $\rho r_0^3 = 0.23$. The step function description of the cluster density, which is referred to as the liquid drop model (LDM), constitutes a crude approximation, while in real life there is a finite-size, smooth cluster surface density profile, which is due to the large zero-point motion of the atoms [84, 106, 112, 113]. $\rho(r)$ decreases from ρ to zero over the distance scale of $(2$–$4)r_0 \simeq 6$–10 Å. The thickness of the surface region is approximately $\sim 8\pi r_0^3 N^{2/3}$, while the ratio between the surface and volume regions is $\sim 6N^{-1/3}$. The atom–atom scattering length is $a \simeq 2.6$ Å [106], so that $\rho a^3 \simeq 0.4$ for this strongly interacting boson fluid cluster.

The trapped condensate in a magnetic and/or optical trap is not bound, but rather produced and kept in an external harmonic confining potential $V_{ext} = m\omega_{HO}^2 r^2/2$ [14, 106], which acts on each atom of mass m and where ω_{HO} is the trap characteristic frequency. For noninteracting atoms in a trap of frequency ω_{HO} the condensate corresponds to the lowest single-particle state in a harmonic potential, with a harmonic width $A_{HO} = (\hbar/m\omega_{HO})^{1/2}$. A common situation for an atomic condensate (e.g., ^7Li [49], ^{23}Na [48, 114], or ^{87}Rb [47, 115, 116]) involves $N = 10^5$–10^6 atoms. The characteristic trap frequency is $\omega_{HO}/2\pi \sim 10$–100 Hz, while the width of the condensate is $A_{HO} \sim 10$–100 μm. The atom–atom scattering length is $a \simeq 10$–100 Å, so that $a/A_{HO} \sim 10^{-4}$–10^{-3}. The average interatomic distance is $r_0 \sim 10^{-5}$ cm, while the central density is

TABLE I
Length Scales for Finite Ultracold Systems

System	$a\,(\text{Å})^a$	$r_0\,(\text{Å})^b$	$r_M\,(\text{Å})^c$	$R_0\,(\text{Å})^d$	$\rho\,(\text{Å}^{-3})^e$	ρr_0^3	ρa^3	$\varphi\,(\text{Å})^f$
$(^4\text{He})_N$ cluster $N = 10^4$	$\sim$2.6	2.2	3.0	50	2.2×10^{-2}	0.25	0.40	-1
^{87}Rb Bose–Einstein condensate $N \sim 10^5$ $\omega_{HO}/2\pi = 100$ Hz $A_{HO} = 10^4$ Å	$\sim$50	10^3–10^4	3.1^g	5×10^4	10^{-11}–10^{-9}	10^{-2}–1	10^{-6}–10^{-5}	$\sim 10^3$

[a] Atom–atom scattering length.
[b] Interatomic radius.
[c] Radius of interatomic potential well.
[d] Radius of ultracold system $R_0 = N^{1/3} r_0$
[e] Average density.
[f] Healing length.
[g] Approximated by the equilibrium radius of an Na_2 molecule (G. Herzberg, *Spectra of Diatomic Molecules*, Van Nostrand, New York, 1958).

$\rho \sim 10^{13}$–$10^{15}\,\text{cm}^{-3}$ (i.e., 10^{-11}–$10^{-9}\,\text{Å}^{-3}$), whereupon $\rho r_0^3 \sim 10^{-2}$–1, while $\rho a^3 \sim 10^{-5}$–10^{-6} for this unbound system. The structural packing parameter ρr_0^3 is similar for $(^4\text{He})_N$ clusters and for the atomic condensate, whereas the structure–interaction parameter ρa^3 is considerably higher for the self-bound $(^4\text{He})_N$ than for the unbound trapped condensate. Another significant parameter is the healing length, which serves as an order parameter for surface recovery processes $\varphi = (8\pi\rho a)^{-1/2}$ [106], and which is considerably smaller for the $(^4\text{He})_N$ cluster than for the atomic condensate. Table I summarizes characteristic length scales and densities for finite ultracold systems, comparing $(^4\text{He})_N$ clusters and Bose–Einstein atomic condensates. From these data we infer the following:

1. The length parameters r_0 and R_0 manifest a common scaling factor of 10^3 between the $(^4\text{He})_N$ cluster and the atomic condensate. These length scales move from the nanometer size domain for $(^4\text{He})_N$ clusters to the micrometer size domain for the condensate.
2. The density scales by a numerical factor of $\sim 10^9$ between the two classes of systems, while the parameter ρr_0^3 is of the same order of magnitude.
3. The atom–atom scattering lengths a in the two systems are of the same order of magnitude. Consequently the dimensionless parameter ρa^3 varies by 4–6 orders of magnitude between the $(^4\text{He})_N$ clusters and the atomic condensates. This marked difference reflects on the deviations of the ideal gas Bose condensation temperatures from the experimental transition temperatures in these strongly interacting clusters and in weakly interacting condensates (Section I.C).

B. Zero-Point Energy and Kinetic Energy Effects in Quantum Clusters and Ultracold Clouds

Zero-point energy effects can be traced to the light masses of the constituents in quantum clusters and to the extremely low temperatures in the optical molasses and condensates. The zero-point energy effects can be described in terms of the ratio Λ of the quantum lengths

$$\Lambda = \frac{\lambda_{\mathrm{DB}}}{\bar{r}_0} \tag{2}$$

between the de Broglie wavelength λ_{DB} for the relative motion of the particles and the characteristic interparticle distance $\bar{r}_0$. For strongly interacting particles in homonuclear and heteronuclear quantum clusters (categories 1–3), the 6–12 interparticle potential is characterized by a well depth of $\in$, a distance of closest approach of σ, and a well position of $r_M = 2^{1/6}\sigma$. Taking $\lambda_{\mathrm{DB}} = h/(m \in)^{1/2}$, where m is the particle mass and $\bar{r}_0 = \sigma$, the de Boer Λ parameter [11, 113] is

$$\Lambda = \frac{h}{(m \in)^{1/2}\sigma} \tag{3}$$

The de Boer parameter, Eq. (3), corresponds to the quantum length ratio, Eq. (2), at temperature $T = \in /3k_B$, where k_B is the Boltzmann factor. For weakly interacting particles in ultracold clouds (categories 4 and 5) the thermal de Broglie wavelength

$$\lambda_{\mathrm{DB}} = \frac{h}{(3mk_BT)^{1/2}} \tag{4}$$

is expressed in terms of the cloud temperature T (Fig. 3), while $\bar{r}_0 \simeq r_0$. In this case

$$\Lambda = \frac{h}{r_0(3mk_BT)^{1/2}} \tag{5}$$

Table II provides insight into quantum effects in clusters and in clouds. For "nearly classical" elemental clusters [e.g., $(\mathrm{Xe})_N$, $(\mathrm{Kr})_N$, $(\mathrm{Ar})_N$ and $(\mathrm{Ne})_N$], $\Lambda < 1$ and quantum effects are moderately small. For quantum clusters [e.g., $(^4\mathrm{He})_N$, $(^3\mathrm{He})_N$, $(\mathrm{H}_2)_N$, and $(\mathrm{D}_2)_N$], $\Lambda > 1$ and pronounced quantum effects set in. For atomic clouds [e.g., optical molasses of Rb atoms (at $T = 10$–$100\,\mu\mathrm{K}$)] and Bose–Einstein condensates [e.g., ^{6}Li or ^{23}Na atoms (at $T = 10$–$100\ nK$)], the quantum lengths ratio parameter is huge [i.e., $\Lambda \simeq 10^2$–10^3 (Table II)], manifesting large zero-point energy quantum effects at ultralow temperatures.

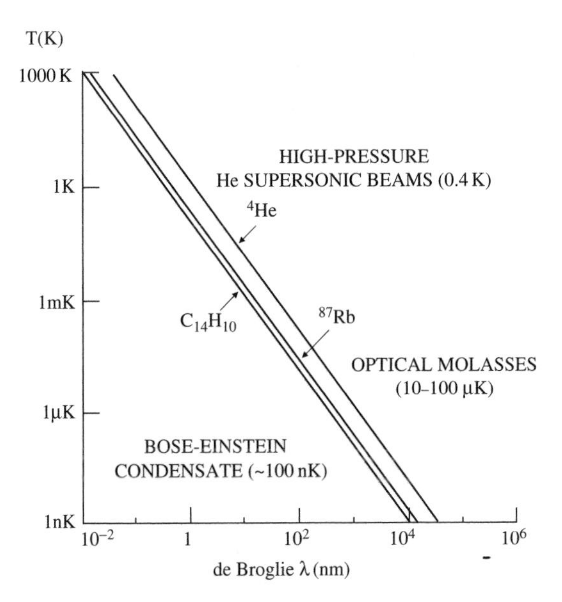

Figure 3. The temperature dependence of the thermal de Broglie wavelengths for several atomic and molecular systems. The relevant temperature domains for Bose–Einstein condensates, optical molasses, and $(^4He)_N$ clusters are marked on the figure.

TABLE II
Quantum Zero-Point Energy Effects for Ultracold Finite Systems

System	$\in$ (K)[a]	$\bar{r}_0$ (Å)[b]	$\Lambda^{(c)}$	Classification
$(Xe)_N$	280	4.4	0.06	"Nearly
$(Kr)_N$	200	4.0	0.11	classical"
$(Ar)_N$	143	3.8	0.20	clusters
$(Ne)_N$	43	3.1	0.60	$\Lambda < 1$
$(H_2)_N$	35	3.4	2.0	Quantum
$(D_2)_N$	35	3.4	1.4	clusters
$(^4He)_N$	11	3.0	2.9	$\Lambda > 1$
$(^3H)_N$	11	3.0	3.3	
Optical molasses of Rb atoms $T \simeq 10^{-5}$ K	—	10^4	150	Ultracold clouds
Finite Bose–Einstein condensates ^7Li, ^{23}Na atoms $T = 10^{-7}$ K	—	10^4	1500	Quantum clouds $\Lambda \sim 10^2$–10^3

[a]Interparticle well depth in clusters.
[b]Interparticle distance of closest approach in clusters and the average interparticle distance in ultracold clouds.
[c]Λ represents the de Boer parameter [Eq. (3)] for clusters and is expressed in terms of the thermal de Broglie wavelength [Eq. (4)] for ultracold clouds.

TABLE III
Increase of the Actual Average Volume Per Particle (v_0) Relative to the Reference Volume (v_c) of the Particle in a Classical solid

System	$v_0/v_c^{a,b}$	Classification
$(Xe)_N$	0.94	"Nearly classical" clusters $v_0/v_c \simeq 1$
$(Ar)_N$	0.98	
$(Ne)_N$	1.06	
$(H_2)_N$	1.3	Quantum cluster $v_0/v_c > 1$
$(^4He)_N$	1.9	
$(^3He)_N$	2.8	
Finite optical molasses ($T \simeq 10^{-5}$ K) and Bose–Einstein condensates ($T \simeq 10^{-7}$ K)	3×10^{11}–4×10^{10}	Ultracold quantum clouds $v_0/v_c \sim 10^{11}$

[a]Data from J. P. Toennies and A. F. Vilesov, *Angew. Chem. Int. Ed.* **43**, 2622 (2004).
[b]$v_c = r_0^3$ for clusters and $v_c \simeq r_M^3$ for ultracold clouds.

The quantum lengths ratio Λ involving the thermal de Broglie wavelength determines the critical temperatures for noninteracting bosons (Section I.C).

Another characterization of zero-point energy effects pertains to the increase in the actual average volume v_0 occupied by a particle in the ultracold system, relative to the reference volume v_c that the particles would occupy in a classical lattice. For "nearly classical" and for quantum clusters we have $v_c = r_0^3$ [11], while for ultracold clouds we have $v_c = r_M^3$ (e.g., $r_M = 1.5\,\text{Å}$ for ^{6}Li and $r_M \simeq 3\,\text{Å}$ for ^{87}Rb [106]). From the data for v_0/v_c assembled in Table III, it is apparent that for "nearly classical" elemental clusters we have $v_0/r_0^3 \sim 1$, for quantum clusters we have $v_0/r_0^3 > 1$, falling in the range $v_0/r_0^3 = 1.3$–2.8, while for ultracold clouds $v_0\ /\ r_M^3$ is huge, being in the range of $\sim 10^{12}$. From these results we conclude that the increase of the ratio v_0/r_0^3 above unity for quantum clusters is due to a marked increase in the zero-point energy. For $(^4He)_N$ and $(^3He)_N$ quantum clusters the ratio v_0/r_0^3 assumes the largest values among clusters, providing a qualitative explanation for the attainment of a liquid state down to 0 K in macroscopic helium under its saturated vapor pressure and in helium clusters. The huge values of v_0/r_M^3 for ultracold atomic clouds again manifest the effects of the large de Broglie wavelength and of zero-point energy quantum effects at these extremely low temperatures.

C. Bose–Einstein Condensation

The properties of an ideal Bose gas are entirely controlled by permutation symmetry, and the resulting Bose–Einstein statistics are obeyed by the particles. All complicating effects of interparticle interactions, which play a dominant role in determining the properties of bulk liquid ^{4}He and of $(^4He)_N$ clusters, are neglected [24, 26, 117–119]. The ideal Bose gas consists of N bosons, each of mass m, within a volume V, with a uniform density $\rho = N/V$ at temperature T. For the uniform infinite system in the thermodynamic limit we have $N,V \rightarrow \infty$, with ρ being fixed. This idealized system consists of a collection of non-interacting bosons in a three-dimensional box [119] whose eigenstates are plane waves with a well-defined momentum p. The statistical description of the bosons in a box involves the specification of the number of particles occupying each of the single-particle eigenstates. The distribution function corresponds to the single-particle distribution, $n(p)$, specifying the number of particles in a state of momentum p. The particles in the gas can be characterized by the thermal de Broglie wavelength λ_{DB}, Eq. (4). When $\lambda_{\mathrm{DB}} \ll r_e$ quantum effects are negligible and the particles behave classically, while when λ_{DB} approaches r_e, there will be an overlap between the wavefunctions for the particles and quantum effects set in (Section I.B).

At high temperatures, when $\lambda_{\mathrm{DB}} \ll r_M$, the properties of the ideal Bose gas are dominated by the thermal motion of the particles. The momenta of these bosons will be distributed according to the classical Maxwell–Boltzmann distribution. The momentum distribution $n(p)$ is then a Gaussian with a width that is proportional to T. At lower temperatures, λ_{DB} approaches the mean interparticle distance r_e, and the effects of Bose quantum statistics result in the deviations of $n(p)$ from the classical Gaussian shape. For bosons without any restriction on the occupation of any state, the quantum effects are manifested in the increase of the occupation of states with small values of p. Boson permutation effects allow for the reduction of the energy of the system by many-particle occupation of lower energy states. The momentum distribution $n(p)$ becomes peaked at small values of p, exhibiting a non-Gaussian shape.

As the temperature is lowered further, the quantum statistical effects dominate the properties of the system, and a transition to the Bose–Einstein condensed phase is manifested. The Bose–Einstein transition temperature T_c^0 in the infinite, noninteracting, uniform boson system is

$$T_c^0 = \left(\frac{2\pi\hbar^2}{mk_B}\right)\left(\frac{\rho}{\xi(3/2)}\right)^{2/3} \tag{6}$$

where $\xi(\cdot)$ is the Rieman zeta function—for example, $\xi(3/2) = 2.61$. The transition temperature [Eq. (6)] is determined by the condition that the thermal de Broglie wavelength is given by some appropriate measure of the interparticle spacing. Defining the thermal wavelength $\lambda_{\mathrm{BE}} = h/(2\pi m k_B T_c^0)^{1/2}$, which is proportional to the de Broglie wavelength at T_c^0 [Eq. (4)], the transition temperature is determined from the condition

$$\rho\lambda_{\mathrm{BE}}^3 = \xi(3/2) \tag{7}$$

At a temperature below T_c^0 a macroscopic occupation of the ground state with $p = 0$ develops. The macroscopic occupation implies that the number $N(0)$ of bosons at the ground state is of the order of the total number N of particles in the system. This is the celebrated "Bose–Einstein condensation."

The transition is characterized by the appearance of a sharp delta-function-type spike in $n(p)$ at $p = 0$, which is proportional to the delta function $\delta(p)$, with a weight that is given by the condensate fraction $N(0)/N$—that is, the fraction of bosons residing in the ground state. The condensate fraction is given by [24, 26, 117–119]

$$\frac{N(0)}{N} = 1 - \left(\frac{T}{T_c^0}\right)^{3/2} \tag{8}$$

At $T = 0$ all the bosons reside in the ground state and $N(0)/N = 1$.

In 1938 London proposed [120] that Bose–Einstein condensation provides a microscopic explanation for superfluidity in liquid ^{4}He. When Eq. (6) is naively applied to bulk ^{4}He, a rather reasonable estimate of $T_c^0 \simeq 3$ K is obtained, which is close to the experimental result for the lambda point temperature of $T_\lambda = 2.17$ K. However, this apparent agreement between the properties of the ideal Bose gas and liquid ^{4}He is unsatisfactory because, due to interatomic interactions, the properties of liquid ^{4}He significantly differ from those of an ideal Bose gas. While in liquid helium a Bose condensate exists, the interatomic interactions will significantly reduce the condensate fraction expected for the ideal Bose gas, Eq. (8) [26]. The effects of interactions in liquid helium are manifested in its phase diagram [121], where there is no low-density phase at a temperature where Bose condensation might take place [26]. Another marked difference between liquid ^{4}He and the ideal Bose gas is the density dependence of the transition temperature. In the ideal gas, T_c^0 increases with increasing density (i.e., $T_c^0 \propto \rho^{2/3}$) according to Eq. (6). On the other hand, the superfluid transition temperature in liquid ^{4}He decreases with increasing density, as experimentally observed from pressure effects on T_λ [122, 123]. Thus the "strong" interactions in liquid helium seem to decrease the condensate fraction. From the point of view of methodology, it should be noted that, in contrast with the ideal gas, the momentum is no longer a "good" quantum number for the self-interacting liquid helium. However, the concept of a condensate fraction is still valid, because one can characterize a finite fraction of atoms with zero momentum in liquid ^{4}He.

The basic concept of the existence of a critical temperature for the onset of macroscopic occupation of a single quantum ground state of a boson system is applicable both for liquid ^{4}He and for weakly interacting low-density atomic vapors. The phenomenon of Bose–Einstein condensation is not limited to an ideal Bose gas and prevails also in a strongly interacting boson system. The bridging between Bose–Einstein condensation in the low-density, weak interaction region (i.e., low-density vapors) and in the high density, strong interaction region (i.e., liquid ^{4}He) was provided by quantum path integral Monte Carlo simulations for a hard-sphere many-boson system [124, 125]. The interparticle interactions are characterized by the hard-core sphere diameter σ, and the bulk system is characterized by the density ρ. The effective dimensionless interaction parameter $\rho\sigma^3$ was varied in the range 10^{-6}–1, with the variation of the critical temperature T_c, normalized to T_c^0 [Eq. (6)], exhibiting three domains (Fig. 4): (i) The low-density, weak interaction region where $\rho\sigma^3 = 10^{-6}$–10^{-2} and where T_c/T_c^0 increases with increasing $\rho\sigma^3$ in the range $T_c/T_c^0 = 1$–1.07. In this region the variation of the critical temperature can be fit by the empirical relation $T_c/T_c^0 \simeq [1 + \alpha(\rho\sigma^3)^{1/3}] \simeq [1 + \alpha(\sigma/r_0)]$, where α ($\sim$2) is a numerical factor. (ii) The maximal value of $T_c/T_c^0 \simeq 1.07$ is reached at $\rho\sigma^3 \sim 10^{-2}$. (iii) The high-density, strong interaction region, where T_c/T_c^0 drops in the range $T_c/T_c^0 = 1.07$–0.55. The high-density, strong interaction range $\rho\sigma^3 = 0.2$–0.4, where T_c/T_c^0 decreases from $\sim$0.70 to $\sim$0.55, faithfully represents the decrease of T_λ with increasing the density in liquid ^{4}He (Fig. 4). The striking result emerging from these quantum simulations is the moderately weak deviation of the critical temperature in strongly interacting boson systems, from the critical temperature T_c^0 [Eq. (6)] for the ideal Bose gas. The large domain of the density-interaction parameter $\rho\sigma^3 = 10^{-6}$–$0.5\, T_c/T_c^0$ varies only in the narrow range of 1.07–0.55.

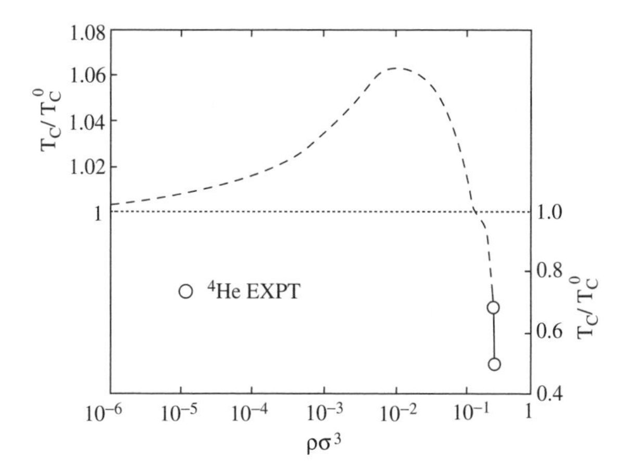

Figure 4. The bridging between Bose–Einstein condensation in the low-density, weak interaction region and in the high-density, "strong" interaction region [124, 125]. Data for T_c/T_c^0, where T_c is the critical temperature and T_c^0 is the critical temperature in an ideal Bose–Einstein gas, were calculated from quantum path integral Monte-Carlo simulations for a hard-sphere many-boson model [124, 125]. The effective dimensionless interaction parameter is $\rho\sigma^3$, where ρ is the density and σ is the hard-core sphere diameter. The two open circles ($\bigcirc$) represent experimental data for bulk liquid ^{4}He.

For the exploration of Bose–Einstein condensation in low-density and ultracold gases, we were concerned with the ideal, spatially uniform Bose gas. This physical situation corresponds to particles in an infinite three-dimensional box. It is interesting to inquire how the critical temperature and condensate fraction are affected by the spatial confinement of the noninteracting bosons. Such a physical situation is of considerable interest for the confinement of ultracold gas in magnetic and optical traps [14]. This is the situation for N noninteracting bosons confined in a harmonic spherical potential $V_{\text{ext}}(r) = (m\omega_{\text{HO}}^2/2)r^2$. The thermodynamic limit is obtained for $N \to \infty$ and $\omega_{\text{HO}} \to 0$ while keeping $N\omega_{\text{HO}}^3$ constant. The critical temperature for the confined noninteracting system is given by [14]

$$T_c^0 = \left(\frac{\hbar\omega_{\text{HO}}}{k_{\text{B}}}\right)\left(\frac{N}{\xi(3)}\right)^{1/3} \tag{9}$$

while the condensate fraction at $T < T_c$ is [14]

$$\frac{N(0)}{N} = 1 - \left(\frac{T}{T_c^0}\right)^3 \tag{10}$$

At $T = 0$ all the particles reside in the lowest state $\in_0 = (3/2)\hbar\omega_{\text{HO}}$. The critical temperature [Eq. (9)] depends on the total number of particles (with a finite value of $\omega_{\text{HO}}N^{1/3}$) and not on the density, as is the case for the homogeneous system [Eq. (6)]. The different temperature dependence for the condensate fraction for the confined boson gas [Eq. (10)] and for the uniform Bose gas [Eq. (8)] can be traced to the higher density of states for the harmonic oscillator relative to that for a particle in a box [14, 24]. Theoretical studies for finite size effects in an ideal finite Bose gas [80, 126] and for a Bose gas trapped in a harmonic potential [14, 127] provided novel information on finite boson systems. These issues will be addressed in Section I.E.

D. Order Parameter and Elementary Excitations

In a Bose fluid the macroscopic complex wavefunction $\Psi(r)$ [130–132] is [128, 133–135]

$$\Psi(\mathbf{r}) = \psi(\mathbf{r})e^{iS(\mathbf{r})} \tag{11}$$

with the modulus $\psi(\mathbf{r})$ representing the condensate density fraction $N(0)$ through the superfluid number density ρ_s,

$$|\psi|^2 = \frac{N(0)}{V} = \rho_s \tag{11a}$$

while the phase $S(\mathbf{r})$ is related to the velocity field $v(\mathbf{r})$ through

$$v(\mathbf{r}) = \left(\frac{\hbar}{m}\right) \nabla S(\mathbf{r}) \tag{11b}$$

For ^{4}He below the lambda point, $v(\mathbf{r})$ represents the velocity of the superfluid. The wavefunction [Eq. (11)] corresponds to a complex local order parameter [130–135] associated with the macroscopic occupation of the Bose–Einstein condensate.

A peculiar feature of Bose condensates and superfluids is that their low-energy excitations correspond to collective modes, which can be described as fluctuations of the order parameter [106]. For uniform dilute boson gases, with an effective interparticle interaction potential $V(r - r') = g\delta(r - r')$, where $g = 4\pi\hbar^2 a/m$ is the coupling constant, the excitations (characterized by energies $\in(k)$) are given by the Bogoliobov spectrum [136]

$$\in(k)^2 = (\hbar^2 k^2/2m)\left[(\hbar^2 k^2/2m) + 2g\rho\right] \tag{12}$$

where k is the wavevector of the excitation. The validity condition for this result is [136]

$$\rho a^3 \ll 1 \tag{12a}$$

which holds for ultracold gases (Section I.A and Table I), but not for liquid ^{4}He bulk or clusters. For large momenta the spectrum corresponds to that of a single free particle $\in(k) = \hbar^2 k^2/2m$, while for low momenta the phonon dispersion curve $\in(k) = \hbar c k$, with c being the sound velocity, is obtained. The crossover between the collective phonon excitations and the single particle excitations occurs when the typical excitation wavelength $\sim k^{-1}$ becomes comparable to the healing length [106] $\varphi = (8\pi\rho a)^{-1/2}$ (Section I.A and Table I). When the local order parameter [Eq. (11)] vanishes at some point (due to impurities, boundary effects, or finite size effects), the healing length is the typical distance for recovering the bulk value of the order parameter. In a finite, nonuniform, dilute condensate obeying Eq. (12a), the low-momentum excitations are not plane waves, but retain their phonon-like character involving the collective motion of the condensate, and manifest a transition to single-particle excitations with increasing the momentum. This theoretical framework was advanced in an attempt to understand the properties of liquid ^{4}He. The advent of this theory [136] took place in 1947, long before the experimental discovery of Bose–Einstein condensation in trapped ultracold gases, for which it is applicable. However, for liquid ^{4}He we obtain $\rho a^3 \sim 1$ (Table I), and this theory is inapplicable.

In liquid ^{4}He the low k collective excitations still involve phonons [137, 138]. However, in this strongly correlated dense fluid the phonon branch does not cross directly to single-particle excitations. The dispersion curve reaches a maximum at $k_{\text{MAX}} \simeq 1\,\text{Å}^{-1}$, and the excitation spectrum forms a minimum with a nearly parabolic dispersion at $k_{\text{MIN}} \simeq 1.9\,\text{Å}^{-1}$ with $\in (k_{\text{MIN}}) = 8.7\,\text{K}$ [137, 138]. The collective excitations near the minimum [17]

$$\in (k) = \in (k_{\text{MIN}}) + \left(\frac{\hbar^2}{2\mu}\right)(k - k_{\text{MIN}})^2 \tag{13}$$

are called rotons [17, 18], whose wavelength is of the order of the interatomic distance and which are related to local order on the atomic scale [17, 18, 137].

The global Hamiltonian for a Bose quantum fluid written in the hydrodynamic form [19] is

$$\hat{H}(\rho) = \left(\frac{m}{2}\right) \int \hat{v}(\mathbf{r})\hat{\rho}(\mathbf{r})\hat{v}(\mathbf{r})d^3r + U[\rho] \tag{14}$$

where $U[\rho]$ is a general functional of the one-particle density ρ, while the first term represents quantum hydrodynamic kinetic energy flow [19]. Quantum mechanical operators with appropriate commutation relations represent the superfluid velocity $\hat{v}$ [Eq. (11b)] and the one-particle density $\hat{\rho}$ [19, 128, 138]. The potential energy in Eq. (14) was expanded to second order in density fluctuations $\delta\rho(r)$, with a plane wave expansion of $\delta\rho$, invoking translational invariance of the bulk local compressibility [138]. This treatment [138] resulted in the Feynman–Bijil spectrum [139, 140] for bulk superfluid helium

$$\in (k) = \frac{\hbar^2 k^2}{2mS(k)} \tag{15}$$

where the structure factor $S(k)$ is the Fourier transform of the two-particle distribution function. This celebrated result established the existence of the phonon spectrum at low k and manifested the interrelationship between the structure and the roton dispersion (with k_{MIN} corresponding to the maximum in $S(k)$) for the quantum fluid. The classical limit for the dispersion relation in the bulk is [141]

$$\in (\text{k}) = \hbar\text{k}\left[\frac{k_B T}{mS(k)}\right]^{1/2} \tag{16}$$

which again manifests a minimum in the k domain, where $S(k)$ reaches its maximal value. From this analysis an iconoclastic conclusion emerges. The appearance of a minimum in the dispersion curve for elementary excitations of a fluid at $k > 1\,\mathring{A}^{-1}$ manifests the existence of collective excitations. However, such a minimum in the dispersion curve [Eq. (13)] does not necessarily mark the existence of quantum effects. Collective excitations with a minimum in the dispersion curve at values of $k \geq 1\,\mathring{A}^{-1}$ were also detected by neutron scattering in classical liquids—for example, liquid ^{4}He above the lambda point, liquid parahydrogen, and liquid D_2O [142–145]. These excitations do not manifest superfluidity, because thermal excitations will dissipate the collective, high-k excitations of a moving particle in a nearly classical fluid. The unique feature of collective roton excitations in ^{4}He, which exist in the domain specified by Eq. (13), is their stability toward dissipation. The same situation will prevail in $(^4\text{He})_N$ clusters. $(^4\text{He})_N$ clusters and other finite Bose–Einstein condensates are characterized by the following finite size effects: (i) *A sharp boundary for the fall-off of the density.* For the LDM, one takes a step function approximation (Section I.A) for the constant density within the cluster radius R_0. (ii) *A discrete spectrum of the excitations.* The excited states for nuclear excitations of a spherical cluster are classified according to the number of radial nodes, n, and the angular momentum quantum numbers ℓ and m.

The discrete level structure is crucial for the lowest energy excitations, whose wavelength is comparable to R_0, and which represent collective excitations of the entire cluster. The eigenmomenta $k_{n\ell}$ are defined by boundary condition (i) for the LDM, with $j_\ell(k_{n\ell}R_0) = 0$, where $j_\ell(\cdot)$ are the spherical Bessel functions. The compressional density fluctuations in a liquid drop give a phonon-like discrete spectrum for all clusters sizes [84, 85, 128]

$$\in_{\ell n}(k_{\ell n}) = \hbar c k_{\ell n} \tag{17}$$

and are determined by the velocity of sound c. The lowest energy breathing mode $(n = 1, \ell = 0)$ is

$$\in_{01} = \frac{\pi\hbar c}{r_0 N^{1/3}} \tag{17a}$$

which was adopted from the theory of nuclear excitations to the realm of dynamic cluster size effects [84, 85]. Equation (17), with a constant value of c (which corresponds to the bulk value), seriously overestimates the energies of the collective breathing mode. A refined liquid drop model (which will be referred to as LDMR) was introduced [128], accounting for the density dependence of c. From the Lee–Yang energy density functional for the imperfect Bose gas the relation between c and the (average) density ρ is [146]

$$c^2 = \left(\frac{(4\pi\hbar^2 a\rho)}{m^2}\right)\left[1 + \left(\frac{240}{15}\right)\left(\frac{\rho a^3}{\pi}\right)^{1/2}\right] \tag{18}$$

TABLE IV
Cluster Size Dependence of the Velocity of Sound c and the Energies $\in_{01}$ of the Lowest Breathing Mode in $(^4\text{He})_N$ Clusters[a]

		$\in_{01}$ (K)		
N	$c(\text{m s}^{-1})$	LDM	LDMR	QLDM
20	154	9.5	5.9	5.0
70	198	6.2	4.8	5.7
240	238	4.0	4.0	3.0
∞	238	0	0	0

[a]Data from M. V. Krishna and K. B. Whaley, *J. Chem. Phys.* **93**, 746 (1990).

The breathing mode energies $\in_{01}$ calculated [128] from the cluster size dependence of the sound velocity (Table IV) are considerably lower than the LDM result for small clusters.

The LDM was extended [128] to incorporate quantum effects advancing a quantum liquid drop model (QLDM). The relation between the discrete dispersion curves for elementary excitation and the density fluctuations in quantum clusters was established in an elegant work [128] based on the hydrodynamic form of the Hamiltonian for a Bose quantum liquid [Eq. (14)]. The procedure [128] was based on the expansion of the Hamiltonian $H(\rho)$ and was second-order in density fluctuations $\delta\rho(\mathbf{r})$ setting the boundary conditions $\delta\rho(r = R_0) = 0$, and on the expansion of $\delta\rho(\mathbf{r})$, $v(\mathbf{r})$, and $\phi(\mathbf{r}, \mathbf{r}') = \delta\rho(\mathbf{r})\delta\rho(\mathbf{r}')$ in spherical Bessel functions. The discrete dispersion curves for $\in_{\ell mn}(k_{\ell n})$ were obtained in the form [128]

$$\in_{\ell mn}(k_{\ell n}) = \frac{\hbar^2 k_{\ell n}^2}{2mS_{\ell m}(k_{\ell n})\nu_{\ell n}} \tag{19}$$

where

$$\nu_{\ell n} = \int_0^{R_0} r^2 [\gamma_\ell(k_{\ell n} r)]^2 \tag{19a}$$

The cluster structure function is [128]

$$\begin{aligned} S_{\ell n}(k_{\ell n})\nu_{\ell n} = {} & \left(\frac{1}{\rho_0 \nu_{\ell n}^2}\right) \iint d^3\mathbf{r}_1 d^3\mathbf{r}_2 \\ & \otimes j_\ell(k_{\ell n} r_1) j_\ell(k_{\ell n} r_2) Y_{\ell m}^*\left(\frac{\mathbf{r}_1}{r_1}\right) Y_m\left(\frac{\mathbf{r}_2}{r_2}\right) \\ & \otimes \langle \delta\rho(\mathbf{r}_1)\delta\rho(\mathbf{r}_2)\rangle \end{aligned} \tag{20}$$

which is the Fourier–Bessel transform of the ground-state fluctuation-density correlation function $\delta\rho(\mathbf{r}_1)\delta\rho(\mathbf{r}_2)$ of the cluster, which can be expressed in the form

$$\langle\delta\rho(\mathbf{r}_1)\delta\rho(\mathbf{r}_2)\rangle = \rho(\mathbf{r}_1)\delta^3(\mathbf{r}_1, \mathbf{r}_2) + \rho(\mathbf{r}_1)\rho(\mathbf{r}_2)[g_2(\mathbf{r}_1, \mathbf{r}_2) - 1] \qquad (20a)$$

where $\rho(\mathbf{r}_j)(j = 1, 2)$ are the one-particle densities and $g_2(\mathbf{r}_1, \mathbf{r}_2)$ is the two-particle distribution function. Equations (19) and (20) constitute the finite cluster analogue of the Feynman–Bijil relation [139, 140, 147], Eq. (15).

From the point of view of general methodology, several comments are in order. First, the appearance of the Fourier–Bessel transform in the structure function [Eq. (20)] reflects on the breakdown of translational invariance, which is prevalent in the case of the bulk. Second, the different symmetries of spherically projected structure functions for the finite system and of plane wave structures for the bulk system are crucial for a proper representation of the cluster excitations. Third, the discrete eigenvectors $k_{\ell n}$ are determined by the boundary conditions. Fourth, the energies $\in_{\ell mn}$ $(k_{\ell n})$ are discrete. However, the complete spectrum for a fixed value of n containing $l = 0, 1, 2, \ldots$ branches would form a continuous smooth curve.

Detailed numerical calculations [128] (for $m = 0$), which are shown in Fig. 5, revealed that the structure function $S_{\ell=0,m=0}(k_{\ell=0,n})$ for $(^4\mathrm{He})_N$ cluster sizes $N = 20, 70, 240$ peaks at $k \simeq \text{Å}^{-1}$, being similar in form to the structure factor $S(k)$ for the bulk liquid [128]. The excitation spectrum $\in_{\ell=0,m=0,n}$ for $N = 240$ and $N = 70$ clearly reveals a maximum around $k^{\mathrm{MAX}}_{\ell=0,n} \simeq 1.2\,\text{Å}^{-1}$ and a minimum around $k^{\mathrm{MIN}}_{\ell=0,n} \simeq 1.8\,\text{Å}^{-1}$ [128]. These $k_{\ell=0,n}$ values are close to the $k_{\mathrm{MAX}} = 1\,\text{Å}^{-1}$ and $k_{\mathrm{MIN}} = 1.9\,\text{Å}^{-1}$ values for the bulk fluid. The low range of $k_{\ell=0,n}$ in the size domain $N = 240, 70$ exhibits a linear $\in_{\ell=0,m=0,n}$ versus $k_{\ell=0,n}$ dependence, exhibiting a phonon spectrum with a slope close to that of the bulk sound velocity [128]. Thus phonon–roton collective excitations exist in $(^4\mathrm{He})_N$ clusters for $N \geq 70$. For the smaller cluster with $N = 20$ the initial slope of $\in_{\ell=0,m=0,n}$ versus $k_{\ell=0,n}$ is linear and is lower than that for the larger clusters, in accord with the cluster size dependence of the sound velocity (Table IV). In the range of $k_{\ell=0,n} \simeq 1.8\,\text{Å}^{-1}$ the dispersion curve for $N = 20$ reveals an inflection point, while no roton minimum is exhibited as is the case for larger clusters. These interesting results [128] pertain to cluster size effects for the realization of collective roton excitations in finite quantum systems. The onset of the roton dispersion curve and the onset of superfluidity in $(^4\mathrm{He})_N$ boson clusters occurs in the cluster size domain of $N \lesssim 70$.

The classical limit of the QLDM [Eqs. (19) and (20)] results in the relation for the LDM for a classical liquid drop [128, 141]

$$\in^{\mathrm{CL}}_{\ell mn} = \hbar k_{\ell n}\left[\frac{k_B T}{\mathrm{m S_{\ell m}(k_{\ell n})\nu_{\ell n}}}\right]^{1/2} \qquad (21)$$

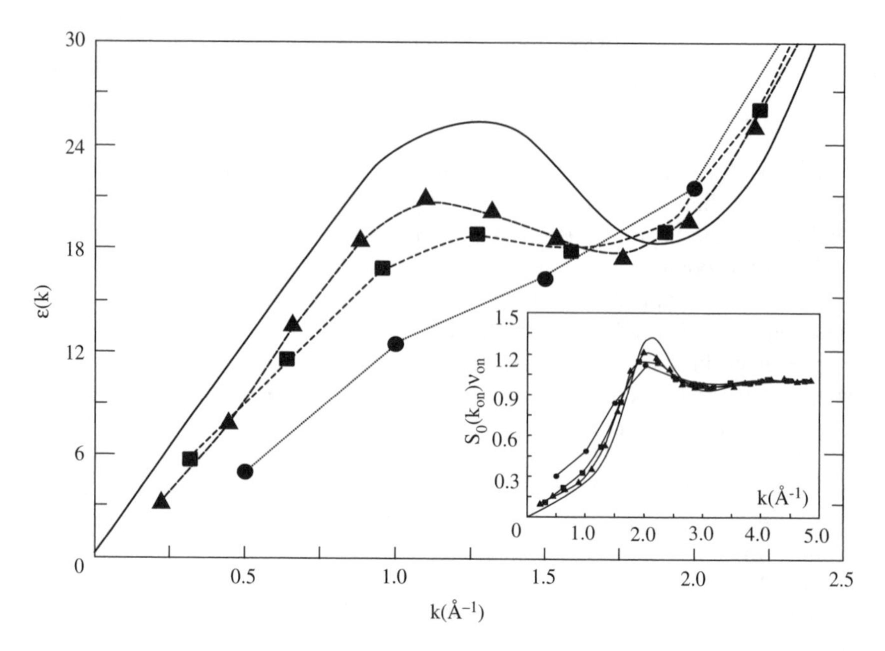

Figure 5. Structure factor and excitation spectrum of $(^4He)_N$ clusters adopted from R. Krishna and K. B. Whaley [128]. The excitation spectrum $E_{\ell n}(k_{\ell n})$ for $\ell = 0$ was calculated by the QLDM [128] for $N = 20(\bullet)$, $N = 70(\blacksquare)$, and $N = 240(\blacktriangle)$. The solid line represents the continuous excitation spectrum from the Feynman–Bijil equation for bulk superfluid ^{4}He. The spherically projected structure factor $S_{\ell n}(k_{\ell n})\nu_{\ell n}$ for $\ell = 0$ is presented in the insert for $(^4He)_N$ clusters with $N = 20(\bullet)$, $N = 70(\blacksquare)$, and $N = 240(\blacktriangle)$, while the solid line represents the structure factor calculated for bulk liquid ^{4}He [128].

This classical result, which manifests a minimum value of $\in_{\ell mn}$ where the maximal value of $S_{\ell m}$ is attained, does not correspond to elementary excitations in a quantum cluster. Indeed, collective, large $k \gtrsim \mathring{A}^{-1}$ cluster excitations will also be manifested in classical clusters. As in the case of the bulk systems discussed above, such collective excitations in classical clusters will be dissipated, while the roton cluster excitations will be robust with respect to dissipation into lower energy excitation (e.g., phonons).

E. Energetics, Thermodynamics, Response, and Dynamics of Ultracold Finite Systems

This review will focus on some facets of the bridging between cluster science and chemical physics of ultracold finite systems—that is, quantum clusters and finite ultracold gases. The following problems will be addressed:

1. *From Finite Ultracold Systems to the Bulk.* The general features of packing, structure and thermodynamic properties, as well as the elementary excitation spectrum of large finite systems, converge to those of the corresponding bulk material for sufficiently large quantum clusters and ultracold clouds.

The convergence of the packing and structural properties of finite systems to those of the bulk manifests some of the simplest predictions of cluster size equations [84–86]. At this stage, two comments are in order: First, the convergence of a specific property of large finite systems to that of the corresponding bulk is general (within some restrictions specified in Section 3 below), but not universal [84]. The "critical" size, or the number N_{MAX} of constituents, of the system for the attainment of the bulk property (within some margin of deviation [84]) is property-dependent [84–86]. Second, in the size domain $N < N_{\text{MAX}}$, the specific property of the finite quantum system manifests the unique structural, thermodynamic, and response characteristics—for example, the state of aggregation, radial distribution function onsets of Bose–Einstein condensation, superfluidity, and elementary excitations. When the size of the quantum system further decreases, specific size effects set in [84] (Section 4 below). Only for rather small system sizes of $N \ll N_{\text{MAX}}$, the unique quantum size effects are expected to be eroded for $N \leq N_{\text{MIN}}$, with the onset cluster size N_{MIN} again being property dependent (Section 5). Regarding the structure and packing of quantum clusters, the radial pair distribution functions for $(^4\text{He})_N (N = 20\text{–}1000)$ [52, 128, 129] clusters converge to those for bulk ^{4}He with increasing the cluster size. The pair distribution function approaches the bulk value, so one can set $N_{\text{MAX}} \gtrsim 10^3$ for this property, with the average structure and density of $(^4\text{He})_N$ clusters converging to that of liquid ^{4}He for $N \geq 10^3$. Regarding the response of $(^4\text{He})_N$ clusters, the dispersion curve for elementary excitations in $(^4\text{He})_N$ clusters [128], which manifests roton-type excitations in the energy domain of 10–20 K at a finite value of the (discrete) momentum, becomes similar to the Feynman–Bijil dispersion curve for phonons and rotons in macroscopic liquid ^{4}He [139, 140, 147] with increasing the cluster size in the range $N \gtrsim 240$ [128], whereupon $N_{\text{MAX}} \gtrsim 240$ for this central property of a boson quantum cluster. For finite, ultracold clouds, structural information is not available, while finite size effects on the critical temperature T_c for Bose–Einstein condensation infer that $N_{\text{MAX}} \simeq 10^3$ (for the attainment of T_c within less than 1% of the bulk value [14]). Regarding the bridging between the structure of the surface of quantum clusters and of the bulk surface, the notable existence of quite large surface profiles (with a width of $t \simeq 6 - 10\,\text{Å}$), which originate from kinetic energy effects (Section I.A), is prevalent both in finite $(^4\text{He})_N$ clusters with $N > 10^3$ and in liquid helium. Similarly, zero-point energy and kinetic energy effects in quantum clusters (Section I.B) bear a close analogy between the large finite ultracold systems and the corresponding bulk systems. In this context, it is important to emphasize that in an analogy between macroscopic liquid helium under its equilibrium vapor pressure, which constitutes the only liquid condensed phase down to $T = 0\,\mathrm{K}$, $(^4\mathrm{He})_N$ and $(^3\mathrm{He})_N$ clusters and droplets are the only clusters that remain liquid down to 0 K [6–11, 51–60, 106, 127–129]. The lowering of the "melting"—that is, the order–disorder (broadened) structural transition temperature in a cluster relative to the corresponding infinite system [148, 149]—may be beneficial for the attainment of lower-temperature liquid quantum clusters [149]. A notable example pertains to the search for a new Bose-condensed finite system in molecular para-H_2 [66, 123, 149]. This perspective depends on the expected value of the critical temperature for Bose–Einstein condensation in molecular hydrogen, which crystallizes at $T_m \sim 14\,\mathrm{K}$ in the bulk, while for clusters the "melting" temperature is expected to be lower [149]. For Bose–Einstein condensation in an ideal uniform H_2 gas, the fictitious liquid density results in the critical temperature [Eq. (6)] of $T_c^0 = 6$–$8\,\mathrm{K}$ [123, 149, 150], which is consistent with early theoretical estimates [150], although a lower value of $T_c = 2.1\,\mathrm{K}$ was inferred from a later analysis [123]. The estimate of $T_c = 6$–8 K, based on Eq. (6), is also consistent with quantum path integral Monte Carlo simulations for finite para-H_2 clusters [66, 151, 152]. The quest for Bose–Einstein condensation and superfluidity in molecular clusters (e.g., para-H_2 and ortho-D_2) is of considerable interest. In this context the exploration of Bose–Einstein condensation and of superfluidity in many-boson systems cannot be accomplished in the bulk but may be feasible for finite quantum clusters.

2. *Bose–Einstein Condensation and Superfluidity in Finite Systems.* The two most important properties for the finite boson $(^4\mathrm{He})_N$ or $(\mathrm{p}\text{-}\mathrm{H}_2)_N$ systems (which are well established in the corresponding homogeneous bulk system of liquid ^{4}He) are superfluidity and Bose–Einstein condensation [6–11, 23, 50, 65–78, 128]. Superfluidity pertains to the hydrodynamic effects of the response to a slow movement of the system's boundaries [6–11, 23, 50, 65–78], while Bose–Einstein condensation manifests off-diagonal long-range order, with the occupation number of the ground state becoming proportional to the number density of the atoms [23, 24, 65–78]. While the properties of superfluidity and of Bose–Einstein condensation are distinct, both phenomena manifest the implications of boson permutation symmetry and are characterized by the same transition temperature, at least in the infinite system [23, 66–69]. In a finite boson system a wealth of novel finite size effects will be manifested, involving surface and boundary effects, spatial inhomogeneity, and the breakdown of translational symmetry. These finite size effects will unveil new facets of elementary excitations, Bose–Einstein condensation, and superfluidity in ultracold finite systems.

The conventional theory and quantum simulations of Bose–Einstein condensation [21–23, 65, 66, 151, 152] reviewed above (Section I.C) rest on the thermodynamic limit of an infinite boson system. Bose condensation in finite systems was originally invoked in the exploration of the properties of nuclei, in which pairs of nucleons bind via the strong force to produce effective bosonic degrees of freedom [153, 154]. These concepts are applicable for nuclei, which constitute finite systems far from the thermodynamic limit. For weakly interacting low-density systems, finite size effects on Bose–Einstein condensation in noninteracting uniform Bose gas [80, 126] and in noninteracting confined Bose gas in an electromagnetic trap [14, 127] were explored. For high-density, strongly interacting ^{4}He finite systems, quantum simulations of the lambda transition in small $(^4\mathrm{He})_N$ clusters ($N = 8$–128) [65, 155] and experimental studies of the specific heat of ^{4}He in confined geometries [156–162] provide information on Bose–Einstein condensation in these finite boson systems. Another important issue in this context of broadened phase changes in finite systems pertains to the inequivalence of canonical and microcanonical ensembles and the role of fluctuations in finite systems. This issue came out for phase changes in nearly classical clusters [163, 164] where the caloric curves for the canonical and microcanonical ensembles were found to be qualitatively different, with the microcanonical ensemble manifesting a negative specific heat in the transition region, which originates from fluctuations. Experimental evidence for S-shaped caloric curves of nearly classical clusters was reported [165]. Recent theoretical studies addressed the role of the specific statistical ensemble and the effects of fluctuations in finite boson systems [166–170].

3. *Size Effects in Ultracold Systems.* Another aspect of the bridging between cluster science and ultracold large finite systems involves size effects [81–89]. Central issues in the broad, interdisciplinary research area of cluster science pertain to the energetics, thermodynamics, spectroscopy, dynamics, and response by the utilization of cluster size equations as scaling laws for the nuclear–electronic response of finite systems [84–87]. When is such size-scaling partial and incomplete? Several examples come to mind in the context of energetics, nuclear dynamics, and cooperative effects. First, specific cluster size effects, involving self-selection and the existence of "magic numbers" for moderately sized clusters, manifest an irregular variation of structure and energetics, which is not amenable to size scaling, with a large abundance of some sizes being due to enhanced energetic stability [81–83, 88, 89]. An interesting recent development in the realm of specific size effects in small $(^4\mathrm{He})_N$ clusters involves "magic numbers." These are not due to the relative energetic stability of the clusters, but rather to effects of interior cluster collective compression modes on the growth kinetics of these clusters [105]. Second, structural characterization and specification of distinct phase-like forms—for example, solid (rigid) and liquid (nonrigid), or solid (rigid) and solid (rigid) configurations—and "smeared" (rounded-off) phase changes between them in clusters and nanoparticles may

differ from the corresponding feature in bulk matter [163, 164]. A related and interesting issue pertains to rounded-off λ phase changes and Bose–Einstein condensation in finite systems of ^{4}He clusters.

4. *Specific Size Effects*. An interesting question in the realm of quantum size effects pertains to the issue of the minimal cluster size that will manifest energetic stability, or to a specific electron level structure. This question concerning the threshold size, which involves specific cluster size effects and changes in the electronic level structure, was addressed in the field of the energetics and response of molecular and metal clusters. Pertinent questions in this context are: What is the minimal cluster size of a metal–atom cluster (Hg_N, Mg_N, Zn_N, or Sr_N) which will exhibit the metal-nonmetal transition [171–173]? What is the minimal cluster size of elemental, ionic, or molecular clusters to support a quasi-free electron state in Xe_N [174, 175], an F center in $(NaCl)_N$ [176], or a solvated electron state in $(NH_3)_N$ [84]? Related questions, such as the following, were raised in the context of cluster reactivity: What is the minimal (or maximal) size of a metal cluster (e.g., Au_N) to induce specific catalytic reactions [177]? This issue of threshold cluster size effects is of considerable interest for quantum clusters. Energetic and dynamic stability will govern the threshold size of ultracold finite systems. A well-known problem involves isotope effects on the minimal cluster size of $(^4He)_N$ and $(^3He)_N$ clusters [51–54]. While the ^4He_N diatomic molecule is stable—that is, the $(^4He)_N$ "cluster" is formed for $N = 2$ [107, 108]—the minimal stable cluster size of $(^3He)_N$ is obtained for $N \simeq 25$ [51–54], manifesting kinetic energy effects. Another interesting problem involves the minimal size of a $(^4He)_N$ cluster that will support an excess electron surface state [178–180] or an interior excess electron state, which is localized in a bubble in $(^4He)_N$ and $(^3He)_N$ clusters. The energetic stability of an interior excess electron state in helium clusters has to be supplemented by the dynamic stability, because dynamic effects involving electron tunneling of the excess electron may result in the depletion of the energetically stable state on the experimental time scale for the interrogation of $(He)_N^-$ clusters [99].

5. *Dynamics*. Cluster dynamics constitutes a rich field, which focused on nuclear dynamics on the time scale of nuclear motion—for example, dissociation dynamics [181], transition state spectroscopy [177, 181, 182], and vibrational energy redistribution [182]. Recent developments pertained to cluster electron dynamics [183], which involved electron–hole coherence of Wannier excitons and exciton wavepacket dynamics in semiconductor clusters and quantum dots [183], ultrafast electron-surface scattering in metallic clusters [184], and the dissipation of plasmons into compression nuclear modes in metal clusters [185]. Another interesting facet of electron dynamics focused on nanoplasma formation and response in extremely highly ionized molecular clusters coupled to an

ultraintense laser field [186]. Perspectives for nuclear dynamics of ultracold quantum clusters involved the description of bulk and surface elementary excitations [85, 86, 128, 129]. Time-resolved nuclear dynamics on the time scale of nuclear motion in quantum clusters involves the extreme medium dilation accompanying the formation of excess electron bubbles in $(\text{He})_N$ clusters in analogy with bulk helium [187, 188]. Electron dynamics in quantum clusters involves electron tunneling from an excess electron bubble [99]. Another interesting development in the realm of cluster dynamics pertained to the energetic and dynamic instability of multicharged clusters. This research area, which was pioneered in 1882 by Lord Rayleigh [189] for the fission of multicharged droplets and in 1939 by Bohr and Wheeler [190] for nuclear fission, was recently extended to encompass the phenomenon of Coulomb explosion of extremely charged elemental and molecular clusters [94]. Coulomb explosion of highly charged molecular clusters—for example, $(\text{Xe}^{+q})_n$ clusters with $q = 1\text{–}36$ and $n = 2000$ [93, 191], or $(\text{D}^+)_n$ clusters with $n = 100\text{–}2000$ [96, 97]—which is induced by extreme multielectron ionization in ultra-intense laser fields (intensity $I = 10^{15}\text{–}10^{20}\ \text{W cm}^{-2}$), being characterized by ultrafast time scales (10–100 fs) and ultrahigh energies for nuclear motion, and with ion energies in the range of 1 keV to 1 meV, corresponding to the energy domain of nuclear physics. The analogy between cluster dynamics and the nuclear dynamics of finite ultracold systems is of interest. Recently, such an analogy was established between cluster Coulomb explosion and the expansion of optical molasses [79], which provides a bridge between nuclear dynamics of clusters and of laser-irradiated ultracold finite clouds.

These general problems outlined above set the cornerstones for this chapter, which will address the energetics, thermodynamics, response, and dynamics of ultracold finite systems. A selective review of the following issues will be presented:

1. What information is available on the (rounded-off) phase changes that characterize Bose–Einstein condensation and superfluidity in $(^4\text{He})_N$ clusters and in ultracold clouds?
2. What are the size effects and scaling laws for the temperature of the λ transition to superfluidity/Bose–Einstein condensation in finite $(^4\text{He})_N$ clusters, and how does the λ temperature relate to the macroscopic condensed phase?
3. What are the size effects and scaling laws for the onset of Bose–Einstein condensation in finite ultracold gases?
4. How can one characterize threshold size effects for superfluidity/Bose–Einstein condensation in small boson systems—for example, what is the minimal size of a $(^4\text{He})_N$ cluster for the attainment of these properties?

5. How can one use macroscopic probes (i.e., excess electron bubbles) to interrogate superfluidity in finite $(^4\mathrm{He})_N$ clusters?
6. How can one establish threshold size effects for the energetic and dynamic stability of excess electron bubbles in $(^4\mathrm{He})_N$ clusters?
7. How can one establish relations and correlations between the nuclear dynamics of molecular clusters and of ultracold gases?

II. SIZE EFFECTS ON THE SUPERFLUID TRANSITION IN $(^4\mathrm{He})_N$ FINITE SYSTEMS

Notable recent developments in the realm of low-temperature large, finite, quantum systems pertain to the exploration of homonuclear molecular clusters (aggregates or nanodroplets) of $(^4\mathrm{He})_N$ where the nuclear dynamics, elementary excitations, and response are dominated by quantum effects and by permutational symmetry [50–68, 155]. Some of the features of the finite ^{4}He boson systems [50, 65, 66, 155–162, 192] are:

1. *The Onset of the Superfluid Transition in the Finite System [50, 65, 66, 155].* This transition is referred to as the λ point in the bulk system. What is the analogy in a finite system? Pioneering quantum path integral Monte-Carlo simulations [65, 66] established the appearance of a rounded-off (smeared) λ transition in finite $(\mathrm{He})_N$ ($N = 64$ and 128) clusters. This was manifested by a maximum in the temperature dependence of the specific heat (Fig. 6), which occurs at the temperature T_λ, with $\Delta T_\lambda = T_\lambda^0 - T_\lambda > 0$ where $T_\lambda^0 = 2.172$ K is the temperature of the λ transition in the bulk [23], while experimental values of ΔT_λ of ^{4}He in confined spaces were recorded [157–162, 192] down to $\Delta T_\lambda \geq 2 \times 10^{-4}$ K. Early experimental studies [156] of the heat capacity of ^{4}He confined in microscopic bubbles (cavities) in Cu foils indicated the occurrence of the superfluid transition with the lowering of T_λ in the confined space; however, pressure effects and size effects on the superfluid transition cannot readily be separated. Relevant in this context of superfluidity in finite systems [157–162, 192] are several experimental studies of the superfluid transition interrogated by the density and specific heat of ^{4}He in confined geometries [157–162, 192]—that is, films [157, 192], cylinders [157, 159, 192] and pores [159, 160]. These confined geometries involve polymer membranes (nucleopore filters), with films of 20–80 Å thickness [157, 192] and cylindrical channels of 10^2–10^3 Å diameter [157, 192] (Fig. 1), porous gold with a pore diameter of 240 Å (Fig. 1) [159], vicor glass involving a highly intercorrelated network of pores of an average diameter of 70 Å [160], and confinement between sheets of Mylar [161] separated by 4600 Å and ^{4}He between Si wafers [162]. These specific heat data manifest the rounding-off of the transition and the shift of its maximum (T_λ) to lower values, that is, $\Delta T_\lambda > 0$. Alternatively, the onset of the appearance of a finite fraction of the

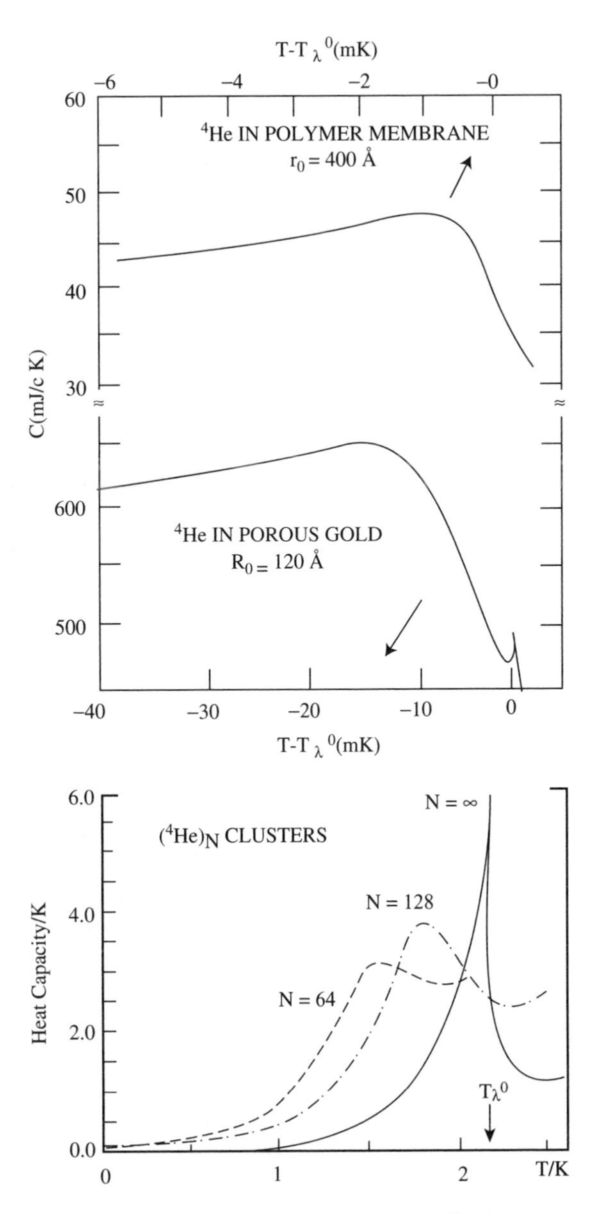

Figure 6. The temperature dependence of the specific heat of $(^4\text{He})_N$ clusters (lower panel for $N = 64$ and $N = 128$) obtained from quantum simulations (Ref. 65) and from experimental data for ^{4}He in porous gold (pore radius $R_0 = 120$ Å, upper panel, Ref. 160) and for ^{4}He in cylindrical channels in polymer membrane (cylinder radius $r_0 = 400$ Å, upper panel, Ref. 159). The bulk infinite system specific heat ($N = \infty$) is presented in the lower panel.

superfluid density can be taken as a measure of the λ transition in the finite system. From the available simulation data for $(^4\text{He})_N$ ($N = 64, 128$) clusters [65] the maximum of the specific heat is manifested at $T_\lambda = 1.58\,\text{K}$ for $N = 64$ and at $T_\lambda = 1.82\,\text{K}$ for $N = 128$, while the onsets of the superfluid density are $T_\lambda = 1.75 \pm 0.10\,\text{K}$ for $N = 64$ and $T_\lambda = 2.0 \pm 0.10\,\text{K}$ for $N = 128$. The superfluid density ρ_s, calculated in conjunction with the specific heat data, starts to increase in the range of the broad maximum of the specific heat [65]. However, the numerical uncertainty in these simulation data precludes definite conclusions. An analysis of ρ_s, based on later quantum simulation data [155] reveals a broad transition region ΔT_ρ (e.g., $\Delta T_\rho \gtrsim 1$ K for $N = 64$), where the superfluid density gradually increases from low values toward unity. These data and their analysis [155] will be discussed later (Section II.E). The experimental data for ^{4}He confined in porous gold [150] and vicor glass [160] also reveal an approximate coincidence of the temperatures corresponding to the maximum of the specific heat and to the onset of the superfluid density. Both observables characterize the rounded-off λ transition in the finite system. "Smeared," rounded-off phase transitions in finite systems may differ from the corresponding features in bulk matter [193–196]. The concept of finite size scaling for phase transitions in a confined system [193] related the lowering of T_λ to the smallest confining dimension L by the Fisher relation

$$\Delta T_\lambda / T_\lambda^0 \sim L^{-1/\nu} \tag{22}$$

where $\nu = 0.67$ is the characteristic exponent for the divergence of the correlation length [134, 135, 155, 195]. Similarly, the region δT_λ of the rounding-off of the specific heat curve is expected to be determined by the relation [193] $\delta T_\lambda \sim L^{-1/\nu}$, whereupon the ratio $\delta T_\lambda / (\Delta T_\lambda / T_\lambda^0)$ is a constant, being size-independent. When these relations for $\Delta T_\lambda / T_\lambda^0$ and δT_λ were originally subjected to an experimental scrutiny [157, 192], it was found that the specific heat data in polymers, films, and cylinders (over a small size domain) obey the Fisher relation [153], Eq. (1); however, the scaling exponent was lower [157, 192] than the value $\nu = 0.67$. A possible resolution of this finite size scaling problem was considered [158] by replacing T_λ^0 by a size-dependent reference temperature. A more elaborate scrutiny of specific heat and superfluid fraction data for finite systems over a larger size domain is called for.

2. *Superfluidity in the Finite Systems.* The quantum path integral simulations [65, 66] for the $(^4\text{He})_N$ ($N = 64$, 128) clusters indicate the onset of the superfluid fraction ϕ_λ in the vicinity of $T \simeq T_\lambda$, with a gradual increase of ϕ_λ with decreasing the temperature, reaching a large finite value ($\phi_\lambda \simeq 0.9$) at $T = 0$. Even more interesting is the use of molecular spectroscopic probes for superfluidity in large $(^4\text{He})_N$ clusters ($N = 10^4$–10^6) at 0.4 K (where $\rho_\lambda \simeq 1$) [71–74]. Another microscopic probe for superfluidity in large $(^4\text{He})_N$ clusters

($N \geq 10^5$) at 0.4 K involves a transport probe—that is, electron tunneling from the electron bubble (Chapter IV)—which provided evidence for vanishingly low viscosity of the superfluid finite system.

3. *Elementary Excitation in the Superfluid Clusters.* The existence of a roton-type collective excitation spectrum in large $(^4\text{He})_N$ clusters ($N = 10^4$–10^6) at 0.4 K was established from electronic spectroscopy of large molecules (e.g., glyoxal [70]) which manifests coupled electronic–roton excitations [70].

While the characteristics of superfluidity and of the elementary excitations in the large, cold ($T = 0.4$ K) $(^4\text{He})_N$ clusters ($N = 10^4$-10^6) were considered in analogy to the properties of the corresponding bulk system [6, 8, 70–78, 102], the interesting problem of size effects on the phenomena of Bose–Einstein condensation and superfluidity in finite boson systems [8, 61, 65–67, 71, 104, 155, 157–162] is not yet fully elucidated. The available information emerges from the path integral Monte-Carlo simulations of $(^4\text{He})_N$ ($N = 64$, 128 [65, 66] and $N = 8$–64 [155]) clusters and from experimental specific heat data of ^{4}He in confined porous systems [157–162, 192]. In this chapter we address the issue of the size scaling of the λ point in finite $(^4\text{He})_N$ clusters. As a starting point, we shall utilize the phenomenological theory of Ginzburg, Pitaevskii, and Sobyanin [134, 135] for the λ transition with proper boundary conditions for free surfaces, to explore the cluster size dependence of T_λ in $(^4\text{He})_N$ clusters. The cluster size scaling theory for superfluidity in $(^4\text{He})_N$ clusters provides a satisfactory semiquantitative account of the results of the path integral Monte-Carlo simulations [65, 66] and of the experimental specific heat data of ^{4}He confined in pores [157–162, 192] for the lowering of T_λ with decreasing the size of the $(^4\text{He})_N$ clusters. The phenomenological theory relates the intensive property (T_λ) of the finite system (of size L) to the correlation length $\xi(T)$ for superfluidity in the corresponding bulk system, with the shift ($T_\lambda^0 - T_\lambda$) depending on the ratio $L/\xi(T)$. This result of the phenomenological model for the size-dependent λ transition is related to the theory of finite size scaling [152, 155, 193–197], which is extensively used to interpret simulations of phase transitions—for example, liquid–vapor critical point [193, 198] and Bose–Einstein condensation in liquid ^{4}He, in a hard-sphere gas [124, 125], and in $(^4\text{He})_N$ clusters [155]. While the finite size scaling theory routinely allows to deduce the transition point for the infinite system from simulations for finite-size samples [155, 193–197, 199], one can invert the argument using finite size scaling for the characterization of the "smeared" λ transition in the finite quantum boson system.

A. A Phenomenological Theory of the Lambda Transition

The Ginzburg–Pitaevskii theory [134] for bulk liquid ^{4}He near the λ point rests on Landau's theory of second-order phase transitions [133]. This theory was extended by Ginzburg and Sobyanin [135] for the treatment of the λ transition in finite systems (e.g., thin films, narrow channels, confined space, and vortices) exploring size effects and confinement on the superfluid transition, which is pertinent for the analysis of the onset of superfluidity in $(^4\text{He})_N$ clusters. This phenomenological theory [134, 135, 199] rests on the introduction of a macroscopic complex wavefunction ψ, which is used as an order parameter for the superfluid transition. The modulus of the complex order parameter ψ, Eq. (11), is related to the superfluid density ρ_s and is normalized in the form

$$\rho_s = m|\psi|^2 \tag{23}$$

where m is the mass of the helium-4 atom. The normal He fluid is considered to be at rest, and the free energy density $f^{(0)}$ (which depends on the pressure p and temperature T) of the homogeneous infinite fluid can be expanded in terms of powers of $|\psi|^2$, while the local free energy density $f(\mathbf{r})$ for an inhomogeneous finite system can be expressed in terms of powers of $|\psi(r)|^2$. Thus for a homogeneous system

$$f^{(0)}(P,T,\psi) = f_1(P,T) + A|\psi|^2 + (B/2)|\psi|^4 + \cdots \tag{24}$$

where f_1 is the free energy density of normal ^{4}He, while the coefficients A and B depend on T and P. From the equilibrium condition for the homogeneous fluid $(\partial f^{(0)}/\partial|\psi|^2)_{P,T} = 0$, Ginzburg and Pitaevskii [134, 135] established the relation $\text{A} + B|\psi|^2 = 0$, which results in an explicit relation between the homogeneous system superfluid density ρ_s and the coefficients A and B, so that $\rho_s = -A/B$. At this stage the phenomenological theory of Ginzburg and Sobyanin can be adopted, representing the bulk order parameter and its superfluid density ρ_s in terms of a critical exponent of the macroscopic system

$$\rho_s/\rho_\lambda = t^\nu \tag{25}$$

with

$$t = (T^0_\lambda - T)/T^0_\lambda \tag{26}$$

where the critical exponent for the superfluid fraction is $\nu = 0.6702$ [200]—that is, manifesting the "2/3 scaling law." T^0_λ is the λ point temperature of the infinite system. Here the superfluid effective density is $\rho_\lambda = 0.351\,\text{g cm}^{-3}$ [200], while $\rho(T_\lambda) = 0.146\,\text{g cm}^{-3}$ is the experimental density at T^0_λ. The equilibrium condition results in $|\psi|^2 = -A/B$, which from Eq. (25) implies that $\rho_s/\rho_\lambda = -A/B \propto t^\nu$. In the temperature range below T^0_λ—that is $(T^0_\lambda - T_\lambda) > 0$—the parameter A is negative [134, 135]. The temperature dependence of the expansion parameters is expressed in the form

$$A = -\alpha t^{2\nu} \qquad (\alpha > 0) \tag{27}$$

$$B = \beta t^{\nu} \tag{28}$$

Eqs. (27) and (28) are consistent with the scaling relations (25). All the terms in the expansion (24) exhibit the same t-dependence, and the parameters α and β are temperature-independent.

The free energy density $f(\mathbf{r})$ of the inhomogeneous finite system with a local order parameter $\psi(\mathbf{r})$ was expressed by adding to $f^{(0)}$, Eq. (24), an even expansion of the gradient term, so that

$$\mathrm{f}(\mathbf{r}) = f_1 + \left(\frac{\hbar^2}{2m}\right)|\nabla\psi(\mathbf{r})|^2 + A|\psi(\mathbf{r})|^2 + \left(\frac{B}{2}\right)|\psi(\mathbf{r})|^4 + \cdots \tag{29}$$

The total free energy $F = \int d^3\mathbf{r}\mathrm{f}(\mathbf{r})$ is minimized with respect to the order parameter. The minimization with respect to ψ^* results in the Schrödinger-type equation

$$-\left(\frac{\hbar^2}{2m}\right)\nabla^2\psi + A\psi + B|\psi|^2\psi + \cdots = 0 \tag{30}$$

At this stage the correlation length $\xi(T)$ for superfluidity in the bulk is introduced:

$$\xi(T) = \left(\frac{\hbar^2}{2m|A|}\right)^{1/2} \tag{31}$$

which, according to Eq. (27), is

$$\xi(T) = \xi_0 t^{-\nu} \tag{32}$$

where

$$\xi_0 = \left(\frac{\hbar^2}{2m|\alpha|}\right)^{1/2} \tag{33}$$

The critical exponent ν for the correlation length, Eq. (32), is identical to that for the superfluid fraction [135, 155, 158, 193–197, 200], Eq. (25). ξ_0, Eq. (33), is the "critical" amplitude for the correlation length. ξ_0 can be related to the superfluid density by the Josephson relation [201]

$$\xi_0 = \frac{k_B T_\lambda^0 m^2}{\hbar^2 \rho_\lambda} \tag{33a}$$

where ρ_λ is the superfluid effective density [200], Eq. (25). Equation (33a) results in $\xi_0 = 3.1$ Å for bulk ^{4}He. Note that this short "critical" amplitude for the correlation length implies that ξ_0 is comparable to the interatomic spacing $r_e = 3.6$ Å in liquid ^{4}He (Section I.A).

The application of Eq. (30) for the order parameter in a finite system (e.g., clusters) requires the introduction of the appropriate boundary condition, with the vanishing of the order parameter—that is, $\psi = 0$ at the boundaries of the cluster. This boundary condition explicitly invokes a step function approximation for the cluster surface profile, while the realistic description of $(^4\text{He})_N$ clusters involves a broadened profile with a FWHM of 6 Å [84, 106]. Introducing the reduced coordinates

$$\mathbf{r}_* = \frac{\mathbf{r}}{\xi(T)} \tag{34}$$

and using Eqs. (31) and (33), Eq. (30) is then expressed in the form

$$-\nabla_*^2\psi + \left[-1 + \left(\frac{B}{A}\right)|\psi|^2 + \cdots\right]\psi = 0 \tag{35}$$

where ∇_*^2 is the Laplacian in the reduced coordinates [Eq. (34)].

Equation (35) was advanced by Ginzburg and Sobyanin for superfluidity in confined finite systems [135]. This theory will be applied herein for the onset of superfluidity of ^{4}He confined in a sphere of radius R_0. Adopting the step function approximation, the boundary condition for the order parameter at the free surface is taken as $\psi(R_0) = 0$. For low values of ψ the first-order linear form of Eq. (30) is

$$\frac{1}{R_*^2}\frac{d}{dR_*}\left(R_*^2\frac{d\psi}{dR_*}\right) + \psi = 0 \tag{36}$$

where

$$R_* = \frac{R}{\xi(T)} \tag{36a}$$

with the lowest solution $\psi(R_*) = \sin R_*/R_*$. The free-surface boundary condition $\psi(R_0/\xi) = 0$ results in $R_0/\xi(T) = \pi$, so that $R_0 = \pi\xi_0 t^{-\nu}$ with t given by Eq. (26). This result implies that the order parameter in the finite system vanishes at the boundary, marking the onset of the superfluid transition in the cluster at the temperature T_λ when

$$\frac{T_\lambda^0 - T_\lambda}{T_\lambda^0} = \left(\frac{\pi\xi_0}{R_0}\right)^{1/\nu} \tag{37}$$

Equation (37) implies that the lowering of the λ temperature T_λ in the finite clusters is given by the relation

$$\frac{T_\lambda^0 - T_\lambda}{T_\lambda^0} = \frac{(\pi\xi_0)^{1/\nu}}{R_0^{1/\nu}} \tag{38a}$$

Setting $R_0 = r_e N^{1/3}$, where N is the number of the He atoms and $r_e (= [m/\rho(T_\lambda)^{1/3}])$ is the constituent radius (the average interatomic distance $r_e = 3.5$ Å), results in

$$\frac{T_\lambda^0 - T_\lambda}{T_\lambda^0} = \frac{(\pi\xi_0/r_e)^{1/\nu}}{N^{1/3\nu}} \tag{38b}$$

Equations (38a) and (38b), together with $\nu \simeq 2/3$, provide the size scaling of the λ point in clusters.

In the original Ginzburg–Sobyanin [135] analysis of ^{4}He superfluidity in confined spaces, the λ transition in a cylinder of radius d_0 and length $\ell(\ell \gg r)$ was considered by making use of Eq. (35) together with the appropriate boundary conditions $\psi(d_0) = 0$ and $(d\psi/dr)_{d_0} = 0$. This treatment results in [135]

$$\frac{T_\lambda^0 - T_\lambda}{T_\lambda^0} = \left(\frac{\alpha\xi_0}{d_0}\right)^{1/\nu} \tag{39}$$

where $\alpha = 2.405$ is the first root of the Bessel function. Equations (38a) and (39) provide explicit expressions [with appropriate numerical coefficients π for spherical clusters, Eqs. (38a) and (38b), and $\alpha = 2.41$ for cylinders, Eq. (39)] for the relation $(T_\lambda^0 - T_\lambda) \propto L^{-1/\nu}$, where $L = R_0$ or $L = d_0$, in accord with Eq. (22).

B. Size Scaling of the λ Point in Clusters and Confined Systems

From the preceding analysis we infer that the depression $\Delta T_\lambda = T_\lambda^0 - T_\lambda$ of the λ point in $(^4\text{He})_N$ clusters, Eqs. (38a) and (38b), size scales as $\Delta T_\lambda/T_\lambda^0 \propto R_0^{-1/\nu} \approx R_0^{-3/2}$, and similarly in cylindrically confined systems, Eq. (39) (of radius d_0), $\Delta T_\lambda/T_\lambda^0 \propto (d_0)^{-1/\nu} \approx (d_0)^{-3/2}$. For $(^4\text{He})_N$ clusters the dependence

TABLE V
Specific Heat of $(^4He)_N$ Finite Systems

System	N	R_0Å	$\Delta T_\lambda/T_\lambda^0$	$\delta(f)$	$\xi_0(g)$Å	δt (h) (K)	$\delta t/(\Delta T_\lambda/T_\lambda^0)$ (K)
Isolated cluster							
Quantum simulations [a]	64	14.3	0.271	2.17	1.9	0.372	1.4
Isolated cluster							
Quantum simulations [a]	128	18.0	0.161	1.82	1.7	0.286	1.8
Porous gold							
Experiment [b]	4.03×10^4	120	6.45×10^{-3}	1.29	1.3	1.8×10^{-2}	2.8
Pores in vycor glass							
Experiment [c]	10^3	35	6.5×10^{-2}	2.06	1.8	5×10^{-3}	0.1
Polymer membranes							
Cylindrical channels	—	$d_0 = 400$Å	3.7×10^{-4}	—	0.9^i	1.5×10^{-3}	4.1
Experiment[d]							
Nucleopore		$d_0 = 150$Å	$2.9 \times 10^{-3} \pm 0.5 \times 10^{-3}$	—	1.23 ± 0.13^i		
Filters	—	$d_0 = 400$Å	$5.5 \times 10^{-4} \pm 0.5 \times 10^{-4}$	—	1.09 ± 0.06^i		
Cylindrical channels		$d_0 = 500$Å	$3.7 \times 10^{-4} \pm 0.5 \times 10^{-4}$	—	1.04 ± 0.09^i		
Experiment[e]		$d_0 = 1000$Å	$1.1 \times 10^{-4} \pm 0.15 \times 10^{-4}$	—	0.90 ± 0.08^i		

[a]Reference 65.
[b]Reference 159.
[c]Reference 160.
[d]References 157, 160, 192.
[e]Reference 192.
[f]$\delta = (\Delta T_\lambda/T_\lambda^0)N^{1/2}$, Eq. (40a).
[g]$\xi_0 = R_0\delta^\nu/\pi$, Eq. (38a).
[h]δt determined by FWHM/2 of $C(T)$ in the region $T < T_\lambda$ for $(^4He)_N$ clusters and $T > T_\lambda$ for confined systems.
[i]ξ_0 determined from Eq. (39).

of δT_λ on the number of constituents, Eq. (38b), is given in the form $\Delta T_\lambda / T_\lambda^0 \propto N^{-1/3\nu} \approx N^{-1/2}$. The relative depression of the λ point these in clusters provides a proper cluster size equation, that is,

$$\frac{\Delta T_\lambda}{T_\lambda^0} = \frac{\delta}{N^{1/2}} \tag{40a}$$

$$= \frac{\gamma}{R_0^{3/2}} \tag{40b}$$

where $\delta = (\pi\xi_0/r_e)^{3/2}$ and $\gamma = (\pi\xi_0)^{3/2}$, with $\Delta T_\lambda \to 0$ for R_0 and $N \to \infty$.

The scaling relation, Eqs. (40a) and (40b), with the proper critical exponent ($\nu \simeq 2/3$) will be utilized to establish the validity of this cluster size equation over a large range of finite spherical $(^4\text{He})_N$ systems ($R_0 = 14$–400 Å, $N = 64 - 1.5 \times 10^6$) from isolated clusters [65, 66] to pores in metals [159] and glasses [160] (Table V and Fig. 7). Concurrently, the scaling relation,

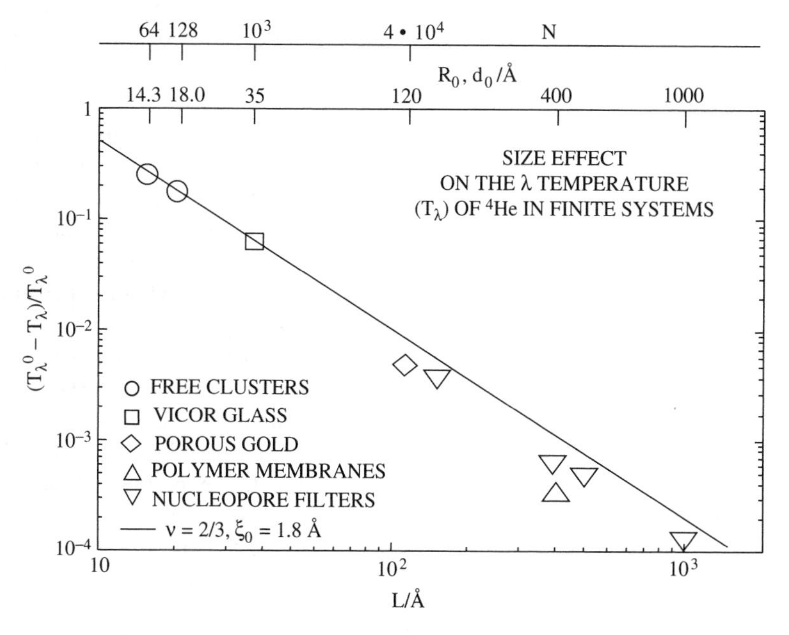

Figure 7. Size scaling of the relative depression $\Delta T_\lambda / T_\lambda^0$ of the λ point of $(^4\text{He})_N$ in finite systems, according to Eqs. (38a) and (39). ○ $(^4\text{He})_N$ clusters of radius R_0 (Ref. 65); □ ^{4}He in vicor glass, pore radius $R_0 = 35$ Å (Ref. 159); ◇ ^{4}He in porous gold, pore radius $R_0 = 120$ Å (Ref. 160); △ ^{4}He confined in cylindrical pores (radius $d_0 = 400$ Å) in polymer membrane (Ref. 159); ▽ ^{4}He in cylindrical pores (radius $d_0 = 150$ Å–1000 Å) in nucleopore filters (Ref. 192). The confining dimension is $L = R_0$ for spherical clusters or pores, or d_0 for cylindrical pores. The solid line corresponds to the size scaling with $\xi_0 = 1.7$ Å and $\nu = 2/3$.

Eq. (39), will be applied with the same critical exponent ($\nu = 2/3$) to account for the depression of the λ point for ^{4}He confined in cylindrical channels of polymer membranes [159] and nucleopore filters [157, 192] (Table V and Fig. 7). Spherical geometry was taken for the isolated clusters [65], for pores in metals [159] and in glasses [160], while cylindrical geometry was taken for the polymers [157, 192]. From this analysis an estimate of the "critical" amplitude ξ_0 for the bulk correlation length will emerge. The fit of the quantum simulations results of Sindzingre, Klein, and Ceperley [65, 66] (Table V and Fig. 7) to Eqs. (40a) and (40b) results in $\delta = 2.17$ and $\xi_0 = 1.9$ Å for $N = 64$, and $\delta = 1.82$ and $\xi_0 = 1.7$ Å for $N = 128$. Thus the finite size scaling law provides a semiquantitative account of the quantum simulation data for small ^{4}He clusters [65, 66]. The cluster size dependence of $\Delta T_\lambda / T_\lambda^0$according to Eqs. (40a) and (40b) was extended over a considerably larger size domain of spherical cavities, whose size was obtained from structural data [159], with the analysis of the experimental specific heat data (Table V and Fig. 7) for ^{4}He in porous gold ($R_0 = 120$ Å) [159] and vicor glass ($R_0 = 35$ Å) [160]. The analysis for these porous spherical systems and, in particular, for the vicor glass, implies complete pore filling [157, 192]. The values of δ, Eq. (40a), obtained for all the finite spherical systems, are nearly constant within a numerical spread of 30% (Table V and Fig. 7), providing evidence for the validity of the cluster size equations, Eqs. (38a), (38b), (40a), and (40b). The values of the "critical" amplitude ξ_0 inferred from this analysis (Table V and Fig. 7) for spherical ^{4}He pores ($R_0 = 35$–120 Å) vary in the range of 1.3–1.8 Å, being close to the values $\xi_0 = 1.7$–1.9 Å obtained for the small clusters ($R_0 = 14$–18 Å). We have also included in Table V and Fig. 7 the experimental specific heat data for ^{4}He in polymer membranes and nucleopore filters [157, 192] with cylindrical channels (with a radius of $d_0 = 150$–1000 Å). Making use of Eq. (39) for the analysis of the experimental data for ^{4}He in cylindrical channels [192], we obtained ξ_0 values in the range $\xi_0 = 1.23 \pm 0.13$ Å for $d_0 = 150$ Å to $\xi_0 = 0.90 \pm 0.08$ Å for $d_0 = 1000$ Å. These values of ξ_0 for the cylindrical channels exhibit a systematic variation of less than 12%, and they are lower by about 50% than the average value of 1.7 ± 0.3 Å evaluated for the experimental data for spherical finite systems ($R_0 = 14$–120 Å). When all these experimental specific heat data are taken together, we infer that $\xi_0 = 1.5 \pm 0.6$ Å. This value of $\xi_0 \approx 1$–2 Å obtained from the analysis of quantum simulation data ($R_0 = 14$–18 Å) and of experimental data for ^{4}He spherical confined systems ($R_0 = 35$–120 Å) and in cylindrical channels ($r_0 = 150$–1000 Å) is lower by a numerical factor of $\sim$1.5–3.0 than the value of $\xi_0 = 3.1$ Å estimated from the Josephson relation, Eq. (33a), for the bulk superfluid. We note in passing that a single value of ξ_0 was used in the analysis of the specific heat data in finite systems. This ξ_0 value corresponds to the infinite fluid. In the experimental papers for porous systems

[159, 160] the values of $\bar{\xi}_0 = 8.4$ Å and 17 Å are given for porous gold and $\bar{\xi}_0 = 93$ Å for vicor glass. These latter $\bar{\xi}_0$ data are based on estimates of the actual superfluid effective densities in the confined systems. In our analysis (Fig. 7) we use for ξ_0 the bulk value.

From the foregoing analysis (Fig. 7) of simulation and experimental data we infer that the size scaling relation $\Delta T_\lambda \propto L^{-3/2}$ (where $L \simeq R_0$ for clusters and nearly spherical confined spaces and $L \simeq r^0$ for cylinders) is obeyed over a wide size domain of $L \simeq 14$–400 Å (i.e., $N = 14$–4×10^4 for clusters and nearly spherical confined spaces), and of $L \simeq 150-1000$ Å for cylindrical channels. This broad range of size domain with the proper critical $\nu \simeq 2/3$ exponent indicates that it is unnecessary to replace T_λ^0 by a size-independent reference temperature, as proposed [158] to account for a lower scaling component reported for (^{4}He) confined in polymer films over a narrow size domain [157, 192].

C. Finite Size Scaling of the Superfluid Transition Temperature and Density

The relation $\Delta T_\lambda \propto R_0^{-1/\nu}$ obtained from the Ginzburg–Pitaevskii–Sobyanin theory for a finite system is related to the theory of second-order phase transitions with the experimental critical parameter, $\nu = 0.67$, for the superfluid fraction and for the correlation length scaling near the critical point of infinite systems [155, 193–197, 199]. This theory implies that the intensive properties of a system of size $L(= R_0)$ depend on the ratio $L/\xi(T) \sim Lt^\nu$, where $\xi(T) = \xi_0 t^{-\nu}$ is the bulk correlation length.

At this stage finite-size scaling theory [155, 193–197, 199] is applicable for the description of the specific heat maximum and of the onset of the superfluid density (see the beginning of Section II), which characterize the rounded-off λ transition. The singular free energy density, f, of the finite system (in the absence of external fields) can be described in terms of a universal function $(Y(\,))$ in the form [194–197] $f = L^{-d}Y(KtL^{1/\nu})$, where K is a metric factor, which contains all the system-dependent aspects of the critical behavior and d is the dimensionality. Defining the parameter $y = KL^{1/\nu}t$, the free energy $f = L^{-d}Y(y)$ yields the specific heat $C = T\,(\partial^2 f/\partial T^2)$. Accordingly, $C \propto Y^{(2)}(y)$, being determined by the second derivative, $Y^{(2)}$, of Y. The maximum of the specific heat $(\partial C/\partial y = 0)$, which characterizes the smeared-out λ transition at T_λ, is manifested at $y = y_{\mathrm{MAX}}$, being exhibited at $Y^{(3)}(y_{\mathrm{MAX}}) = 0$, where $Y^{(3)}$ is the third derivative of Y. Accordingly, the rounded-off λ transition specified by the maximum of the specific heat is exhibited for $t_{\mathrm{MAX}} = y_{\mathrm{MAX}}K^{-1}L^{-1/\nu}$, with $t_{\mathrm{MAX}} = (T_\lambda^0 - T_\lambda)/T_\lambda^0$, in accord with the results of the order parameter analysis of Section II.B. For the sake of generality we shall rewrite this result, setting $T_\lambda^0 = T_c$, where T_c is the critical temperature for the transition in the bulk fluid, so that $t \equiv t_{\mathrm{MAX}}$ is given by

$$t = \frac{T_c - T_\lambda}{T_c} \tag{41}$$

The peak temperature for the specific heat, marking the lambda transition, is [155, 193]

$$T_\lambda = T_c - aL^{-1/\nu} \tag{42}$$

where

$$a = T_c Y_{\mathrm{MAX}} K^{-1} \tag{42a}$$

Equations (41) and (42) are in accord with the results of the order parameter analysis (Section II.A), which was used for the analysis of the experimental results (Table V and Figure 7).

An important issue pertains to the broadening of the specific heat curve $C(T)$ in finite systems (Fig. 6). The region δT_λ of the rounding-off of the specific heat curve was determined from the available quantum simulation [65, 66] and experimental data [157–160, 192] by the (1/2)(FWHM) of $C(T)$ for the range $T < T_\lambda$ for clusters and $T > T_\lambda$ for confined spaces (Table V), which are exhibited at $C(T^{(1)})/C(T_\lambda) = 1/2$. The broadening of the specific heat curve, characterized by the 1/2(FWHM) of C (at $T^{(1)} < T_\lambda$) is given by the 1/2(FWHM) of the $Y^{(2)}(y)$ function. This results in the relative width of the specific heat curve

$$\delta T = \frac{T^{(1)} - T_\lambda}{T_c} \propto L^{-1/\nu} \tag{43}$$

From Eqs. (41) and (43) one infers the same finite size scaling of δT and t, in accord with Fisher's analysis [193]. From this analysis, one concludes that $\delta T/t =$ constant, being size-independent. Indeed, this relation is reasonably well obeyed (within a numerical factor of 3 over the range $R_0 = 14$–400 Å) for the quantum simulations for small clusters, for porous gold, and for the membrane polymer (Table V). However, a marked (one order of magnitude) deviation from this relation is exhibited for ^{4}He confined in vicor glass (Table V), which may be attributed to constrained randomness effects [203, 204] and which calls for further scrutiny.

Finite size scaling of the order parameter $|\psi|^2 = \rho_s/\rho$, Eqs. (11) and (23), for the superfluid transition provides significant information on the size dependence of the superfluid density ρ_s/ρ over the temperature range around T_c. The finite size scaling is [155, 195, 196]

$$\frac{\rho_s}{\rho} = L^{-1}Q(L^{1/\nu}t) \tag{44}$$

where $t = (T - T_c)/T_c$, ν is the correlation length exponent, and $Q(\cdot)$ is an (unknown) analytic function of a finite argument. The linearization of $Q(\cdot)$ results in

$$Q(L^{1/\nu}t) = \alpha + \beta(L^{1/\nu}t) \tag{45}$$

where α and β are numerical constants. The characteristic length scale for the clusters is taken as $L \simeq R_0$, i.e., $L \simeq r_e N^{1/3}$, according to Eq. (1), with $r_e = 3.6$ Å (Section II). Equation (45) assumed the form

$$N^{1/3}\left(\frac{\rho_s}{\rho}\right) = A + B(N^{1/3}t) \tag{46}$$

with

$$A = \alpha/r_e \tag{46a}$$

$$B = \beta r_e^{(1/\nu-1)} \tag{46b}$$

Superfluid density data in $(^4\text{He})_N$ clusters ($N = 8$–64), obtained from quantum simulation data with periodic boundary conditions [155] (Fig. 8), obey Eq. (46),

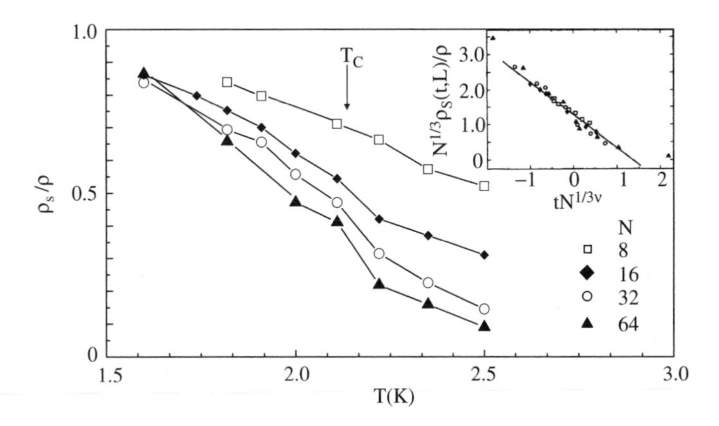

Figure 8. The cluster size and temperature dependence of the superfluid relative density ρ_s/ρ for $(^4\text{He})_N$ clusters ($N = 8, 16, 32, 64$) [155]. Data obtained from quantum path integral Monte-Carlo simulations [155]. Finite size scaling of ρ_s/ρ, according to Eq. (46), is presented in the insert.

as is apparent from the linear plot of $N^{1/3}(\rho_s/\rho)$ vs $N^{1/3}t$, adopted from reference 155 (insert to Fig. 8). From the insert to Fig. 8 we estimate $A = 1.20$ and $B = 1.20$–0.86. The linear fit resulted [155] in the size-independent parameters $\nu = 0.72 \pm 0.1$ and $T_c = 2.17 \pm 0.05$ K. The function $Q(0) = \alpha = Ar_e$ [Eq. (46a)] assumes the value $Q(0) = 4.3$ Å. The universal constant [155, 205, 206]

$$\tilde{X} = (\hbar^2\rho/mk_BT_c)Q(0) \tag{47}$$

is then given by $\tilde{X} = 0.54 \pm 0.05$. This result should be compared with the value $\tilde{X} = 0.49 \pm 0.01$ for a 3*DXY* model of a classical spin system [205, 206]. The good agreement between the $\tilde{X}$ values for the classical system [205, 206] and for the quantum $(^3\text{He})_N$ system implies that both systems belong to the same universality class [155].

From the analysis of the cluster size dependence of the superfluid density (or order parameter) the following conclusions emerge:

(i) The finite size scaling provides a confirmation of the critical exponent ν, with value $\nu = 0.72 \pm 0.1$ obtained from the analysis [155], which is in agreement with the experimental value of $\nu = 0.6702$ [200]. ν is cluster size independent.

(ii) The transition temperature $T_c = 2.17 \pm 0.05$ K is cluster size dependent. This conclusion contradicts a previous conjecture [158] regarding the size dependence of T_c.

(iii) The value of $T_c = 2.17 \pm 0.05$ K, obtained for the superfluid density data [155] (Fig. 8), is very close to the experimental value of $T_\lambda^0 = 2.172$ K [29] for the lambda-specific heat transition in infinite ^{4}He.

(iv) The phase transition in the finite system is "rounded off." The finite size scaling theory accounts for the spread of the ρ_s data over a broad temperature domain, which spans the temperature range $T > T_c$ as well. The broadening of the specific heat curve near T_c is also accounted for.

(v) The size dependence of the specific heat data for T_λ in a finite system [Eq. (42)] and the superfluid density data for ρ_s/ρ [Eq. (46)] are described by the same, size-independent, critical temperature T_c. This is an important conclusion of the finite size scaling theory, which bridges between superfluidity and thermodynamics of finite $(^4\text{He})_N$ systems.

(vi) Superfluidity features are manifested for small $(^4\text{He})_N$ clusters with low N $(N = 8)$ [155]. An interesting open problem in this context is that the quantum simulations were performed for periodic boundary conditions [155]. The role of the boundary conditions on the superfluid and thermodynamic properties of finite $(^4\text{He})_N$ clusters and other boson systems requires further scrutiny. A central issue in the field of thermodynamics, superfluidity, and elementary excitations in finite $(^4\text{He})_N$ systems is the characterization of the minimal cluster size, which will manifest Bose–Einstein condensation and superfluidity.

D. Threshold Size Effects for Superfluidity

A surprising result emerging from the quantum simulations [65, 66] of small $(^4\text{He})_N$ clusters and the analyses in Sections II.B and II.C is the manifestation of a well-characterized, broadened, high-order phase transition for small $(^4\text{He})_N$ clusters (i.e., $N = 8$) for the superfluid density [155], and $N = 32$ for the appearance of the lambda transition [65]. An open question pertains to the threshold size of these equations: What is the system's smallest size for the exhibition of superfluidity and what are the corresponding phase transitions?

The short correlation length for superfluidity in bulk ^4He implies that threshold cluster sizes are small—that is, of the order of interatomic distance. A simple-minded argument will imply that the minimal cluster size R_0^{MIN} for the realization of a superfluidity transition is $R_0^{\text{MIN}} > \xi_0$. Using Eq. (40b), a lower limit for R_0^{MIN} will be manifested for $T_\lambda \to 0$, whereupon $(\pi\xi_0/R_0^{\text{MIN}})^{3/2} \sim 1$ and $R_0^{\text{MIN}} \sim \pi\xi_0$. Taking the short correlation length $\xi_0 \sim 2$ Å, we roughly estimate that $R_0^{\text{MIN}} \sim 6$ Å, so that the smallest ^4He cluster will consist of a central atom and its first coordination layer. Thus the threshold size domain for the realization of the lambda transition is $N_{\text{MIN}} \sim 5$–13. Such a low value of N_{MIN} is consistent with the value $N_{\text{MIN}} \leq 8$ for the exhibition of the superfluid density in finite systems [155]. Finally, the threshold size for the appearance of rotons in the elementary excitation spectra of $(^4\text{He})_N$ clusters [128] is realized for $20 < N_{\text{MIN}} < 70$ (Section I.D).

In Table VI we assemble the data for N_{MIN} for the manifestation of superfluidity in $(^4\text{He})_N$ clusters. These values of N_{MIN} for different physical attributes of small finite systems (Table VI) may be property-dependent. The quantification of the size dependence of T_λ and of ρ_s (Section II.C) implies that

TABLE VI
Threshold Size of $(^4\text{He})_N$ Clusters for the λ Transition, Bose–Einstein Condensation and Superfluidity in $(^4\text{He})_N$ Finite Systems

Property	N_{MIN}	Method
Thermodynamic lambda point	5–13	$R_0^{\text{MIN}} \sim \pi\xi_0$ (Section II.B)
Superfluid density	≤ 8	Quantum path integral Monte-Carlo simulations ([155] and Section II.C)
Rotons, collective excitations	$20 < N_{\text{MIN}} < 70$	Calculations of cluster structure function ([128] and Section I.D)

the critical temperature T_c and the critical exponent ν of the correlation function are cluster size independent (Section II.C). The results of the finite size scaling theory provide guidelines for the description of these "smeared out" thermodynamic properties in small systems, establishing a thermodynamic–superfluidity relation, which implies the identity of the broadening of the 'high–order transition' for both T_λ and ρ_s. On the other hand, the interrelationship between the onset of roton elementary excitations and the superfluid density was not established, and threshold sizes may be different for these two classes of observables.

III. ENERGETICS OF ELECTRON BUBBLES IN $(^4He)_N$ CLUSTERS

A. Excess Electron Localization on and in Bulk Liquid Helium and Helium Clusters

The use of electron bubbles as microscopic probes for superfluidity in bulk ^{4}He dates back to the pioneering 1960 studies of Meyer and Reif [207], who determined the Landau–Feynman roton energy from the temperature dependence of the electron mobility. In view of the fundamental importance of probing collective excitations in finite, interacting boson quantum systems, we shall present in this chapter a theoretical study of excess electron bubbles in $(^4He)_N$ clusters [8, 208, 209]. We shall address the structure, energetics, and energetic stability of the electron bubble in $(^4He)_N$ clusters. This structural and energetic information will be utilized in Section IV for the evaluation of electron tunneling times from electron bubbles in these clusters, elucidating the dynamic stability of the electron bubble. The bubble transport in $(^4He)_N$ clusters is qualitatively and quantitatively distinct in superfluid clusters, due to their vanishingly small viscosity, as compared to viscous normal helium clusters (Section IV), whereupon the dynamics of electron tunneling from bubbles in $(^4He)_N$ clusters will provide microscopic probing of superfluidity in these finite quantum systems.

The pseudopotential between an electron and a helium atom is strongly short-range repulsive, with a very weak long-range attractive polarization interaction [210–213]. Accordingly, the conduction band energy for an excess quasifree electron in structurally unperturbed bulk liquid He is large and positive—that is, $V_0 = 1.06$ eV for ^{4}He [210–222] and $V_0 \sim 0.9$ eV for ^{3}He (at $p = 1$ atm) [212, 214, 215]—with the conduction band lying above the vacuum level (Fig. 9). The direct implication of these high positive energies of the quasi-free electron state is the exterior and interior localization of the excess electron. Two distinct types of excess electron states in and on bulk liquid He are manifested (Fig. 9), involving the electron exterior surface state

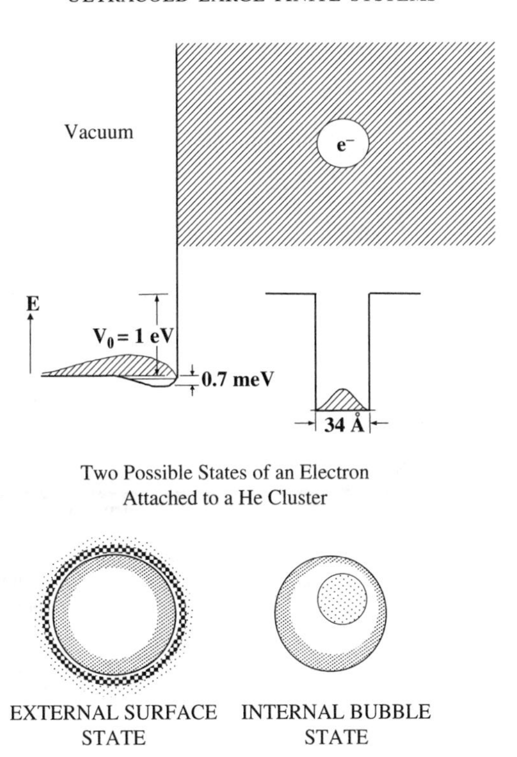

Figure 9. A schematic representation of the energetics of an excess electron interacting with bulk liquid helium, where it can reside either in a surface state with a binding energy $E_s = -0.7$ meV or in an interior bubble state with a radius $R_b = 17$ Å and a total binding energy of 0.36 eV (i.e., 0.70 eV below the conduction band energy V_0). A sufficiently large $(\mathrm{He})_N$ cluster can attach an excess electron in an external surface state or in an internal bubble state.

[215, 223–231] and the electron interior bubble state [213, 214, 232–240]. The excess electron surface state is stabilized by a weak image potential, which results in an electron localized within a one-dimensional Coulomb potential with a large barrier (Fig. 9) [215, 223–231]. The excess electron bubble state involves local fluid dilation, leading to a localized, energetically stable state of the electron confined in a cavity, pertaining to electron localization accompanied by large configurational changes in bulk liquid helium (Fig. 9) [213, 214, 232–240].

The excess electron surface state and the electron bubble state constitute two distinct ground states and two electronic manifolds of bound electronic states, with the surface states converging to the vacuum level, while the bubble states converge to the liquid conduction band (Fig. 9). The two electronic manifolds

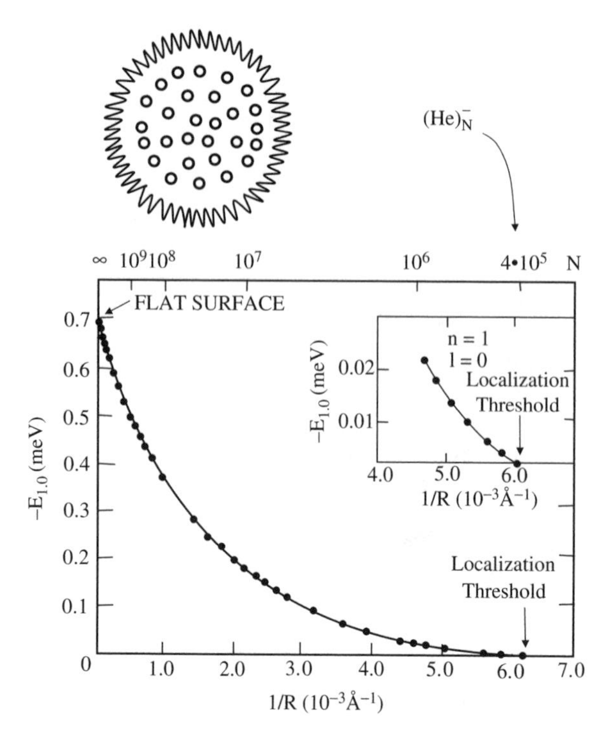

Figure 10. Cluster size dependence of the binding energy of an electron in a surface state ($n = 1$, $l = 0$) on $(^4He)_N$ clusters. The localization threshold is manifested at $N_c = 5.7 \times 10^5$ and the binding energy increases with increasing the cluster radius R, according to the scaling law $[E_s(R) - E_s(\infty) \propto (R - R_c)^2$, reaching the flat surface binding energy $E_s(\infty) = -0.7$ meV [Ref. 178–180].

are separated by the large energy barrier V_0, with weak electron tunneling from interior bubble states located in the vicinity of the surface [239]. A similar physical situation prevails for excess electron localization on and in $(He)_N$ clusters. The excess electron external surface state was predicted to be realized [178–180] above a threshold cluster size N_c and a cluster radius R_c ($N_c = 3 \times 10^5$ for $(^4He)_N$ and $N_c = 5.7 \times 10^5$ for $(^3He)_N$), above which the image potential is sufficiently strong to support a bound ground state. The binding energy $E_s(R)$ on a cluster of radius R (Fig. 10) is described by a threshold cluster size equation with the scaling law $[E_s(R) - E_s(\infty)] \propto (R - R_c)^2$, converging to the bulk value $E_s(\infty) = -0.74$ meV [178]. The huge mean radius $\langle r \rangle$ of this "halo state" diverges when $R \rightarrow R_c$ (Fig. 11). The internal electron bubble state was predicted to be realized in sufficiently large He clusters [208, 209]. The experimental genesis of this field rested on

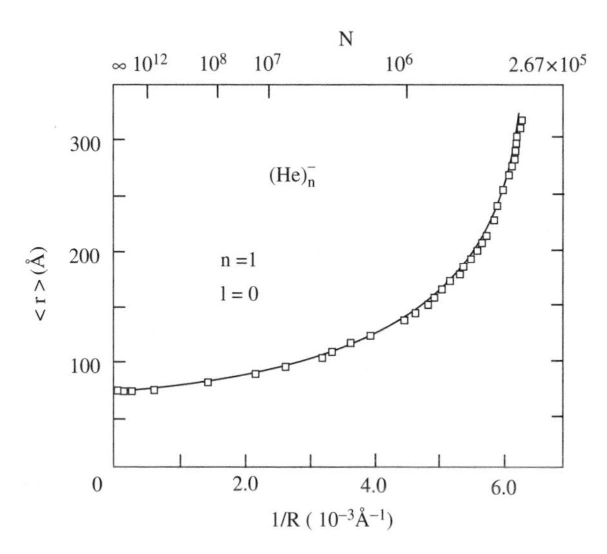

Figure 11. The mean radius $\langle r \rangle$ of the charge distribution of the "halo" ground surface state of an excess electron on $(^4\mathrm{He})_N$ clusters. $\langle r \rangle$ diverges at the localization onset $N_c = 5.7 \times 10^5$ [Ref. 178–180].

the metastable excitation of large helium clusters by electron impact, as well as on the observation of electron attachment to helium clusters [241]. Extensive experimental studies [8, 208, 209, 242] used electron capture to determine the size distributions of very large $(^4\mathrm{He})_N$ clusters with an average size of $\bar{N} = 10^5$–10^8. The significant observation [208, 209] that the negative $(\mathrm{He})_N^-$ cluster ions do not field ionize in electric fields of 10^3 V/cm on a time scale of 50 μs, seems to rule out the formation of excess electron surface states on these clusters, under current experimental conditions. On the basis of these experimental observations, it was proposed [209] that electron bombardment of $(\mathrm{He})_N$ clusters results in the formation of interior electron bubbles. Further experimental evidence for the formation of internal electron bubbles via electron attachment to large clusters ($N = 10^5$–10^8) was reported [243]. In important experiments [99, 244, 245], dramatic differences were observed for the time scale for the detachment of electrons from $(^4\mathrm{He})_N^-$ clusters at 0.37 K and from $(^3\mathrm{He})_N^-$ clusters at 0.15 K. Electron detachment from $(^4\mathrm{He})_N^-$ clusters in the size domain of $N = 10^5$–10^7 was characterized by lifetimes in the range of 10^{-2}s to 3×10^{-1}s [99, 242–246], with the cluster size dependence of these lifetimes being established [245]. These, lifetimes are shortened by the presence of heavy rare gas impurities [244]. On the other hand, considerably longer lifetimes were observed for electron detachment from $(^3\mathrm{He})_N^-$ clusters [99]. This observation was interpreted in terms of the dynamics for the motion of the

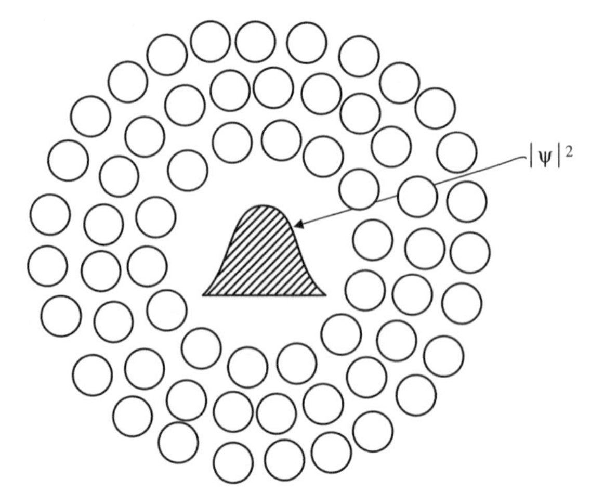

Figure 12. Artist's view of the excess electron bubble localized state in a $(He)_N$ cluster.

electron bubble in superfluid $(^4He)_N^-$ clusters, in contrast with the viscous bubble dynamics in the normal $(^3He)_N^-$ clusters [99, 243–245]. We shall now proceed to explore structure–dynamics–function relations for electron bubbles (Fig. 12) in helium clusters.

B. Bubbles in Helium Clusters

We consider a cluster of $N\,^4$He atoms of mass m and radius r_0, together with a single excess electron. The subsystem of the helium atoms will be treated by the density functional formalism [113, 247]. The excess electron will be treated quantum mechanically. The energetics and charge distribution of the electron were calculated within the framework of the adiabatic approximation for each fixed nuclear configuration.

We shall first treat the structure and energetics of an empty bubble in the center of a large neutral $(^4He)_N$ cluster $(N = 10^3–10^7)$ using a phenomenological density functional approach [247]. We express the internal energy E of the nonuniform system by a functional of the number density $n(r)$

$$E[n(r)] = \int \in (n(r))\, d^3r \tag{48}$$

where $\in (n(r))$ is the energy density of a uniform helium system, neglecting effective interactions between different portions within the cluster, which arise from nonuniformity. Thus zero-point renormalization effects in the density functional, originating from the nonuniformity of the system, are disregarded

[247]. The energy density is represented as a power series in the local density, $n(r)$, in the form

$$\in (n(r)) = A_1 n^2 + A_2 n^3 + A_3 n^4 \tag{49}$$

The coefficients $A_i (i = 1\text{–}3)$ are determined by the condition that the energy density, along with the chemical potential and compressibility from Eq. (48) in the bulk limit ($N \to \infty$), corresponds to these properties for the macroscopic liquid helium at zero temperature and pressure [20, 232, 248].

The density corresponding to the ground state of the system minimizes its total energy and can be obtained from the Euler equation

$$\delta \left\{ E[n(r)] - \mu \int n(r)\, d^3 r \right\} = 0 \tag{50}$$

where μ is the chemical potential [247]. For a spherical helium droplet of radius R we obtain from Eqs. (48)–(50)

$$\mu = -\left(\frac{\hbar^2}{2m}\right)\left(\frac{\nabla^2 n(r)^{1/2}}{n(r)^{1/2}}\right) + 2A_1 n(r) + 3A_2 n^2(r) + 4A_3 n^3(r) \tag{51}$$

This result can be expressed in a dimensionless form:

$$\nabla^2 g(x) = g(x) B(x) \tag{52}$$

where

$$B(x) = \left(a_1 g^2(x) + a_2 g^4(x) + a_3 g^6(x) - \frac{\mu}{E_v} \right) \tag{53}$$

The function $g(x)$ is defined in terms of the normalized local density:

$$g(x) = n^{1/2}(x) / n_0^{1/2}(x) \tag{53a}$$

where x is the normalized radius

$$x = \frac{r}{r_f} \tag{53b}$$

with

$$r_f = \left(\frac{\hbar^2}{2mE_v}\right)^{1/2} \tag{53c}$$

and the coefficients a_i are expressed in terms of the parameters A_i [Eq. (49)], which are given by

$$a_i = (i+1)A_i n_0^i / E_v, \qquad i = 1\text{–}3 \tag{53d}$$

n_0 is the average number density in the bulk at zero temperature and pressure and E_v is the binding energy per atom in the bulk, which was taken from the experimental data [106, 232, 248] as $E_v = 0.616$ meV. The coefficients a_i, Eq. (53d), are $a_1 = -2.2$, $a_2 = -2.4$ and $a_3 = 3.6$.

The internal cluster energy E_c and the number of atoms N in the cluster are given by

$$E_c = 4\pi \int \in (n(r)) r^2 \, dr \tag{54}$$

$$N = 4\pi \int n(r) r^2 \, dr \tag{55}$$

Equations (54) and (55) are applicable both for an ordinary cluster and for a cluster with a bubble. To characterize the density profile for the cluster with a bubble, we choose the helium atom density function in the form of a void at $r < R_b - t_1/2$, a rising profile toward a constant density with increasing r beyond the void boundary at $r > R_b - t_1/2$, and an onset of the cluster exterior decreasing density profile for $r > R - t_2/2$. Here R_b is the bubble radius, R is the cluster radius, t_1 is an effective thickness parameter for the density profile of the bubble wall, and t_2 is the thickness of the cluster surface density profile. The explicit form of the helium density profile was taken as

$$n(r) = 0, \qquad 0 < r < R_b - \frac{t_1}{2} \tag{56a}$$

$$n(r) = n_0[1 - (1 + br)\exp(-b^3 r^3)]^3, \qquad R_b - \frac{t_1}{2} < r < R - \frac{t_2}{2} \tag{56b}$$

$$n(r) = (c)\arctan\left\{\left[\sinh\left(\frac{2r}{t_2}\right)\right]^{-1}\right\}, \qquad r > R - \frac{t_2}{2} \tag{56c}$$

The parameter b in Eq. (56b) specifies spatial saturation taking $b = [R_b - t_1/2]^{-1}$. For sufficiently large clusters the density in the interior of the cluster [Eq. (56b)] converges to the bulk value n_0. The parameter c in Eq. (56c) is taken as $(c) = (2n_0/\pi)$. Equation (56b) was advanced on the basis of previous work on nonuniform ^{4}He near a hard wall [247]. Equation (56c) represents the surface density profile of the cluster with a bubble in the form of the gudermannian function [178–180].

The density functional approach used above for the energetics of the cluster was applied by us for the cluster with a bubble. It is assumed that t_1, $t_2 < R_b$ and t_1, $t_2 \ll R$, so that nonuniformity effects created by the bubble formation are small. We employed the trial function for the density [Eqs. (56a)–(56c)] and for the calculations of $g(x)$ [Eq. (53a)] to compute $B(x)$ [Eq. (53)] and then to solve Eq. (51) numerically. The new density $n(r)$ thus obtained was used to calculate $B(x)$ in a self-consistent procedure. Equations (54) and (55) were then used to calculate the cluster internal energy $E_c(R)$ and the number of particles N for the cluster with a bubble. Calculations of the cluster energy with a bubble $E_c(R_b, R, N)$ [Eq. (54)] were performed for several, fixed bubble radii R_b with a constant number N of particles. The cluster energies also depend on the density profile thicknesses t_1 and t_2, which were varied in the calculations in the range 6–10 Å. The energy of a cluster without a bubble $E_c(R_b = 0, R, N)$ was calculated for $R_b = 0$ and $t_1 = 0$, with varying the exterior density profile thickness. The cluster reorganization energy $E_d(R_b, R, N)$ of the cluster upon the formation of a bubble of radius R_b at constant N is given by

$$E_d(R_b, R, N) = E_c(R_b, R, N) - E_c(R_b = 0, R, N) \tag{57}$$

Calculations of the energetics of bubble formation over a range of cluster sizes $(N = 6.5 \times 10^3$ to $2 \times 10^5)$ were performed. Figure 13 portrays the calculated binding energies E_c/N per atom for a ^{4}He cluster without a bubble. The cluster size dependence of E_c/N per atom for ordinary $(^4\text{He})_N$ clusters in the larger size domain $(N = 6.5 \times 10^3$ to $2 \times 10^5)$ is portrayed in Fig. 13. These energies obey the cluster size equation for the LDM [51–53, 84, 106]

$$\frac{E_c}{N} = E_v + E_s\left(\frac{r_0}{R}\right) \tag{58}$$

$E_v = -0.610$ meV is the volume energy per atom and $E_s = 1.60$ meV is the surface energy per atom. These energetic parameters are in agreement with the experimental value [232] $E_v = -0.616$ meV for the atom binding energy in bulk ^{4}He and with the surface energy $E_s = 1.603$ meV inferred from previous theoretical results [106] for smaller clusters $(N = 128$–$728)$. An additional contribution to E_c/N involves the cluster curvature energy $E_u(r_0/R)^2$ with $E_u = 1.034$ meV [51–54, 106, 248]. The curvature energy term makes only a small contribution to the large clusters studied by us; that is, for $N = 6.5 \times 10^3$ the relative contributions of the curvature energy to the surface energy $(E_u/E_s)(r_0/R)$, is 3%. Our results for the larger clusters $(N = 6.5 \times 10^3$ to $2 \times 10^5)$ are in accord with previous quantum mechanical and density functional [51–54, 106, 128, 129] calculations for smaller clusters $(N = 128$–$728)$ for

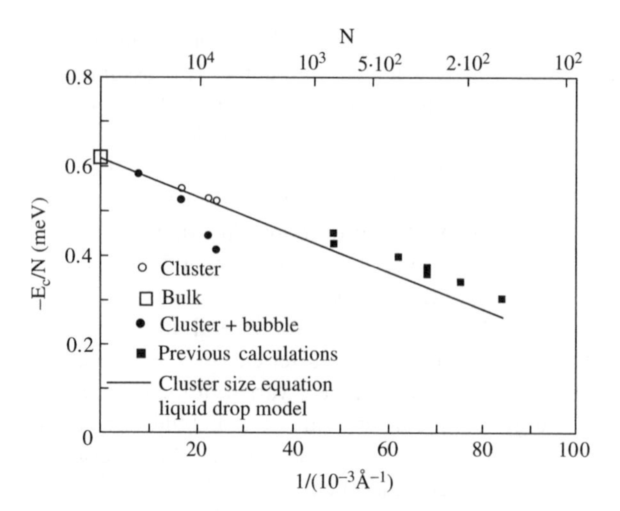

Figure 13. The cluster size dependence of the calculated binding energies per atom for a $(^4\text{He})_N$ cluster ($N = 6.5 \times 10^3$ to 1.88×10^5) of radius R without a bubble (marked as cluster) and for a cluster with a bubble at the equilibrium electron bubble radius R_b (marked as cluster + bubble). The experimental binding energy per atom in the bulk [232, 248], $E_c/N = -0.616$ meV ($R, N = \infty$), is presented (marked as bulk). Previous computational results for the lower size domain $N = 128$–728 [51–54, 106, 128, 129] are also included. The calculated data for the large ($N = 10^5 - 10^7$) clusters ($N = 6.5 \times 10^3$ to 1.88×10^5), as well as the bulk value of E_c/N without a bubble, follow a linear dependence versus $1/R$ and are represented by the liquid drop model, with the cluster size equation [Eq. (58)] (solid line). The dashed curve connecting the E_c/N data with a bubble was drawn to guide the eye. The calculated data for the smaller clusters ($N = 128$) manifest systematic positive deviations from the liquid drop model, caused by the curvature term, which was neglected.

which the positive deviation (Fig. 13) from Eq. (58) originates from the cluster curvature energy in this size domain.

In Fig. 13 we also present the energetics of the $(^4\text{He})_N$ cluster with a bubble at the equilibrium electron bubble radius, with R_b inferred (Section III.C) from the electron bubble. These results manifest the marked increase of E_c/N upon bubble formation, which is due to cluster deformation. Data were obtained on the bubble radius R_b, the cluster deformation energy per atom E_d/N [Eq. (57)], the cluster mean density n, and the cluster radius R for $(^4\text{He})_N$ clusters. These results reflect on the energetic implications (i.e., the increase of E_d/N) and on the structural manifestations (i.e., cluster expansion with increasing the bubble radius).

The density profiles for several clusters at different bubble radii are portrayed in Fig. 14. These density profiles reflect on the formation of a "helium balloon" with a finite thickness ($\delta R \simeq R - R_b$) in the cluster. The profile thicknesses for the bubble and for the cluster surface obtained from this model are $t_1, t_2 \cong 6$–10 Å.

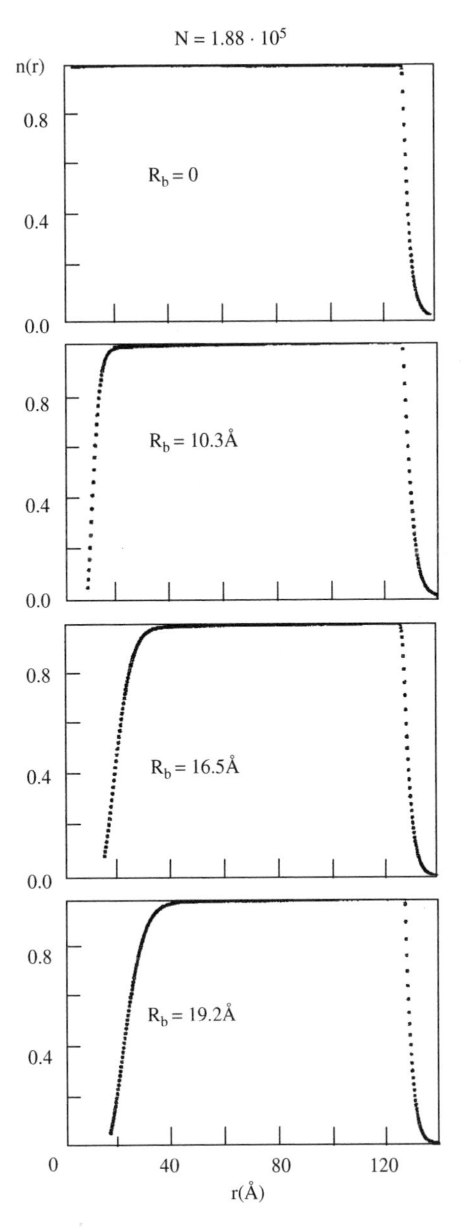

Figure 14. The density profiles at different bubble radii for $(^4\text{He})_N$ clusters with $N = 1.88 \times 10^5$ reflecting on the formation of a "helium balloon" with a finite thickness ($\Delta R = R - R_b$) in the cluster. The profile thicknesses for the bubble and for the cluster surface obtained from this simple model are t_1, $t_2 \cong$ 6–10 Å (see text). The exterior surface profile of the cluster was characterized by the 90–10% fall-off width w_2, while the interior bubble profile was characterized by a 10–90% rise width w_1. For $N = 1.88 \times 10^5$ clusters $w_1 = 6.2$ Å for $R_b = 0$ (no bubble), $w_2 = 7.8$ Å and $w_1 = 6.2$ Å for $R_b = 10.3$ Å, while $w_2 = 6.8$ Å and $w_1 = 12.3$ Å for $R_b = 19.2$ Å. These results demonstrate that the cluster surface profile width w_2 remains nearly independent of the bubble size, while w_1 increases with increasing the bubble radius.

These values of t_2 for the finite large $(^4\text{He})_N$ clusters studied herein are in accord with the results of previous calculations for macroscopic $(^4\text{He})_N$ clusters [55, 58, 59], as well as for experimental data for macroscopic liquid ^{4}He.

The energetics of the formation of a "helium balloon" (i.e., a helium cluster with a bubble at its center) reveals high reorganization energies, which for $R_b = 14.4$ Å (corresponding to the value of R_b^e for $N = 6500$) fall in the range $E_d = 0.72$ eV for $N = 6.5 \times 10^3$ to $E_d = 0.26$ eV for $N = 1.88 \times 10^5$, and increase with decreasing N at a fixed value of R_b. These E_d values increase with increasing the bubble radius R_b for clusters with a fixed value of N. It is also instructive to note that for the cluster size domain studied herein the E_d values are considerably higher than the bubble formation energy in the bulk $E_d(\infty) \simeq 4\pi\gamma R_b^2$, where γ is the surface tension. While the reorganization energy in the bulk is dominated by the bubble surface energy, the reorganization energy for bubble formation in the cluster is determined by three contributions: the interior bubble surface energy $E_b(R_b)$, the exterior cluster surface energy $E_c(R)$, and the cluster energy increase due to density changes $\Delta(n(r); N)$. All these three energy contributions are cluster size-dependent.

For a rough estimate of the surface energy contributions we shall use a step function density profile, so that $E_b(R_b) = 4\pi R_b^2\gamma$ and $E_c(R) = 8\pi\gamma R\Delta R$, where $\Delta R(\ll R)$ is the expansion of the cluster radius upon the formation of the bubble, that is, $\Delta R = [R(R_b) - R(R_b = 0)]$. Within the framework of this approximate relation, we have

$$E_d(R_b, R, N) = 4\pi R_b^2\gamma + 8\pi\gamma R\Delta R + \Delta(n(r); N) \tag{59}$$

The surface term contributions to E_d in Eq. (59) are moderately small. Thus for $N = 6.5 \times 10^3 (R = 43.7$ Å) at the equilibrium bubble radius $R_b = 14.4$ Å, we find from the complete simulations that $E_d = 0.72$ eV, while $\Delta R = 2.7$ Å. Thus $E_b(R_b) = 5.7 \times 10^{-2}$ eV and $E_c(R) = 6.5 \times 10^{-2}$ eV, with $E_b + E_c = 0.122$ eV providing a contribution of $\sim$16% to the reorganization energy. The dominating contribution to E_d [Eq. (59)] for the cluster size domain studied herein originates from the contribution of the density changes—that is, the third term in Eq. (59). With increasing the cluster size toward the bulk $(N \to \infty)$, we have $E_e(R) \to 0$ and $\Delta(n(r); N) \to 0$, with $E_d(R_b, R \to \infty, N \to \infty) \to E_b(R_b)$.

C. The Electron Bubble

We now introduce an excess electron into the bubble, which is located in the center of the helium cluster at a fixed nuclear configuration of the "helium balloon." The electronic energy of the excess electron will be calculated within the Born–Oppenheimer separability approximation. We modified the nonlocal effective potential developed by us for surface excess electron states on helium clusters [178–180] for the case of an excess electron in a bubble of radius R_b

located in a cluster of radius R. This potential $V(r)$ at distance r from the center of the bubble (and of the cluster) will be subdivided into interior and exterior contributions in the form

$$V(r) = V_{<}(r), \qquad r \leq R_b - \frac{t_1}{2} \tag{60a}$$

$$V(r) = V_{>}(r), \qquad r \geq R_b - \frac{t_1}{2} \tag{60b}$$

where the thickness density profile of the bubble wall is defined by Eq. (56) and r is the distance from the center of the cluster.

The exterior contribution $V_{>}(r)$ to the potential in Eq. (60b) is determined by the energy of the quasi-free electron in the finite system, being given by [213, 215, 222]

$$V_{>}(r) = T + V_p(r) \tag{61}$$

where the repulsive short-range contribution T is represented by the Wigner–Seitz model with a hard-core pseudo-potential with radius a, which is taken as the e–He scattering length [210, 213, 215, 233, 249, 250]. The attractive contribution V_p is given as the polarization energy of the cluster, which is induced by the electron within the Wigner–Seitz cell [212, 213, 215, 222]. The cluster polarization energy is expressed as the sum of the contribution U_p^{in} of the atom inside the Wigner–Seitz cell, the contribution U_p^{out} of the atoms outside the Wigner–Seitz cell in an infinite medium, and the correction term V_p^c to the polarization energy for the finite size of the cluster, due to the excluded volume effect.

$$V_p(r) = U_p^{\text{in}} + U_p^{\text{out}} + V_p^c(r, R) \tag{62}$$

where

$$U_p^{\text{in}} = \left(\frac{2\pi\hbar^2}{2m_e}\right)\bar{n}a_p \tag{63a}$$

$$U_p^{\text{out}} = -2\pi\left(\frac{4\pi}{3}\right)^{1/3}\alpha e^2\bar{n}^{4/3}\left(\frac{1+8\pi\bar{n}a}{3}\right)^{-1} \tag{63b}$$

and

$$V_p^c(r, R) = \left(\frac{e^2}{2R}\right)(1-\varepsilon^{-1})\sum_{j=0}^{\infty}\frac{j+1}{(\varepsilon j + j + 1)(r/R)^{2j}} \tag{63c}$$

Here a_p is the e–He scattering length due to the polarization potential, which was taken as [215] $a_p = -0.1$ Å, α is the atomic polarizability, and $\bar{n}$ is the average helium density.

The interior contribution $V_<(r)$ to the potential [Eq. (60a)] is given by the superposition of electron–atom pseudopotentials exerted on the electron by the helium atoms within the surface density profile of the bubble walls and by the electronic polarization potential $V_i(r)$ induced within the region of the bubble, which is represented in terms of a cluster image potential

$$V_<(r) = \int_{R_b - t_1/2}^{R_b + t_1/2} d^3 r' \mathrm{v}_{\mathrm{ps}}(r' - r) n(r') + V_i(r) \tag{64}$$

where v_{ps} is the electron–He-atom pseudopotential [210–213, 249, 250] and $n(r)$ is the bubble surface density profile [Eq. (56b)]. The first term in Eq. (64) is the contribution of the polarization potential from the density profile of the bubble. The second term, $V_i(r)$, is the polarization potential induced within the rest of the cluster outside the bubble, which is given by

$$V_i(r) = V_i(r, R) - V_i(r, R_b) \tag{65}$$

where $V_i(r, R)$ is the image potential for a helium cluster of radius R and $V_i(r, R_b)$ is the image potential for the cluster region occupied by a bubble. Equation (65) assumes the form

$$\begin{aligned} V_i(y) = \left(\frac{e^2}{4R}\right)\left(\frac{1}{\varepsilon - 1}\right)/(1+\varepsilon)\Bigg[\frac{2\beta}{\beta^2 - y^2} - \frac{2}{1 - y^2} \\ + \left(\frac{1}{y}\right)\left(\ln\left|\frac{\beta + y}{\beta - y}\right| - \varepsilon \ln\left|\frac{1+y}{1-y}\right|\right)\Bigg] \end{aligned} \tag{66}$$

where ε is the dielectric function (taken as that for macroscopic helium), $y = r/R$, and $\beta = R_b/R$.

The potential $V(r)$ is given by the interior contribution $V_<(r)$ [Eqs. (60a), (64–66)] and by the exterior contribution $V_>(r)$ [Eqs. (60b), (61), (63a–63c)]. To obtain the ground-state electronic energy E_e of the bound excess electron in the bubble, we solved numerically the one-electron Schrödinger equation

$$\left[-\left(\frac{\hbar^2}{2m_e}\right)\nabla^2 + V(r) - E_e\right]\psi(r) = 0 \tag{67}$$

where m_e is the electron mass. The total energy $E_t(R_b, R, N)$ of the electron bubble states in a helium cluster is expressed in the form

$$E_t(R_b, R, N) = E_e(R_b, R, N) + E_d(R_b, R, N) \tag{68}$$

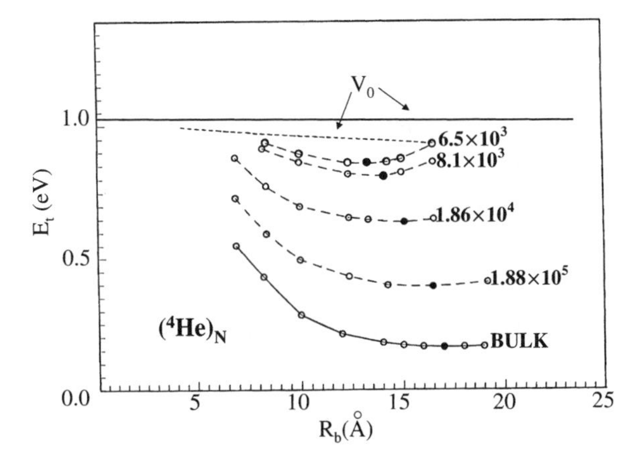

Figure 15. The potential energy surfaces for the excess electron bubble states in $(^4\text{He})_N$ clusters portraying the total energy $E_t(R_b, R, N)$ versus the bubble radius R_b for fixed values of N marked on the curves. The open and full points represent the results of the computations for the clusters using the density functional method for $E_d(R_b, R, N)$ and the quantum mechanical treatment for $E_e(R_b, R, N)$, while for the bulk we took $E_d(R_b, R \to \infty, N \to \infty) = 4\pi\gamma R^2$. The black point (•) on each configurational diagram represents the equilibrium bubble radius. The R_b-dependence of the energy of the quasi-free electron state $V_0(R_b, R, N)$ in the cluster of the smallest size of $N = 6.5 \times 10^3$ (dashed line) and the bulk value of V_0 (solid line) are also presented. The V_0 values for each R_b for $N = 8.1 \times 10^3$ to 1.88×10^5 fall between these two nearly straight lines.

where the cluster reorganization energy $E_d(R_b, R, N)$ is given by Eq. (57). The energies $E_t(R_b, R, N)$, $E_e(R_b, R, N)$ and $E_d(R_b, R, N)$ in Eq. (68) are determined by the bubble radius R_b and the cluster radius R, as well as by the density profile parameters t_1 and t_2, and by the number of atoms N. The potential energy surfaces for the excess electron bubble states in ^{4}He clusters in the ground electronic state are portrayed in Fig. 15, where we display $E_t(R_b, R, N)$ versus R_b for fixed values of N. These energetic configurational diagrams exhibit the most stable configuration at their minimal energies at $R_b = R_b^e$. The equilibrium electron bubble radii R_b^e and total energies E_t^e, corresponding to the minima of these potential curves, are summarized in Fig. 16. The equilibrium bubble radius increased from $R_b^e = 13.4\,\text{Å}$ at $N = 6.5 \times 10^3$ to $R_b^e = 16.6\,\text{Å}$ for $N = 1.88 \times 10^5$, while the total energy E_t^e at the equilibrium configuration decreases nearly linearly from $E_t^e = 0.86\,\text{eV}$ for $N = 6.5 \times 10^3$ to $E_t^e = 0.38\,\text{eV}$ for $N = 1.88 \times 10^5$ (Fig. 16). The electronic energies are $E_e = 0.160\,\text{eV}$ for $N = 6.5 \times 10^3$, $E_e = 0.126\,\text{eV}$ for $N = 1.86 \times 10^4$ and $E_e = 0.102\,\text{eV}$ for $N = 1.88 \times 10^5$. The increase of E_e with decreasing the cluster size is due to the increase of R_b^e with increasing N. To complete the presentation of the energetic parameters, we also present in Fig. 15 the R_b dependence of the energy of the quasi-free electron state $V_0(R_b^e, R, N)$ in clusters of different sizes, which were

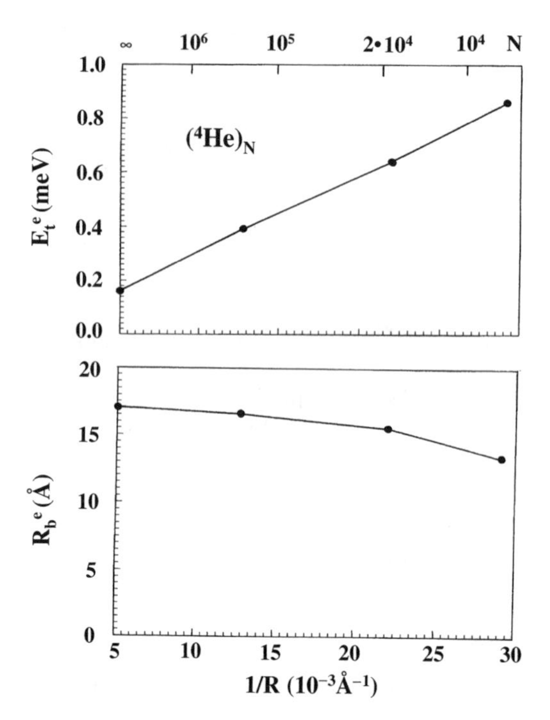

Figure 16. The cluster size dependence of the equilibrium electron bubble radii R_b^e and the total ground-state energies E_t^e, corresponding to the minima of the potential curves of Fig. 15.

obtained from the solution of Eq. (67), with the exterior potential given by Eqs. (61) and (63a–63c). These V_0 values in clusters are reduced by less than 10% relative to the bulk value. For the smallest cluster with $N = 6.5 \times 10^3$ studied herein $V_0 = 0.95\,\text{eV}$ and for $N = 1.88 \times 10^5$ we have $V_0 = 1.02\,\text{eV}$, while the bulk value is $V_0 = 1.06\,\text{eV}$ (Fig. 15). These energetic data will subsequently be utilized for the energetic stability of the electron bubble.

D. Energetic Stability of the Electron Bubble

The energy of the excess electron bubble in the ground electronic state at its equilibrium bubble radius R_b^e, with the corresponding cluster radius R^e, is determined by the contributions of the electronic energy and the cluster reorganization energy, with $E_t(R_b^e, R^e, N)$ [Eq. (68)] being positive relative to the vacuum level, while for a broad range of cluster sizes this energy is lower than the cluster conduction band energy. The equilibrium energy of an electron bubble increases with decreasing N; at some value of N it will become higher than V_0, marking the onset of the energetic instability of the electron bubble. A central question is: What is the minimal cluster size for which the electron bubble

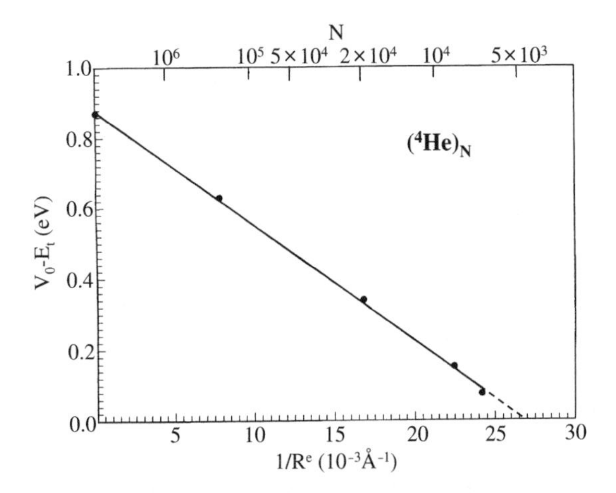

Figure 17. The dependence of the energy gap $(V_0 - E_t^e)$ between the quasi-free electron energy and the total ground-state energy at the equilibrium configuration of the electron bubble on the reciprocal value of the cluster radius at this equilibrium configuration $1/R^e$ for clusters in the range $N = 6.5 \times 10^3$ to 1.86×10^4 and for the bulk. A crude extrapolation of this linear dependence of $V_o - E_t^e$ to zero leads to a localization threshold at $R \leq 39$ Å, which corresponds to $N \cong 5 \times 10^3$.

is energetically stable? The energetic stability condition for the excess electron bubble state (i.e., an electron in a "helium balloon") is given by

$$E_t^e\left(R_b^e, R^e, N\right) \leq V_0\left(R_b^e, R^e, N\right) \tag{69}$$

In Fig. 17 we present the plot of $V_0(R_b^e, R^e, N) - E_t^e(R_b^e, R^e, N)$ versus $1/R^e$. An extrapolation of this linear dependence to $V_0 - E_t^e = 0$ (Fig. 17) results in the energetic localization threshold at the cluster equilibrium radius of $R^e \leq 39$ Å, of a cluster which contains an electron bubble. For such a cluster, the energetic localization threshold is $N = \left[(R^e)^3 - (R_b^e)^3\right]/r_0^3$, where R_b^e is the equilibrium radius of the electron bubble, which assumes the value $R_b^e \simeq 13.5$ Å in this cluster size domain. Accordingly, we estimate $N \simeq 5200$ for the minimal cluster size for which the electron bubble is energetically stable. This energetic localization threshold constitutes an upper limit for the cluster size, which allows for the existence of the electron bubble state. Dynamic effects, due to electronic tunneling of the excess electron from the bubble to the vacuum, may result in the depletion of the energetically stable excess electron bubble state on the experimental time scale for the interrogation of $(\mathrm{He})_N^-$ clusters (1–10^{-6} [243–245]). Accordingly, the dynamic stability of the excess electron bubble state in $(^4\mathrm{He})_N$ clusters on the experimental time scale may be realized only for cluster sizes that exceed those dictated by the energetic stability. We now proceed to explore the facets of dynamic stability of the excess electron bubble.

IV. ELECTRON BUBBLES AS PROBES FOR SUPERFLUIDITY IN $(^4He)_N$ CLUSTERS

It is of considerable interest to use the electron bubble as a probe for elementary excitations in finite boson quantum systems—that is, $(^4He)_N$ clusters [99, 128, 208, 209, 243–245]. These clusters are definitely liquid down to 0 K [46–49] and, on the basis of quantum path integral simulations [65, 155], were theoretically predicted (see Chapter II) to undergo a rounded-off superfluid phase transition already at surprising small cluster sizes [i.e., $N_{MIN} = 8$–70 (Table VI)], where the threshold size for superfluidity and/or Bose–Einstein condensation can be property-dependent (Section II.D). The size of the $(^4He)_N$ clusters employed in the experiments of Toennies and co-workers [242–246] and of Northby and co-workers [208, 209] (i.e., $N \simeq 10^4$–10^7) are considerably larger than N_{MIN}. In this large cluster size domain the λ point temperature depression is small [199], that is, $(T_\lambda - T_\lambda^0)/T_\lambda^0 \simeq 2 \times 10^{-2} - 2 \times 10^{-3}$ for $N = 10^4$–10^7. Thus for the current experimentally accessible temperature of 0.4 K, the large $(^4He)_N^-$ clusters ($N = 10^4$–10^7) studied by Toennies and co-workers [99, 242–246] are superfluid.

Electron tunneling dynamics from electron bubbles in helium clusters strongly depends on the transport dynamics of the electron bubble within the cluster. In normal fluid $(^4He)_N$ and $(^3He)_N$ clusters the electron bubble motion is damped, while in $(^4He)_N$ superfluid clusters this motion is nondissipative [99]. Accordingly, bubble transport dynamics in $(^4He)_N$ clusters dominates the time scale for electron tunneling from the bubble, providing a benchmark for superfluidity in finite boson systems [245, 251]. In this chapter we address (a) the dynamics of electron tunneling from bubbles in $(^4He)_N$ and $(^3He)_N$ clusters [99, 209, 242–245, 251] and (b) the role of intracluster bubble transport on the lifetime of the bubble states. Our analysis provides semiquantitative information on electron bubbles in $(^4He)_N$ clusters as microscopic nanoprobes for superfluidity in finite quantum systems, in accord with the ideas underlying the work of Toennies and co-workers [99, 242–245].

A. Dynamic Processes of Electrons in Helium Clusters

Following the injection of an excess electron into a helium cluster, a sequence of dynamic processes is realized which involve:

1. Quasi-free electron thermalization, which for electrons in the energy range of a kiloelectronvolt (keV) in macroscopic liquid He is characterized by a time scale of 0.3–0.5 ps, with a characteristic spatial range of ~50–60 Å at 1.4 K [252].
2. Localization of the quasi-free electron. The dynamics of the transition from the quasi-free electron state to the localized bubble state in the

cluster involves electron localization accompanied by large configurational dilation. The electron bubble expansion time τ_b located in the center of the cluster was estimated on the basis of our previous calculations for electron localization in bulk liquid helium [187, 188]. For the finite size of the cluster, with the bubble energy being taken as the cluster reorganization energy, we obtained explicit expressions for the lifetime τ_b for the formation of the excess electron bubble. From the continuity condition $r^2\hat{V} = R_b^2\hat{U}$, where $\hat{U}$ is the velocity of the cavity boundary and $\hat{V}$ is the local radial velocity [187, 188]. In the absence of energy dissipation [187, 188] the boundary kinetic energy equals the total change of the free energy

$$\Delta F = \left(\frac{\bar{n}}{2}\right) \int dr \frac{4\pi R_b^4 U^2}{r^2} \tag{70}$$

where $\bar{n}$ is the average cluster density. ΔF can be expressed in the form

$$\Delta F = 2\pi\bar{n}R_b^3 U^2(1 - \bar{\beta}(R_b, R)) \tag{71}$$

where $\bar{\beta} = R_b/R$. Following our previous procedure [187, 188] we obtain the bubble formation time in the cluster from the relation

$$\tau_b = (2\pi\bar{n})^{1/2} \int_{R_0}^{R_b} dr \left[\frac{r^3[1 - \bar{\beta}(r, R)]}{V_0(R_b) - E_t(r)}\right]^{1/2} \tag{72}$$

where $\bar{\beta}(r, R) = r/R$ is a correction factor for the finite cluster size, $V_0(R_b)$ is the quasi-free electron energy at the incipient cavity radius R_b exhibited at the crossing of the potential energy surfaces for the quasi-free electron state $V_0(R_b)$, and for the localized state $E_t(R_b)$ [Eq. (68)] at $R_b < R_b^e$ [187]. On the basis of Eq. (72) we estimate the following values of bubble formation times for the $(^4\mathrm{He})_N$ cluster: $N = 1.88 \times 10^5$ at $T = 0.4$ K, $\tau_b = 3.6$ ps without dissipation and $\tau_b = 7.8$ ps when medium dissipation was taken into account [188]. These dissipation effects are negligible in superfluids [188], due to the vanishing viscosity. For a $(^3\mathrm{He})_N$ cluster with the same N the relaxation time was calculated to be $\tau_b = 4.4$ ps without dissipation and $\tau_b = 9.0$ ps with dissipation.

3. Electron tunneling from the bubble.
4. The motion of the electron bubble in the field of the image potential within the cluster.

Ultrafast processes 1 and 2 will not further be considered; rather they will be used to set a temporal lower limit of $t > 10$ ps for the electron tunneling dynamics from the bubble. In what follows we shall consider the dynamics of electron tunneling in conjunction with the bubble motion in the cluster. This problem is of considerable interest, because electron tunneling is expected to be extremely sensitive to the spatial hydrodynamic motion of the bubble, providing a microscopic nanoprobe for superfluidity in ^{4}He clusters [245, 251], as experimentally demonstrated by Northby and co-workers [208, 209] and by Toennies and co-workers [99, 242–245].

B. Electron Tunneling from Bubbles in $(^4\text{He})_N$ and $(^3\text{He})_N$ Clusters

Electron tunneling rates from the ground state of electron bubbles through the surface of macroscopic liquid helium were previously calculated [239]. In what follows, an extension of these results will be provided for electron tunneling from electron bubbles of radius R_b in helium clusters of radius R, where the center of the bubble is located at a distance r from the cluster center, with the shortest distance, $d = (R - r)$, between the bubble center and the cluster surface. The tunneling process is characterized by a barrier height of $V_0 - E_e$, which is given by the energy gap between the quasi-free electron energy V_0 and the electronic energy of the ground electronic state E_e at the bubble equilibrium configuration, and by a barrier width $(X - R_b)$, where the distance X from the center of the bubbles to some point on the cluster surface is $X \geq d$. The tunneling probability $F(X)$ is approximated by the WKB expression

$$F(X) = \nu \exp(-2\alpha X) \tag{73}$$

where

$$\alpha = \left[(2m_e/\hbar^2)(V_0 - E_e)\right]^{1/2} \tag{74}$$

and the tunneling frequency is

$$\nu = \frac{(2V_0/m_e)^{1/2}}{2R_b} \tag{75}$$

The tunneling transition rate through a solid angle $d\Omega$ is $F(X(\Omega))d\Omega/4\pi$, where $X(\Omega)$ depends on the angular coordinates, while the total tunneling transition rate is

$$\phi(d) = \nu \frac{\int_\Omega F(X(\Omega))\, d\Omega}{4\pi} \tag{76}$$

The general form of the total transition rate is expected to be

$$\phi(d) = A \exp(-\beta d) \tag{77}$$

where β is the exponential parameter and A is the preexponential factor. Equation (77) manifests the common exponential distance (d)-dependence of electron tunneling processes [253].

Two exact results emerge from this analysis:

1. For electron tunneling from a bubble with a perpendicular (shortest) distance d from an infinite plane surface, Schoepe and Raydfield [239] showed that the parameters in Eq. (77) are

$$\beta = 2\alpha \tag{78a}$$

and

$$A = \left(\frac{V_0}{2m_e R_b^2}\right)^{1/2} \exp(2\alpha R_b) \frac{\exp(-1/\alpha d)}{(4\alpha d)} \tag{78b}$$

The preexponential factor exhibits a weak algebraic d-dependence.

2. For electron tunneling from a "helium balloon"—i.e., from a bubble located at the center of the cluster, where $X(\Omega) = (R - R_b)$ for all values of Ω—Eqs. (74)–(76) result in the simple form of the parameters in Eq. (77)

$$\beta = 2\alpha \tag{79a}$$

and

$$A = \left(\frac{V_0}{2m_e R_b^2}\right)^{1/2} \exp(2\alpha R_b) \tag{79b}$$

while

$$d = R \tag{79c}$$

The exponential distance dependence of the total tunneling transition rate for the two configurations, described by cases 1 and 2 above, is identical, while the preexponential factor for the bubble near a plane surface is smaller by a numerical factor of $\exp(-1/\alpha d)/4\alpha d$ than for the bubble located in the center of the cluster. For $2\alpha \simeq 1\,\text{Å}^{-1}$, inferred from Eq. (74), one estimates that $4\alpha d \gg 1$, and for the inifinite plane this reduction factor is $\sim 1/4\alpha d$. We compare electron tunneling rates from a bubble located in the center of the cluster with a bubble whose center is displaced from the cluster center, that is, $d < R$. Both cases will be characterized by the same exponential parameter $\beta = 2\alpha$. The preexponential factor $\bar{A}$ will be reduced relative to case 2, [Eq. (79b)]. Thus for the general case of tunneling from an electron bubble in a cluster we expect that

$$\beta = 2\alpha$$
$$\bar{A} = \varphi(d,R)\left(V_0/2m_e R_b^2\right)^{1/2}\exp(2\alpha R_b) \tag{80}$$

where $\varphi(d, R)$ is the correction factor for the displacement of the bubble center from the cluster center, with $\varphi(R, R) = 1$ (case 2), and $\varphi(d, \infty) = \exp(-1/\alpha d)/4\alpha d$ (case 1).

We now consider electron tunneling rates from a bubble in a cluster. The origin of the coordinate axes will be taken at the cluster center and the center of the bubble is taken at $r = R - d$ on the z axis. The distance from the center of the bubble to a point specified by the polar coordinates (R, θ, ϕ) on the cluster surface is given by

$$X(\theta) = \left[d^2 + 2(R^2 - Rd)(1 - \cos\theta)\right]^{1/2} \tag{81}$$

The tunneling rate [Eq. (76)] is

$$\phi(d,R) = \left(\frac{\nu}{2}\right)\int_0^{\pi} d\theta\,\sin\theta\,\exp[-\beta(X(\theta) - R_b)] \tag{82}$$

The integration in Eq. (82) with $X(\theta)$ given by Eq. (81) results in

$$\begin{aligned}\phi(d,R) = {} & \nu\exp(\beta R_b)[1/\beta^2(R^2 - Rd)]\{\exp(-\beta d)(\beta d + 1) \\ & - \exp[-\beta(d^2 + 4R^2 - 4Rd)^{1/2}][\beta(d^2 + 4R^2 - 4Rd)^{1/2} + 1]\}\end{aligned} \tag{83}$$

where $\beta = 2\alpha$ [Eqs. (74) and (78a)]. In the limit when the bubble approaches the cluster center, (i.e., $\delta = R - d \ll R$), we define a parameter

$$\bar{A} = 2\left[\left(\frac{R}{d}\right)^2 - \left(\frac{R}{d}\right)\right] \tag{84}$$

for the expansion near $d \sim R$, which in this limit corresponds to $\bar{A} \simeq 2(\delta/R) \ll 1$. Equation (83) assumes the form

$$\phi(d,\ R) = \nu \exp(\beta R_b)\left(\frac{1}{\beta^2 d^2 \bar{A}}\right)\{[\exp(-\beta d)(\beta d + 1)] - \exp[-\beta d(2\bar{A}+1)^{1/2}][\beta d(2\bar{A}+1)^{1/2}+1]\} \quad (85a)$$

The expansion of Eq. (85a) in powers of $\bar{A}$ results in

$$\phi(d,R) = \nu \exp(\beta R_b)\exp(-\beta R) + O(\bar{A}^2) \quad (85b)$$

which converges to the expression for electron tunneling from a bubble located at the center of the cluster [Eqs. (77), (79a)–(79c)], corresponding to case 2 above. Larger clusters ($N = 1.88 \times 10^5 - 10^7$ with $R = 127$ Å–477 Å) are of interest in the context of electron tunneling from the bubble at $d \leq 50$ Å [252], so that the shortest barrier width is $d - R_b \leq 35$ Å, which is lower than the cluster radius; that is, $(d - R_b) \ll R$ and $d/R = 0.2$–0.5. In this limit for the small d/R expansion, it will be convenient to express Eq. (83) in the alternative form

$$\phi(d,R) = \nu\exp(\beta R_b)\left\{\beta^2 R^2\left[1-\left(\frac{d}{R}\right)\right]\right\}^{-1}\left\{\exp(-\beta d)(\beta d+1)-\exp\left[-\beta R\left[\left(\frac{d^2}{R^2}\right)+4-4\left(\frac{d}{R}\right)\right]^{1/2}\right]\left[\beta R\left[\left(\frac{d^2}{R^2}\right)+4-4\left(\frac{d}{R}\right)\right]^{1/2}+1\right]\right\} \quad (86)$$

For the range of $d/R \leq 0.5$ and $\beta d \gg 1$, Eq. (86) reduces to

$$\phi(d,R) = \exp(-\beta d)\ \exp(\beta R_b)\left(\frac{1}{\beta d}\right)\left(\frac{d}{R}\right)^2\Big/\left(1-\frac{d}{R}\right) \quad (86a)$$

The correction factor, $\varphi(R,\ d)$ [Eq. (80)] for the preexponential factor in the tunneling process accompanying the displacement of the bubble center from the cluster center toward the cluster surface—that is, $d \sim (0.1$–$0.5)R$—is

$$\varphi(d,\ R) = \left(\frac{1}{\beta d}\right)\left(\frac{d}{R}\right)^2\Big/\left(1-\frac{d}{R}\right) \quad (86b)$$

From this analysis, two conclusions emerge. First, the displacement of the center of the bubble results in a reduction of the preexponential frequency factor by a numerical factor given by Eq. (86). Taking $\beta \simeq 1\ \text{Å}^{-1}$ and $R = 127$ Å, $\varphi(d,\ R) = 3.4 \times 10^{-3}$. So, moving the bubble center to the cluster center will increase the preexponential factor in the tunneling probability by 2–3 orders of magnitude. Concurrently, in view of the exponential decrease of $\phi(d, R)$ with increasing d [Eq. (77)], the tunneling transition rate will exponentially decrease, overwhelming the weaker algebraic-type increase of the preexponential factor. Consequently, $\phi(d, R)$ will decrease by moving the bubble to the bubble center. Second, for tunneling from an electron bubble in a cluster, the correction factor $\varphi(d, R)$ is smaller by a numerical factor of $(d/R)^2/[1-(d/R)]$, relative to that for tunneling from a bubble located near a flat surface at the same value of d. This effect on the preexponential factor in the tunneling probability is modest, because the major contribution to the variations of $\phi(d, R)$ originates from its exponential distance dependence. Accordingly, the contribution of the cluster surface curvature to the transition rate (at a fixed value of d) is not large and does not exceed one order of magnitude.

C. Electron Tunneling Times

The electron tunneling times τ from a bubble located at the radial distance d in a fixed spatial configuration within the cluster

$$\tau = \frac{1}{\phi(d, R)} \tag{87}$$

were calculated from Eqs. (77), (80) and (86) for the range $d = 33$–60 Å in clusters of sizes $N = 1.86 \times 10^4$ ($R = 58.1$ Å) and 1.88×10^5 ($R = 127$ Å), and $N = 10^7$ ($R = 488$ Å). At a fixed value of d the tunneling time exhibits a weak cluster size dependence; that is, for $d = 39$ Å we found that τ, [Eq. (87)] increases by a numerical factor of 4 between the cluster sizes $N = 1.86 \times 10^4$ and $N = 10^7$. This weak variation of τ originates mainly from the dependence of the exponential parameter $\beta = 2\alpha$ [Eqs. (74), (80) and (86)] on the energy gap $(V_0 - E_t)$, which slightly decreases with increasing the cluster size (Section III.C). The electron tunneling lifetimes are very sensitive to the shortest radial distance d of the bubble center from the cluster boundary, exhibiting an exponential distance dependence (Fig. 18). The phenomenological description of the distance dependence of the rate of the tunneling probability [Eq. (77)] for $N = 1.88 \times 10^5$ (Fig. 18) results in the exponential parameter $\beta = 1.01$ Å^{-1}, which was calculated from Eqs. (74) and (80), with $V_0 - E_e = 0.92$ eV (Section III.C). The preexponential factor was estimated to be $A = 4.1 \times 10^{20}$ s^{-1}.

The relevant spatial range of the d values for the initial location of the bubble can be inferred from the characteristic spatial range $L \simeq 50$ Å for thermalization of electrons in macroscopic liquid helium [252], which constitutes an upper limit for d. The lower limit for d is due to the experimental limitations on the time scale for electron detection of the He_N^- ions [104, 244, 245], which fall in the range of $t > 10^{-6}$ s. The corresponding distance for electron tunneling of

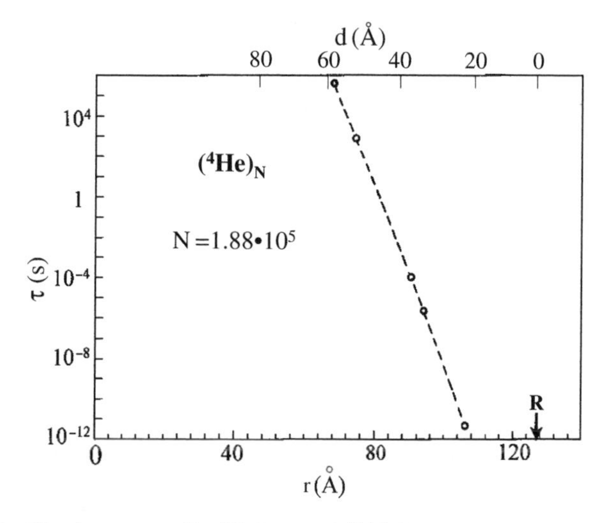

Figure 18. The electron tunneling lifetime $\tau = 1/\phi(d)$ from the electron bubble whose center is located at distance $d = R - r$ from the cluster surface for $N = 1.88 \times 10^5$ and $R = 127$ Å. The electron tunneling times from the bubble exhibit an exponential distance dependence [Eq. (77)].

$\tau = 10^{-6}$ s inferred from Fig. 18 is $d_{\text{MIN}} = 29$ Å. We thus expect the d values for the interrogation of tunneling from electron bubbles to fall in the physically acceptable region $d_{\text{MIN}} = 29\,\text{Å} \leq d \leq L \simeq 50\,\text{Å}$. The exponential distance dependence of the electron tunneling times from the bubble implies that very different physical situations will be encountered for dissipative bubble motion in normal $(^4\text{He})_N$ and $(^3\text{He})_N$ clusters and in the absence of dissipative bubble motion in superfluid $(^4\text{He})_N$ clusters.

D. Motion of the Electron Bubble in the Image Potential in Superfluuid and Normal Fluid $(\text{He})_N$ Clusters

The electron bubble motion within the cluster is described to occur in the image potential well $V_{\text{IM}}(r)$ [209], which is given from Eq. (66) in the form

$$V_{\text{IM}}(d;R) = V_i(r) - V_i(0)$$
$$= \left(\frac{2\Sigma}{R}\right)\left\{\left(1 - \left(\frac{r}{R}\right)^2\right)^{-1} + \left(\frac{R}{2r}\right) \ln\left[(R+r)/(R-r)\right] - 1\right\} \qquad (88)$$

where $r = (R - d)$ and $\Sigma = e^2(\varepsilon - 1)/4\varepsilon(\varepsilon + 1)$. For the electron bubble located in the center of the cluster (i.e., $d = R$), we have $V_{\text{IM}}(d = R; R) = 0$.

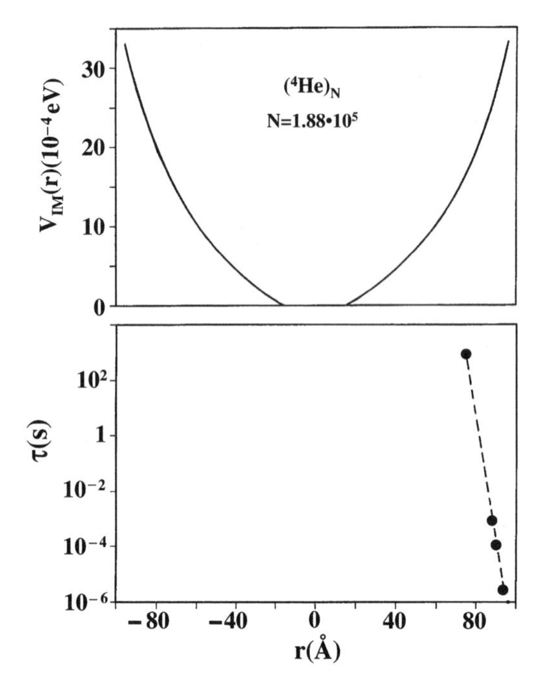

Figure 19. The image potential $V_{\mathrm{IM}}(r)$ [Eq. (88)] for the motion of the electron bubble in an $(^4\mathrm{He})_N$ cluster with $N = 1.88 \times 10^5$ and $R = 127$ Å (upper panel). The lower panel shows the exponential dependence of the electron tunneling times [Eq. (87)] near the cluster boundary versus the distance $r = R - d$ of the centers of the bubble and of the cluster. The τ data calculated from Eqs. (80) and (86) are presented in the range $\tau = 10^{-2}$–10^{-6} s.

In Fig. 19 we portray the image potential dependence for a $(^4\mathrm{He})_N$ cluster with $N = 1.88 \times 10^5$ on the initial distance r from the cluster center. Approximating the image potential by a one-dimensional harmonic potential $V_{\mathrm{IM}}(r, R) \simeq kr^2/2$, we estimate that $k \simeq 5 \times 10^{-3}$ erg cm^{-2}. At the localization length $d = R - r$, where the electron bubble is initially formed, $\phi(R, d)$ exhibits an exponential d-dependence. In Fig. 19 we also include the electron tunneling times at these distances. In our treatment it will be implicitly assumed that the ground-state electronic energy, the cluster deformation energy accompanying bubble formation, and the equilibrium bubble radius exhibit a weak dependence on d. Then the energetic changes accompanying the displacement of the bubble center within the cluster solely result from the image potential. Following initial localization the electron bubble will move in the image potential well. Two limiting cases involving distinct modes of the motion of the electron bubble, which is dissipative in normal $(^4\mathrm{He})_N$ and in $(^3\mathrm{He})_N$ clusters and nondissipative in superfluid $(^4\mathrm{He})_N$ clusters, will now be considered.

In normal $(^4He)_N$ clusters at temperature $T > T_\lambda$ above the λ point, as well as in $(^3He)_N$ clusters at $T > 0.03\,K$, the bubble motion within the image potential well is dissipative. On the other hand, the bubble motion in superfluid $(^4He)_N$ clusters will be nearly undamped. The bubble translational dynamics at distance r from the cluster center is described by an equation of motion, which for the one-dimensional case is

$$M_b\ddot{r} + \frac{dV_{\rm IM}(r)}{dr} + 4\pi\eta R_b\dot{r} = 0 \tag{89}$$

where M_b is the effective mass of the electron bubble, which involves the first external layer of the atoms being taken as $M_b \simeq 200$ m. k is the force constant of the image potential [Eq. (88)], which is approximated as a harmonic potential with a restoring force constant k, and ν is the viscosity of the fluid, which varies dramatically between the normal fluid ($\eta = 200$ μpoise) [248] and the superfluid ($\eta = 10^{-9}$–10^{-11} μpoise) [254]. The second term on the left-hand side of Eq. (89) represents the restoring force $\partial V_1(r)/\partial r = kr$. The third term on the left-hand side of Eq. (89) represents the drag force. In the normal fluid the drag force dominates over the restoring force (i.e., $4\pi\eta R_b\dot{r} \gg kr$) and the time evolution is

$$r(t) = \exp\left(-\frac{t}{\tau_D}\right) \tag{89a}$$

with the characteristic damping time for the bubble motion, inferred from Stokes' law [99], being

$$\tau_D = \frac{M_b}{4\pi\eta R_b} \tag{89b}$$

A rough estimate for the normal fluid results in $\tau_D = 4 \times 10^{-12}$ s [99]. The bubble oscillation time τ_0 in the well of the image potential can be inferred from Eq. (87), with the third term on the left-hand side of this equation being neglected, resulting in

$$R(t) \propto \exp\left(\frac{it}{\tau_0}\right) \tag{89c}$$

with the oscillation time being

$$\tau_0 = \left(\frac{M_b}{k}\right)^{1/2} \tag{89d}$$

Using $k = 5 \times 10^{-3}\ \text{erg cm}^{-2}$, estimated from Eq. (88), results in $\tau_0 \simeq 5 \times 10^{-10}$ s. In the normal fluid $\tau_D \ll \tau_0$ and the motion of the bubble will be overdamped, being described by Eq. (89a). The bubble will then relax to the cluster center on the time scale of τ_D, which is short relative to the electron tunneling time from the normal fluid cluster. Accordingly, electron tunneling from bubbles in normal liquid clusters of helium will occur from a thermally equilibrated position of the bubble.

For the bubble motion in the image potential with the superfluid, one infers that for the experimentally relevant temperature of 0.4 K [6–11, 99, 242–245] only the superfluid component is expected to prevail in the cluster size domain ($N > 10^3$), which is of interest to us. The drag force acting on the electron bubble in superfluid $(^4\text{He})_N$ clusters is expected to be vanishingly small due to the negligible viscosity (i.e., $\eta < 10^{-9}$–10^{-11} micropoise [254]), whereupon the damping time [Eq. (89b)] is $\tau_D \simeq 10$ s. τ_D is larger by 10 orders of magnitude than the bubble oscillation time [Eq. (89d)], $\tau_0 = 5 \times 10^{-10}$ s. The exceedingly long damping time marks the negligible dissipation motion of the bubble, which will undergo oscillatory motion. As $\tau_0 \ll \tau_D$, the separation of time scales between fast oscillation and ultralong damping enables us to consider electron tunneling from the bubble, which oscillates within the cluster.

E. Ultraslow Electron Tunneling Rates from Normal Fluid Helium Clusters

From the foregoing analysis in Section IV.D we concluded that in normal fluid $(^4\text{He})_N$ or $(^3\text{He})_N$ clusters the electron bubble will relax to the cluster center on the time scale of the damping time $\tau_D \sim 4 \times 10^{-12}$ s. Furthermore, when the electron tunneling times are longer than the damping times (i.e., $\tau > \tau_D$), which is realized for the initial bubble distance of $d > 26$ Å (see Figs. 18 and 19), the bubble relaxes to the minimum image potential being located in the vicinity of $r = 0$, without electron escape during the translational relaxation. This initial distance of $d = 26$ Å for the attainment of configurational damping to the cluster center is shorter than the lower limit $d_{\text{MIN}} = 29$ Å for the physically acceptable d region for experimental interrogation of the electron bubble [99] and for the upper limit of $L \simeq 50$ Å (Section IV.C). For the physically acceptable d domain (Section IV.C) the motion of the electron bubble in the normal cluster fluid is overdamped toward the equilibrium configuration in the center of the cluster, and electron tunneling from the centrally located electron bubble, (i.e., from an electron in the "helium balloon") will prevail. The electron bubble configurational distribution $p_s(r)$ in the image potential within the cluster prior to electron tunneling was taken in the form of an equilibrium Boltzmann distribution in the absence of tunneling, which is described by the static approximation [208, 209]

$$p_s(r) = \left(\frac{1}{2}\right)\exp\left(-\frac{V_{\text{IM}}(r)}{k_B T}\right) \tag{90}$$

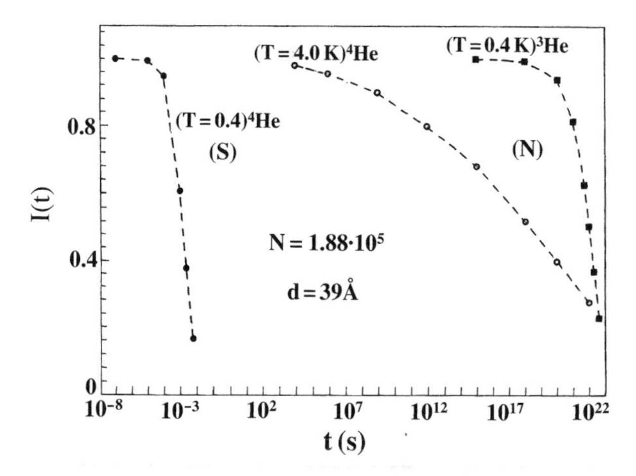

Figure 20. The dependence of $I(t)$ on t [Eq. (91)] for normal $(^3\text{He})_N$ ($T = 0.4$ K) and normal $(^4\text{He})_N$ ($T = 4.0$ K) clusters ($N = 1.88 \times 10^5$). These data are compared with the $I(t)$ dependence on t [Eqs. (92) and (93)] for a superfluid $(^4\text{He})_N$ ($N = 1.88 \times 10^5$) cluster at $T = 0.4$ K. All calculations for $d = 39$ Å. (N) represents normal fluid clusters while (S) represents a superfluid cluster.

The time-dependent electron current function $I(t)$ is given by

$$I(\text{t}) = \frac{\int_0^R p_s(r)\exp(-\phi(r)t)\,dr}{\int_0^R p_s(r)\,dr} \tag{91}$$

where $\phi(r)$ is given by Eqs. (80) and (86) with $r = R - d$. In Fig. 20 we portray typical $I(t)$ versus t curves calculated from Eqs. (90) and (91) for normal fluid $(^4\text{He})_N$ clusters at $T = 2.5$ K and 4.0 K and for normal fluid $(^3\text{He})_N$ clusters at $T = 0.4$ K, 1.25 K and 4.0 K, as well as for superfluid $(^4\text{He})_N$ clusters at $T = 0.4$ K (Section IV.F). The electron tunneling lifetime τ_{TUN} was estimated from $I(\tau_{\text{TUN}})/I(0) = 1/e$. The electron tunneling lifetimes from the normal fluid $(^4\text{He})_N$ and $(^3\text{He})_N$ clusters involve electron escape from a bubble which is overdamped toward an equilibrium configuration in the center of the cluster. The $I(t)$ versus t curves are exponential for $(^3\text{He})_N$ clusters at $T = 0.4$ K, while for higher temperatures of $T = 3$–4 K the $I(t)$ versus t curves are nonexponential, being of the form of stretched exponentials. The marked temperature dependence of the electron current function arises from the contribution of the equilibrium Boltzmann distribution [Eq. (90)]. A central result of this analysis involves the extremely long values of the electron tunneling times in normal $(^4\text{He})_N$ (at $T > T_\lambda$) clusters with $\tau_{\text{TUN}} = 10^{22}-10^{23}$ s and in $(^3\text{He})_N$ clusters with $\tau_{\text{TUN}} = 10^{20}-10^{22}$ s (Table VII). These dramatic results for extremely long

TABLE VII
Calculated Tunneling Lifetimes τ_{TUN} of an Electron from an Electron Bubble in Normal Fluid $(^4He)_N$ Clusters (at $T = 2.5$–4 K), in $(^3He)_N$ Clusters (at $T = 0.4$–2.5 K), and in Superfluid $(^4He)_N$ Clusters (at $T = 0.4$ K)[a]

Cluster	N	T (K)	τ_{TUN} (s) $d = 39$ Å	τ_{TUN}(s) $d = 38$ Å	Quantum State
$(^4He)_N$	10^7	0.4	0.26[c]	0.090[c]	Superfluid
$(^4He)_N$	10^6	0.4	0.040[c]	0.018[c]	Superfluid
$(^4He)_N$	1.88×10^5	0.4	0.018[c]	0.006[c]	Superfluid
$(He)_N$	1.88×10^5	0.4	0.008[c]		Superfluid
$(^4He)_N$	1.88×10^5	4.0	10^{22}[b]		Normal
$(^4He)_N$	1.88×10^5	2.5	10^{23}[b]		Normal
$(^3He)_N$	1.88×10^5	2.5	10^{20}[b]		Normal
$(^3He)_N$	1.88×10^5	1.3	10^{20}[b]		Normal
$(^3He)_N$	1.88×10^5	0.4	10^{22}[b]		Normal

[a]The electron bubble's structural dissipation lifetime in the image potential is $\tau_D = 4 \times 10^{-12}$ s for the normal fluid and $\tau_D \sim 10$ s for the superfluid cluster. The initial position of the bubble is $d = 39$ Å.
[b] τ_{TUN} defined by the relation $I(\tau_{TUN})/I(0) = e^{-1}$.
[c]τ_{TUN} from Eq. (94).

electron escape times from normal helium clusters are in qualitative agreement with the previous analysis of the static model [208, 209] and with recent work [245]. These ultralong lifetimes for electron tunneling from a centrally located electron bubble in normal $(^4He)_N$ and $(^3He)_N$ clusters approach the lifetime of the universe. Of course, these estimates of the ultralong tunneling lifetimes from a configurationally damped electron bubble in the normal fluid clusters are grossly oversimplified as other dynamic processes will result in the annihilation of the negative helium cluster on such unphysical time scales. The results of these lifetime calculations constitute proof that electron tunneling from a bubble in a normal fluid cluster is so long that it is not amenable to experimental observation under any real-life conditions.

F. Fast Electron Tunneling Rates from Superfluid Helium Clusters

The negligible dissipation effects for the bubble motion in superfluid clusters, which is characterized by exceedingly long damping times of $\tau_D \simeq 10$ s (Section IV.D), induce a free oscillatory bubble motion in the image potential with a short oscillation lifetime of $\tau_0 \sim 5 \times 10^{-10}$ s. Furthermore, the long damping time τ_D is comparable to the electron tunneling times (Fig. 18) only for the distance of $d \simeq 48$ Å, which is exceedingly close to the upper limit $L \simeq 50$ Å for the physically acceptable d region introduced in Section IV.C. Electron bubbles initially produced in this physically acceptable region in the superfluid cluster will manifest tunneling during the oscillatory translational motion in the image potential, without damping of the bubble motion. This physical picture was advanced by Farnik and Toennies [245] and by the present authors [251]. We shall now provide model calculations for this interesting problem.

For the motion of an electron bubble within the superfluid cluster the static, harmonic approximation (Section IV.E) breaks down. Consider an electron bubble initially located at the radial distance d from the cluster boundary. The dynamic spatial distribution $p_f(r)$ of the electron bubble in the image potential V_{IM} [Eq. (88)] in the absence of dissipation can be described by the probability of the bubble location at distance $-d \leq r \leq d$ from the cluster surface, where the bubble moves back and forth from $-d$ up to d in the image potential.

The dynamic spatial distribution (per unit length) of the electron bubble

$$p_f(r) = \left|\frac{Y}{v(r)}\right| \tag{92}$$

where

$$\mathrm{v}(r) = \left\{\frac{2[E - V_{\mathrm{IM}}(r)]}{M_B}\right\}^{1/2} \tag{92a}$$

and

$$Y = \int dr \left\{\frac{2[E - V_{\mathrm{IM}}(r)]}{M_B}\right\}^{-1/2} \tag{92b}$$

E is the initial potential energy in the image potential, $\mathrm{v}(r)$ is the velocity of the bubble motion with its center at distance r from the cluster center, Y is the period of the bubble oscillation, and $V_{\mathrm{IM}}(r)$ is the image potential, [Eq. (88)]. In this case the electron current $I(t)$ was calculated in the form

$$I(t) = \int_{-d}^{d} p_f(r) \exp(-\phi(r)t)\, dr \tag{93}$$

where again $\phi(r)$ is given by Eqs. (77), (78a), (80) and (86). Moreover, the major contribution to the electron escape is from the narrow range $d \ldots (d + \Delta d)$, where $\Delta d/d \ll 1$. On the basis of the exponential distance dependence of $\phi(d)$ [Eq. (77)] we (arbitrarily but physically) choose Δd from the reduction of ϕ by two orders of magnitude—that is, $\phi(d + \Delta d)/\phi(d) = 0.01$—so that $\Delta d = 4.4/\beta \approx 4.4$ Å. We then calculate the integral [Eq. (90)] in the limits $d \ldots d + \Delta d$

$$I(t) = \int_{d}^{d+\Delta d} p_f(r) \exp(-\phi(r)t)\, dr \tag{93a}$$

In Fig. 20 we portray the results of model calculations for the time dependence of $I(t)$, based on Eqs. (86), (92) and (93), for superfluid $(^4\text{He})_N$ clusters with $N = 1.88 \times 10^5$ at $T = 0.4$ K. These calculations were performed for $d = 39$ Å, which falls well in the physically acceptable region of d. This $I(t)$ versus t curve exhibits a near exponential decay with time, with the characteristic lifetime τ_{TUN} for electron tunneling. Using the simple relation $I(\tau_{\text{TUN}})/I(0) = e^{-1}$ we obtained from Fig. 20 that $\tau_{\text{TUN}} \simeq 8 \times 10^{-3}$s for $N = 1.88 \times 10^5$, which is included in Table VII.

In Fig. 20 we compare the current decay curve $I(t)$ versus t for the $(^4\text{He})_N$ superfluid cluster ($N = 1.88 \times 10^5$) with $I(t)$ versus t curves for normal fluid clusters of the same size, i.e., $(^3\text{He})_N$ at $t = 0.4$ K and $(^4\text{He})_N$ at $t = 4.0$ K, calculated by the procedure of Section IV.E, where the corresponding decay curves manifest astronomical decay times (Table VII). The huge 22–24 orders of magnitude difference between the electron tunneling times from electron bubbles in the superfluid clusters and in the normal fluid clusters (Fig. 20 and Table VII) demonstrate the role of electron bubbles to interrogate unique differences in the superfluid and normal fluid viscosity, which are manifested in the damping time for the bubble motion, as also noted by Farnik and Toennies [245]. The ultralong tunneling times from electron bubbles predicted herein seem to be consistent with the experimental results of Toennies and co-workers [99, 243–245], who did not detect electron emission from electron bubbles in large normal fluid $(^3\text{He})_N$ clusters. One concludes that detachment of electron bubbles cannot be observed from normal fluid large clusters and is amenable to experimental observation only for superfluid large clusters, in accord with the analysis of Toennies and co-workers [99, 243–245].

Further calculations of τ_{TUN} utilized the more elaborate expression for the tunneling lifetimes

$$\tau_{\text{TUN}} = \left| \frac{I(t)}{(dI(t)/dt)} \right| \quad \text{at} \quad \frac{I(t)}{I(0)} = \frac{1}{e} \tag{94}$$

and provided information on the cluster size dependence of τ_{TUN} in superfluid $(^4\text{He})_N$ clusters, for $d = 38$–39 Å, which are presented in Table VII. The tunneling lifetimes were calculated in the range of $d = 38$–39 Å and for a cluster size domain of $N = 2 \times 10^5$–10^7.

In Fig. 21 we portray the calculated tunneling lifetimes for cluster sizes $N = 2 \times 10^5, 10^6$, and 10^7 over a broad range of the initial distances d. The exponential d dependence of τ_{TUN}, Eq. (77), causes a lengthening of the tunneling lifetime τ_{TUN} with increasing d. τ_{TUN} increases by a numerical factor of 2–3 from $d = 38$ Å to $d = 39$ Å (Table VII).

Most important is the cluster size dependence of the electron tunneling times. The calculations of the dependence of τ_{TUN} on the cluster radius at fixed values

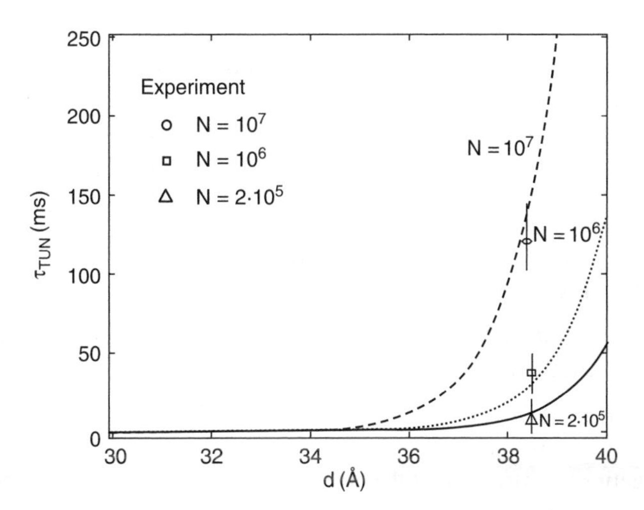

Figure 21. The lifetimes for electron tunneling from the electron bubble versus the initial value of d for superfluid $(^4\text{He})_N$ clusters at $T = 0.4$ K for cluster sizes $N = 1.9 \times 10^5$, 10^6, and 10^7. τ_{TUN} is calculated from Eqs. (77), (80), and (86). These theoretical results for times of electron detachment from superfluid $(^4\text{He})_N^-$ clusters are compared with the experimental data [245], which are marked on the curves. Good agreement between theory and experiment is achieved for $d = 38.5$ Å.

of d (Table VII and Fig. 21) exhibit an increase of the tunneling lifetimes by a numerical factor of $\sim$15 for the cluster size domain $N = 2 \times 10^4$–10^7. This result reflects on the role of the surface curvature of the cluster on τ_{TUN}, which shortens the lifetime in smaller clusters. The electron tunneling times $\tau_{\text{TUN}} = 6 \times 10^{-3}$ s–0.26 s in the range $d = 38$–39 Å calculated from the electron tunneling model advanced herein, are in very good agreement with the experimental results of Farnik and Toennies [245] on electron detachment lifetimes from $(^4\text{He})_N^-$ clusters in the size domain $N = 10^5$–10^7. The tunneling times calculated from $d = 38.5$ Å (i.e., $\tau_{\text{TUN}} = 11$ ms for $N = 2 \times 10^5$, $\tau_{\text{TUN}} = 32$ ms for $N = 10^6$ and $\tau_{\text{TUN}} = 155$ ms for $N = 10^7$) are in good agreement with the experimental results [245], as is apparent from Fig. 21. The agreement between our theory [251], as presented herein, and the experimental reality [245] inspires confidence in the utilization of electron bubbles as a probe for superfluidity of $(^4\text{He})_N$ clusters.

Of course, our model calculations of τ_{TUN} for a superfluid cluster use the initial distance d as a fitting parameter. Further refinements and extensions of the theory of electron tunneling dynamics in superfluid clusters will be of interest. These will involve the following amendments and additions: (i) a treatment of small local changes in energetics, structure, and charge distribution of the electron bubble, when it is dislocated from the cluster center; (ii) examination of the implications of an inhomogeneous distribution of the initial bubble distance d on the current decay curves; (iii) the exploration of the implications of the intermediate thermal relaxation time falling between the slow limit and the fast limit in $(^4He)_N$ clusters below the lambda point where the superfluid fraction is finite but lower than unity, so the viscosity assumes a mean value between the superfluid and normal fluid properties. These refinements and extensions of the theory will be of considerable interest, however, they will not change the general conclusions emerging from the present work.

G. Superfluidity Effects on the Formation Dynamics and on Electron Tunneling of Electron Bubbles in $(^4He)_N$ Clusters

We explored the energetics, formation dynamics and dynamic instability of electron bubbles in large $(^4He)_N$ and $(^3He)_N$ clusters ($N = 6.5 \times 10^3$–10^7). The energetics and structure of the electron bubble (Section III), which pertain to the deformation energy for the bubble formation, the ground-state energy of the localized excess electron, the total energy, and the equilibrium nuclear configuration, are insensitive to the properties of the superfluid, being nearly identical for $(^4He)_N$ normal fluid clusters above the lambda point ($T > T_\lambda$), for normal fluid $(^3He)_N$ clusters and for $(^4He)_N$ superfluid clusters (at $T < T_\lambda$). The localization dynamics of the quasi-free electron state to the electron bubble state in a $(^4He)_N$ cluster (Section IV.D), which corresponds to intra-cluster ultrafast dynamics on the time scale of nuclear motion, exhibits weak effects of superfluidity on the lifetime τ_b for the formation of the equilibrium electron bubble configuration. These superfluidity effects originate from medium dissipation accompanying the electron bubble expansion and depend on the medium viscosity [188], which is drastically different for the normal fluid cluster and for the superfluid cluster [248, 254]. When the dissipation effect is taken into account for $N = 1.88 \times 10^5$ clusters at 0.4 K, then for the $(^4He)_N$ superfluid cluster we have $\tau_b = 9.0$ ps, with a 15% increase of τ_b originating from the reduction of dissipation effects in the latter case. The physical situation regarding superfluidity effects on the electron bubble dynamics is drastically different for electron tunneling from bubbles in $(^4He)_N$ and $(^3He)_N$ clusters. Electron tunneling from bubbles is grossly affected by the distinct mode of motion of the electron bubble (Section IV.F), which is dissipative in normal fluid $(^4He)_N$ and $(^3He)_N$ clusters and nondissipative in superfluid $(^4He)_N$ clusters. The quantitative distinction in the motional damping times of the electron bubble (i.e., $\tau_D \simeq 4 \times 10^{-12}$ s for the normal fluid cluster and $\tau_D \simeq 10$ s for the superfluid cluster) will induce electron tunneling from the electron bubble located in the vicinity of the center of the normal fluid cluster and from the electron bubble located near the bubble boundary of the superfluid cluster. These distinct locations of the bubble during electron tunneling will result in ultralong tunneling lifetimes ($\tau_{TUN} \sim 10^{20}$ s) for normal fluid clusters and short lifetimes ($\tau \sim 10^{-2}$–1 s) for superfluid clusters.

These experimental results for electron detachment from $(He)_N^-$ superfluid clusters [99, 242–245] and the present analysis reflect beautifully on the role of electron bubbles as microscopic probes for superfluidity of finite boson quantum clusters. The classical 1960 studies of Meyer and Reif [207] provided direct information on the roton energy from the interrogation of the temperature dependence of the electron mobility in bulk superfluid helium. Our analysis and the experimental results [242–245] enable the interrogation and theoretical exploration of the electron bubble translational motion in the image potential within normal fluid and superfluid clusters, allowing us to infer on the dramatic effects of superfluidity in large finite boson quantum clusters using the techniques of electron detachment.

H. Dynamic Stability of the Electron Bubbles

In the exploration of the dynamics of electron tunneling from bubbles in superfluid and normal fluid $(^4He)_N$ and $(^3He)_N$ clusters, we focused on large clusters with $N = 10^5$–10^7. An interesting question in the realm of quantum size effects pertains to threshold size effects (Section I.D)—for example: What is the minimal helium cluster size to support an excess electron bubble? From the analysis of the energetic stability of the electron bubble in $(He)_N$ clusters (Section III), we concluded that the minimal cluster for which the electron bubble is energetically stable corresponds to $N \simeq 5200$. We have already pointed out that dynamic effects involving electron tunneling of the electron bubble may result in the depletion of the energetically stable electron bubble state on the experimental time scale for the interrogation of $(He)_N^-$ clusters ($t_{EXP} \simeq 1 - 10^{-6}$ s) [99, 242–245]. Accordingly, the dynamic stability criterion is governed by the experimental conditions for the detachment of an excess electron from $(He)_N^-$ clusters.

The dynamic stability criterion is governed by the condition $\tau_{TUN} > t_{EXP}$, where t_{EXP} is the lowest limit for the time scale of experimental detection of $(He)_N^-$ ions [42–45]. Taking $t_{EXP} \simeq 10^{-6}$ s, we infer that dynamic stability will prevail for $\tau_{TUN} > 10^{-6}$ s. We are concerned here with electron tunneling from moderately small $(He)_N$ clusters. Therefore we shall consider electron tunneling from a centrally located bubble (a "helium balloon") in these moderately small clusters. Using Eqs. (77), (79a), (79b), and (79c) to calculate $\phi(R, R)$, we can estimate $\tau_{TUN} = 1/\phi(R, R)$ for this case. τ_{TUN} can then be expressed by the relation $\tau_{TUN} = A^{-1}\exp(\beta d)$, with $\beta = 1\,\text{Å}^{-1}$, $A^{-1} = 1.7 \times 10^{-24}$ s, and $d = R$. The minimal cluster radius for dynamic localization is inferred from the relation $A^{-1}\exp(\beta d) > 10^{-6}$ s, resulting in $d = R \geq 41$ Å. For this value of R we infer that dynamic stability is ensured for $N = [(R)^3 - (R_b^e)^3]/r_0^3$, where $R_b^e \simeq 13.5$ Å, as appropriate for an 'electron balloon' in small clusters (Section III). The lowest cluster size for dynamic stability of the electron bubble is $N \simeq 6200$. This minimal cluster size, which satisfies the constraints of dynamic stability, should be compared with the onset of energetic stability, is estimated to be $N = 5200$ (Section III). There is near coincidence (within the uncertainty of the estimates) between the onsets of the energetic and dynamics stability. We conclude that the electron bubble is amenable to experimental observation for $N = 5700 \pm 500, R \simeq 40$ Å, and $R_b = 13.5$ Å. This prediction provides an interesting avenue for the experimental search for the lowest cluster size allowing for electron localization in a helium balloon which supports a localized excess electron.

V. EXCURSIONS IN THE WORLD OF FINITE, ULTRACOLD GASES

In the preceding Sections we addressed some features of energetics, thermodynamics, dynamics, and function of "strongly" interacting $(^4He)_N$ and $(^3He)_N$ quantum clusters, and excess electron bubbles in these finite, ultracold quantum systems in the temperature domain $T < T_c$, where T_c corresponds to the critical temperature for the transition to the superfluid state and for Bose–Einstein condensation in these high-density, finite quantum systems. We shall now proceed to review some new facets of energetics, thermodynamics, and dynamics of finite, ultracold gases, which unveil some fascinating aspects of the thermodynamics and energetics of Bose–Einstein condensation and of the dynamics of the expansion of optical molasses in low-density, finite quantum systems.

A. Finite Size Effects on Bose–Einstein Condensation in Confined Systems

The realization of Bose–Einstein condensation in assemblies of ultracold atoms in magnetic traps, optical traps, and microwave traps (briefly reviewed in Section I) was conducted with a maximal sample size of $\sim 10^7$ atoms [14]. Consequently, the thermodynamic limit for the Bose–Einstein condensation is never reached, and the high-order phase transition is broadened (Section II). Finite size effects on the Bose–Einstein condensation temperature and other thermodynamic attributes of the dilute gas open up new horizons in the exploration of high-order phase changes in low-density quantum systems. It will be appropriate to refer to this high-order phase transition in a finite system as a "high-order phase change," in analogy with solid–liquid first-order phase changes in clusters [148].

The theory of Bose–Einstein condensation of N noninteracting atoms confined in a finite, three-dimensional cavity of size L [80, 126] starts from the single-particle wavefunctions for a particle in a box. The density of states in the momentum space is [126]

$$g(p) = \frac{4\pi L^3}{h^3} p^2 - \frac{3\pi L^2}{\hbar^2} p \tag{95}$$

where the first term, proportional to the volume, is the Weyl term [126], which is sufficient to describe the system in the infinite volume limit, while the second term is the finite volume correction. This treatment led to the critical temperature T_c for Bose–Einstein condensation in the ideal finite system, relative to the critical temperature T_c^0 [Eq. (6)] in the infinite system [126]

$$\frac{T_c}{T_c^0} = [1 + ABN^{-1/3} \ln(BN^{-1/3})]^{-2/3} \tag{96}$$

where

$$A = \frac{2}{(3\pi)^{1/2} \zeta(3/2)} \tag{96a}$$

and

$$B = [(3\pi)^{1/2}/2]\zeta(3/2)^{1/3} \tag{96b}$$

The logarithmic term ln $BN^{-/13}$ on the right-hand side of Eq. (96) originates from the boundary corrections to the density of states and is determined by the physical boundary conditions for the vanishing of the wave function at the edges of the box [126]. If periodic boundary conditions would be imposed, the logarithmic term will be absent [126]. This point raises again the specific role of boundary conditions in determining the observables for finite quantum systems (Section II).

For sufficiently large finite systems, Eq. (96) results in

$$\frac{T_c}{T_c^0} = 1 - \left(\frac{2}{3}\right) ABN^{-1/3} \ln(BN^{-1/3}) \tag{97}$$

which constitutes a cluster size equation. For sufficiently large clusters the critical temperature T_c is decreased relative to the Bose–Einstein temperatures T_c^0 [Eq. (6)], assuming the form

$$\frac{T_c^0 - T_c}{T_c^0} = \left(\frac{2}{3}\right) ABN^{-1/3} \ln(BN^{-1/3}) \tag{98}$$

The reduction of T_c is proportional to $N^{-1/3} \propto L$, so that $(T_c^0 - T_c)/T_c^0 \propto L^{-1}$. The convergence to the properties of the infinite systems [i.e., $(T_c^0 - T_c) \to 0$] requires the removal of the logarithmic singularity in Eq. (98) by the imposition of periodic boundary conditions on the system.

Finite size effects on the critical temperature for Bose–Einstein condensation of a noninteracting Bose gas confined in a harmonic trap manifests the reduction of the condensate fraction and the lowering of the transition temperature, as compared to the infinite system [14, 127]. For an N particle condensate, the shift of the critical temperature T_c, relative to that for the $N \to \infty$ limit T_c^0, is given by the cluster size scaling relation [14, 127]

$$\frac{T_c^0 - T_c}{T_c^0} = \left\{ \frac{\bar{\omega}\zeta(2)}{2\omega_{h0}[\xi(3)]^{2/3}} \right\} N^{-1/3} \tag{99}$$

where T_c^0 [Eq. (9)] is now the Bose–Einstein condensation temperature in a harmonic trap. ω_{h0} is the average oscillator frequency, $\bar{\omega} = \omega_{h0}(2 + \lambda)/\lambda^{1/3}$, with λ being the symmetry parameter of the trap and $\xi(j)$ being the Rieman zeta function. Alternatively, one can express the lowering of the critical temperature as $(T_c^0 - T_c) \propto N^{-1/3}$, or $(T_c^0 - T_c) \propto L^{-1}$, where L is the characteristic length of the three-dimensional system. Equation (99) describes a size scaling relation for the critical temperature of the Bose–Einstein condensation of atoms with a zero scattering length, with the convergence to the properties of the infinite system in the thermodynamic limit—that is, $(T_c^0 - T_c)/T_c^0 \to 0$—when $N \to \infty$.

Both a noninteracting Bose gas in a finite volume [Eq. (96)] and a noninteracting Bose gas confined in a harmonic trap [Eq. (99)] exhibit finite size effects for the lowering of the critical temperature relative to the critical temperature for the corresponding case. For the ideal Bose gas, this reduction of T_c scales in both cases as $(T_c^0 - T_c) \propto N^{-1/3} \propto L^{-1}$. These results are different from the reduction of the critical temperature for the superfluid transition and for Bose–Einstein condensation of $(^4He)_N$ clusters (Section II). These scale as $(T_c^0 - T_c) \propto N^{-1/3\nu}$ or $(T_c^0 - T_c) \propto L^{-1/\nu}$ [Eqs. (37), (38a), (38b), and (42)], where $\nu = 0.670$ [200] is the critical exponent for the superfluid fraction and for the correlation function in bulk liquid ^4He. The different size dependence of the critical temperature for the noninteracting, ideal, confined Bose gas and for "strongly" interacting $(^4He)_N$ clusters raises the distinct possibility that these systems may belong to different universality classes, being characterized by distinct critical exponents.

B. A Molecular Description of the Bose-Einstein Condensation

An ultracold dilute gas of bosonic atoms constitutes a many-body system of weakly interacting constituents. An attempt to bridge between the thermodynamic picture of a rounded-off Bose–Einstein phase transition and molecular concepts was advanced [255]. Such a generalized molecular structure allows for the introduction of a collective coordinate and of an effective potential for the many-boson system, which allows for the characterization of the Bose–Einstein condensate and for the assessment of its upper size limit [255]. In what follows we provide a brief review of this work [255].

The Hamiltonian of N identical atoms of mass m, confined in a trap approximated by a spherically symmetric harmonic oscillator of frequency ρ, is given by the Schrödinger equation

$$H = \frac{\hbar^2}{2m}\sum_i \nabla_i^2 + \frac{1}{2}\sum_i m\omega^2 r_i^2 + \sum_{i<j} U_{\rm int}(\mathbf{r}_i - \mathbf{r}_j) \tag{100}$$

where $\mathbf{r}_i$ is the distance vector of the ith atom from the trap center, and $U_{\rm int}$ is the pairwise interatomic potential. In treating a low-density condensate, a contact potential [256] was introduced

$$U_{\rm int}(\mathbf{r}_j - \mathbf{r}_j) = \left(\frac{4\pi\hbar 2a}{m}\right)\delta(\mathbf{r}_i - \mathbf{r}_j) \tag{101}$$

where a is the atom–atom scattering length.

To describe the Bose–Einstein condensate, a coordinate transformation was applied, with the introduction of hyperspherical collective coordinates [255, 257–259], where one of the coordinates is the hyperradius

$$\hat{R} = \left(\frac{1}{N}\sum_i r_i^2\right)^{1/2} \tag{102}$$

This parameterization was widely utilized in nuclear [257, 258] and atomic [259] physics. $\hat{R}$ describes the hyperradius of a $3N$-dimensional space, while the remaining $(3N-1)$ coordinates are given in terms of hyperangles, $\mathbf{\Omega} = \{\Omega_1, \Omega_2, \ldots, \Omega_{3N-1}\}$, which parameterize the hypersphere. The "grand angular momentum" $\mathbf{\Lambda}^2$ for the system has eigenfunctions, which are called "hyperspherical harmonics" [260]. The construction of an approximate solution to the Schrödinger equation [Eq. (100)] rested on the expansion of the many-body wavefunction $\psi(\hat{\boldsymbol{R}}, \mathbf{\Omega})$ into hyperspherical harmonics. This expansion results in a set of coupled differential equations for the hyperradial expansion coefficients. Such a situation is ubiquitous for the Born–Oppenheimer approximation in molecular chemical physics [261]. At this stage, the lowest term of the expansion is chosen, in analogy with the K–harmonic approximation in nuclear theory [262].

The approximate one-dimensional wavefunction $F(\hat{R})$ for the condensate was obtained from the Schödinger equation

$$\left[\left(-\frac{\hbar^2}{2M}\right)\frac{d^2}{d\hat{R}^2} + V_{\text{eff}}(\hat{R})\right]F(\hat{R}) = EF(\hat{R}) \tag{103}$$

where the effective potential is

$$V_{\text{eff}}(\hat{R}) = \frac{\hbar^2}{2M}\frac{(3N-1)(3N-3)}{4\hat{R}^2} + \frac{1}{2}M\omega^2\hat{R}^2 + \zeta\sqrt{\frac{1}{2\pi}}\frac{\hbar^2 a}{M}\frac{N^2(N-1)}{\hat{R}^3} \tag{104}$$

with $M = nm$, emphasizing that Eq. (103) describes the quantum mechanical motion of the system as a whole, and $\zeta = 1.807$. One should note that even when all the atoms of the condensate have zero angular momentum about the center of the trap, there remains an effective centrifugal barrier, representing the term proportional to $1/\hat{R}^2$ in Eq. (104). This term represents the kinetic energy cost of confining all atoms within a small region near the center of the trap. This energy is responsible for stabilizing an atomic condensate with $a < 0$ against collapse.

To assess the stability of these collective systems, it is useful to plot a schematic representation (Fig. 22) for the effective potential, Eq. (104), where $\hat{R}$ is expressed in harmonic oscillator units $(\hbar/m\omega)^{1/2}$ [255]. The effective potential is portrayed for different values of the scattering length. In the

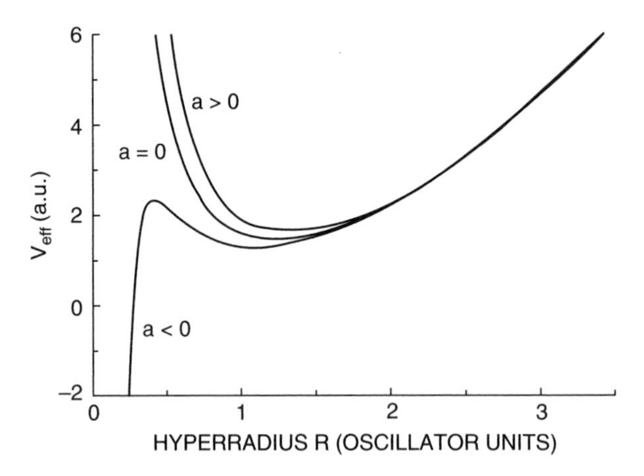

Figure 22. A schematic presentation of the effective potentials $V_{\text{eff}}(R)$ [Eq. (103)] for the collective motion of the N-boson system [255]. The effective potentials are marked on the curves by $a < 0$, $a = 0$, and $a > 0$.

noninteracting limit for a = 0, the system decouples to N harmonic oscillators with the energy levels

$$E = \hbar\omega\left[\sum n_i + \frac{3N}{2}\right], \qquad a = 0 \tag{105}$$

For a nonzero scattering length, V_{eff} is characterized by either repulsive $(a > 0)$ or attractive $(a < 0)$ interaction contributions, as manifested in Fig. 22. We also note that the effective potential [Eq. (104)] depends on the number N of particles. For $a > 0$, V_{eff} is binding for all numbers of particles N. The system is localized within a nonharmonic binding collective potential (Fig. 22) and a "molecular cloud" is formed without any constraints on the number of particles. The situation is drastically different for $a < 0$. The N-dependent effective potential changes its shape for some "critical" number of particles, which will be denoted as N_{CR}. For $N < N_{\text{CR}}$, V_{eff} exhibits a local minimum (Fig. 22), which characterizes the metastable region, where the metastable condensate can exist on a sufficiently long time scale for experimental observation. The metastable region in Eq. (104) is stabilized by the $1/\hat{R}^2$ effective centrifugal repulsion. At smaller $\hat{R}$ values, a potential barrier appears, separating the metastable region from the "collapse region," which is dominated by the $1/\hat{R}^3$ component of V_{eff} [Eq. (104)]. When N increases, this component increases as $\sim N^3$ and at $N \sim N_{\text{CR}}$ the barrier vanishes and so does the metastable region. For $N > N_C$ the potential is purely attractive and the condensate does not exist. The critical size, N_C, of a finite boson system with attractive interactions $(a < 0)$ was estimated from the analysis of the potential (104) in the form [255]

$$N_C \simeq 0.671\left(\frac{\hbar}{m\omega|a|}\right) \tag{106}$$

For ^{7}Li atoms, with $a = -27$ bohr [263] in a harmonic trap with an average frequency of $\omega = 144$ Hz [267], Eq. (106) resulted in $N_C \sim 1450$ [254], in excellent agreement with variational estimates [265].

The foregoing analysis [264] reflects on novel features of collective nuclear dynamics in finite ultracold systems. This leads to the concept of macroscopic tunneling [255, 265, 266]. The WKB approximation used [255, 265] for the macroscopic tunneling rate Γ (expressed in atoms/s) is

$$\Gamma = N\nu\exp(-2\Phi) \tag{107}$$

where

$$\Phi = \int_{R_{\text{in}}}^{R_{\text{out}}} dR\sqrt{\frac{2M}{\hbar^2}[V_{\text{eff}}(R) - E]} \tag{107a}$$

with the integral being taken between the inner (R_{in}) and the outer (R_{out}) turning points, while the frequency is

$$\nu = \frac{1}{2\pi}\left[M^{-1}(d^2V_{eff}/dR^2)R_{MIN}\right]^{1/2} \tag{108}$$

with R_{MIN} being the position of the local minimum of $V_{eff}(R)$. Numerical estimates [255] resulted in the size dependence of the macroscopic tunneling rate $\Gamma \sim 3 \times 10^5 \exp[1.6(N - N_c)]$ atoms/s, with $N_c = 1460$. The rate of macroscopic tunneling exhibits a strong exponential dependence on N. This treatment [255] opens up the exploration of collective, macroscopic dynamics in ultracold, finite boson systems. It will be interesting to extend this formalism to consider wavepacket dynamics of collective states of the macroscopic system. Removing the K approximation will induce coupling between the zero-order K nearly harmonic levels, inducing dynamic effects reminiscent of intramolecular radiationless transitions [267].

C. Nuclear Dynamics of Expansion of Optical Molasses

To establish analogies and relations between the nuclear dynamics of clusters and ultracold gases, we proceed to the nuclear dynamics of optical molasses [79]. Pruvost and her colleagues established an interesting analogy between the expansion dynamics of ultracold ($T = 10$–$100\,\mu K$) optical molasses and Coulomb explosion of multicharged molecular clusters [93–98]. The optical molasses involve a cloud of trapped, laser-irradiated, neutral atoms (e.g., Rb) in a magnetic trap (Fig. 23), which is characterized by a density of $\rho = 10^{11}$–10^{13} atoms cm^{-3}. When the magnetic trap is being suppressed, the cloud expands via the radiative trapping force $\vec{F}$. This radiative trapping force $\vec{F}_{ij}$, which originates from photon emission and reabsorption between a pair of atoms (e.g., i and j) separated by distance r_{ij}, is given by [268]

$$\vec{F}_{ij} = \frac{K}{4\pi r_{ij}^2} \tag{109}$$

The coefficient K is [268]

$$K = \frac{(\sigma_R - \sigma_L)\sigma_L I}{c} \tag{110}$$

where σ_L is the laser absorption cross section, σ_R is the reabsorption cross section, I is the laser intensity, and c is the velocity of light. The radiative trapping force $\vec{F}_{ij}$, [Eqs. (109) and (110)], is proportional to r_{ij}^{-2}, being analogous

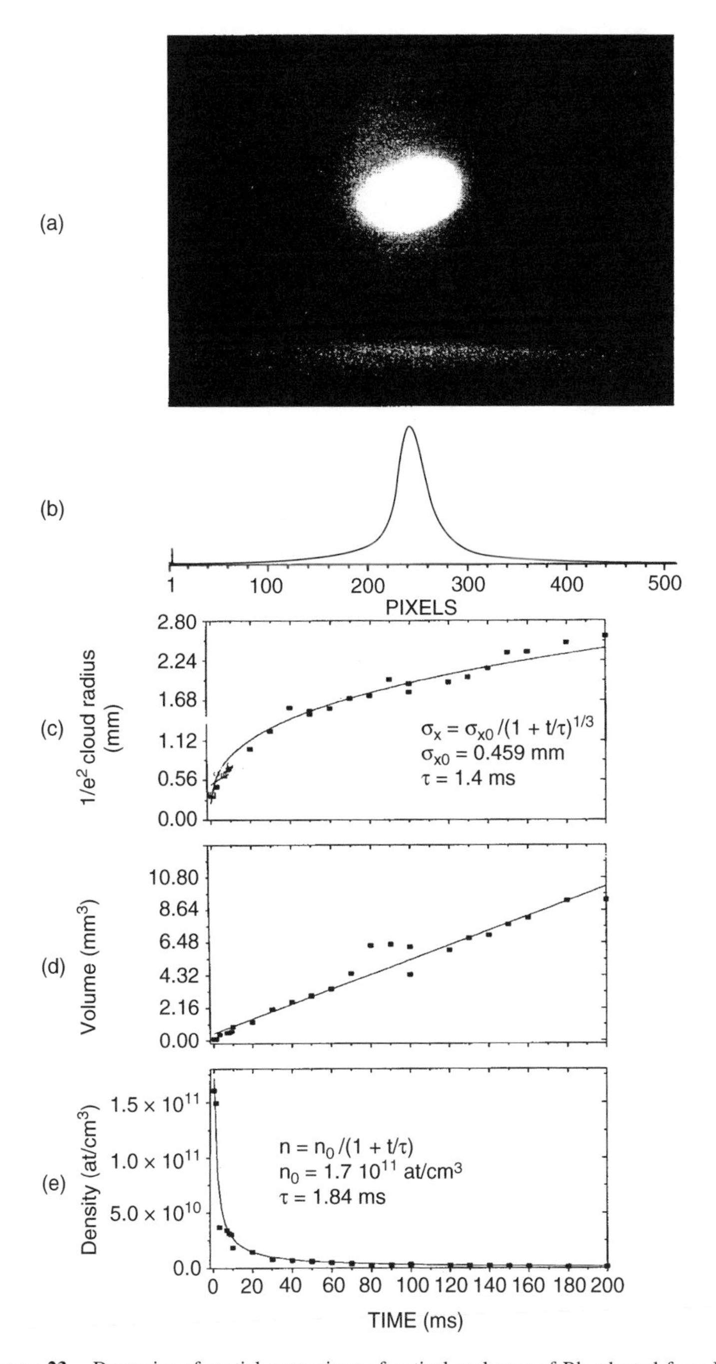

Figure 23. Dynamics of spatial expansions of optical molasses of Rb adapted from Pruvost et al. (data from Ref. 79). (*a*) A photograph of the irradiated cloud at $t = 0$. (*b*) Excited atom distribution in the irradiated cloud at $t = 0$. (*c*) Time dependence of the cloud radius. (*d*) Time dependence of the volume of the irradiated cloud. (*e*) Time dependence of the density of the expanding cloud.

with the Coulomb law for the electrostatic force $\vec{F}_e$ between two ions of charges q, which is given by

$$\vec{F}_{e,ij} = \frac{q^2}{4\pi \in_0 r_{ij}^2} \tag{111}$$

The isomorphism between the radiative trapping force [Eq. (109)] and the electrostatic force [Eq. (111)] allows to characterize the coefficient K [Eq. (110)] in terms of an effective charge for the specification of the radiative trapping force

$$q_{\text{eff}} = \gamma e \tag{112}$$

where e is the electronic charge. The coefficient γ is given by

$$\gamma = \frac{(K \in_0)^{1/2}}{e} \tag{113}$$

Numerical estimates for Rb atoms [79, 268] resulted in the coefficient $\gamma = 3.5 \times 10^{-5}$ and the effective charge $q_{\text{eff}} = 3.5 \times 10^{-5} e$. Accordingly, the radiative trapping force is analogous to Coulomb's law, and the expansion of the irradiated cloud is isomorphous to a Coulomb explosion of a multicharged cluster with charge q on each ion. The time scale for the expansion of the optical molasses of Rb atoms (Fig. 23) is $\tau_M \simeq 1.4$ ms. The translational temperature of those expanding molasses falls in the range $T = 100$–$10\,\mu$K (Fig. 24). The

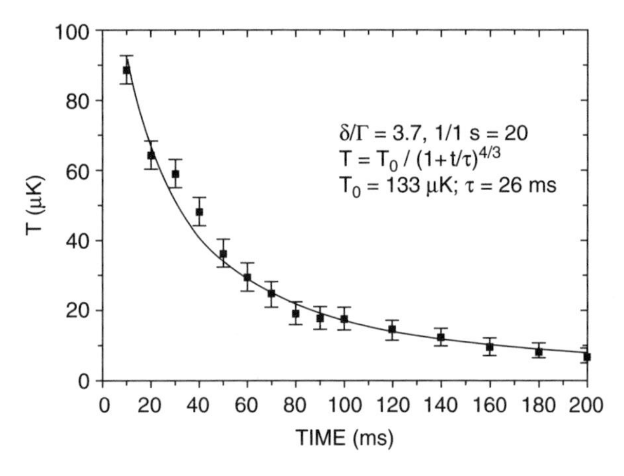

Figure 24. The time-dependence temperature of the expanding optical molasses of Rb (data from Ref. 79).

isomorphism between intracloud radiative trapping and cluster Coulomb explosion allows for the advancement of a theoretical framework for the dynamics of cloud expansion and its time scale [79].

Coulomb explosion of highly charged homonuclear molecular clusters—for example, $(D^+)_N(N = 55\text{–}7 \times 10^4)$ or $(Xe^{q^+})_N(N = 55-10^4$ and $q = 1\text{–}36)$—is induced by extreme multielectron ionization in ultraintense laser fields [95–98, 186, 188]. The two important features of Coulomb explosion dynamics are the ultrashort (fs) time scales for the spatial expansion of the ions and the production of energetic (keV–MeV) ions. When the time scales for the compound cluster electronic ionization processes in the ultraintense laser field are faster than the nuclear processes of Coulomb explosion, the cluster vertical ionization (CVI) approximation becomes valid [96, 97]. The CVI decouples the time scales for the dynamics of heavy particles (ions) from the dynamics of electrons. Under the CVI conditions, Coulomb explosion of the highly charged ionic cluster involves a nuclear dynamic process. The nuclear dynamics of the ion expansion of uniformly expanding homonuclear, multicharged clusters, with a charge q on each atom, is characterized by a time scale [96]

$$\tau_{\text{EX}} = 2.137q^{-1}\left(\frac{m}{\rho}\right)^{1/2} \tag{114}$$

where ρ is the initial ion density—that is, $\rho = \left(4\pi r_0^3/3\right)^{-1}$, with r_0 being the constituent radius and m the ion mass. The following units are used in Eq. (114): (m, amu), (q, e), (r_0, Å), (ρ, Å^{-3}) and (τ_{EX}, fs). τ_{EX} [Eq. (114)] represents the time for the doubling of the initial cluster radius. The characteristic Coulomb explosion times [Eq. (114)] reveal the following features: (i) charge dependence, with $\tau_{\text{EX}} \propto q^{-1}$, in accordance [93, 96, 97] with analyses and simulations (Fig. 25); (ii) mass effect, with $\tau_{\text{EX}} \propto m^{1/2}$, providing predictions for isotope effects on the dynamics; (iii) dependence on the initial structure of the system with $\tau_{\text{EX}} \propto (\rho)^{-1/2} \propto (r_0)^{3/2}$; (iv) lack of cluster size dependence. For sufficiently large clusters, no cluster size dependence of τ_{EX} is expected. The absence of cluster size scaling is expected to prevail for sufficiently large clusters, where the continuum approximation underlying the derivation of Eq. (114) holds. From numerical simulations [93] under CVI conditions of Coulomb explosion of $(Xe^{q+})_N$ clusters ($q = 1\text{–}8$ and $N = 2\text{–}55$), portrayed in Fig. 25, we inferred that the size invariance of τ_{EX} sets in for $N > 35$. The Coulomb explosion lifetime of $(Xe^+)_N(N > 35)$ clusters is $\tau_{\text{EX}} \simeq 100$ fs (Fig. 25), in accord with the analytic results [Eq. (114)].

Both the cluster Coulomb explosion time τ_{EX} and the expansion time of optical molasses τ_M obey the relation [Eq. (114)] $\tau_{\text{EX}}, \tau_M \propto q^{-1}m^{1/2}r_0^{3/2}$. We shall scale the cluster Coulomb explosion time $\tau_{\text{EX}} = 100$ fs for $(Xe^+)_N$ (with $q = $ e, $m_{\text{Xe}} = 127$ amu, and $r_0 = 2.5$ Å) to obtain the time scale τ_M for the

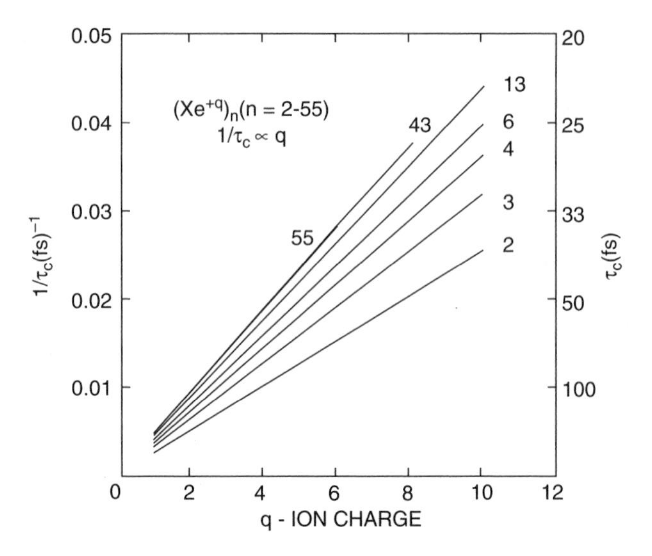

Figure 25. Molecular dynamics simulations of time-resolved Coulomb explosion of $(Xe^{q+})_N$ clusters ($N = 2$–55; $q = 1$–10) (data from Ref. 93). The times τ_c for the doubling of the radius of the initial distribution of the Xe^{q+} ions scales as q^{-1}, according to Eq. (114).

expansion of optical molasses of Rb presented in Fig. 23 (where $q = 3.5 \times 10^{-5}$e, $m_{Rb} = 87$ amu, and $r_0 = 10^4$ Å). Pruvost et al. [79] set

$$\tau_M = \mathfrak{J}\tau_{EX} \tag{115}$$

where the scaling factor, according to Eq. (114), is $\mathfrak{J} = 10^{10}$. This courageous extrapolation results in $\tau_M \simeq 1$ ms. This prediction is in good agreement with the recent experimental data [79] presented in Fig. 23, where the expansion of bright three-dimensional optical molasses of Rb is characterized by the time scale of $\tau_M \simeq 1.4$ ms. This analysis builds bridges between nuclear dynamics of clusters and of ultracold atomic clouds.

VI. EPILOGUE

At this stage it will be appropriate to quote from a "brief" 107-page-long review article on the "Electronic States of Molecular Crystals Under Pressure," which was written in 1965 by Stuart Rice and Joshua Jortner [269]. That field of chemical physics led throughout a time span of 40 years to the exploration of conducting polymers, to the advent of molecular electronics, and to the understanding of exciton migration in the plant photosynthetic reaction center. In the last section of that review [269] the authors addressed the "exhausted and blurry-eyed reader," suggesting some new experiments "which are by no means easy" and finally stating that "with a sigh of relief, we end this review."

We shall adhere to the Rice–Jortner recipe [269] in the concluding remarks of the present review, which addresses structure, thermodynamics, elementary excitations, dynamics, and function of large, finite, ultracold systems. New developments in this field, which bridges between quantum clusters and ultracold finite clouds, will emerge from concurrent theoretical and experimental efforts, some of which will be based on the present review. The experiments we would like to suggest for future developments are by no means easy. However, they appear to provide new insight into collective excitations in quantum clusters, as well as into nuclear and electron dynamics in ultracold finite systems. We would like to suggest the following:

1. *"Supersolid" Clusters*. It was suggested [270–272] that the property of nonclassical rotational inertia possessed by macroscopic superfluid ^{4}He may be shared by solid ^{4}He (under pressure), provided that the solid is Bose-condensed. Such a "supersolid" will then manifest resistance-free flow [273]. Recent probable observation of supersolid ^{4}He, using acoustical waves and heat pulses, was reported [274]. A "soft" solid ^{4}He is expected to possess highly mobile lattice defects, such as vacant sites or dislocations. It is possible that the lattice vacancies (or the atoms hopping among them) will exhibit Bose–Einstein condensation and superfluidity. If ^{4}He exhibits superfluidity in the solid phase, a resistance-free transport of microscopic probes (e.g., electron bubbles) may prevail in the supersolid. Indeed, excess electron localization in a bubble in solid helium was proposed [240], and the theoretical description of electron bubbles in solid ^{4}He is supported by experimental evidence for electron mobility in the "normal" solid [240]. It will be extremely interesting to explore solid ^{4}He clusters produced under high pressure. Excess electron detachment for these solid clusters may provide evidence for "supersolid" behavior.

Another rigid quantum cluster, which may exhibit "supersolid" properties, involves the molecular cluster of solid *para*-hydrogen. The quantum simulations of Sindzingre et al. [151] report that an $(H_2)_{64}$ cluster at 1.0–0.5 K is an amorphous solid. For $N = 33$ and $N = 13$ the radial density profile indicates "liquid"-type behavior above $T_M = 5$ K. Below 5 K, structural sensitivity is manifested with these clusters, revealing specific rigid structure—for example, a pentagonal bipyramidal core for $N = 33$ (temperature not specified) and an icosohedral structure for $N = 13$ (at 3 K). "Superfluidity" was inferred from temperature dependence of the superfluid fraction in the clusters, with the temperature onset (T_λ) of the decrease of the normal fraction being $T_\lambda \simeq 3.5$ K for $N = 13$ and $N = 18$ and, $T_\lambda < 1$ K for $N = 33$. These results imply that $T_M > T_\lambda$, so that ordinary superfluidity cannot be claimed. Accordingly, the original Ginzburg–Sobyanin hypothesis [149] for $(\mathrm{H}_2)_N$ clusters is violated.

Sindzingre et al. [151] proposed that these clusters correspond to a "supersolid." This is an interesting speculation, supported by the finding of permutation cycles in quantum simulations [151]. Again, the use of electron bubbles in $(p\text{-H}_2)_N$ clusters, as possible probes for "supersolid" behavior, will be of considerable interest. The formation of electron bubbles in liquid H_2 was established [212] and electron bubble formation in solid H_2 clusters is feasible, in analogy with electron localization in solid $(^4\text{He})_N$ clusters [240]. If indeed solid $(p\text{-H}_2)_N$ manifests "supersolid" properties, the abundant lattice vacancies in this quantum cluster might drive coherent, nearly friction-free motion of the electron bubble. The search for supersolid properties in solid $(p\text{-H}_2)_N$ (and solid $(^4\text{He})_N$) clusters by electron tunneling from bubbles (Section IV) will be extremely interesting.

2. *Time-Resolved Dynamics of the Bose–Einstein Condensation.* The nuclear dynamics for the development of a coherent population of the global ground state of a Bose–Einstein condensate [275, 276] pertains to collective nuclear dynamics on the time scale of nuclear motion. A fascinating experiment will involve an "instantaneous" temperature decrease (a negative temperature jump) in a gas, with the attainment of a sufficiently low temperature $T < T_c$. Such an experiment can be considered for liquid $(^4\text{He})_N$ clusters, or for finite cold gases, which will be "instantaneously" cooled down below T_c and their nuclear dynamics will be interrogated. Such a compound collective nuclear dynamic process will involve several steps [275, 276], which will, in analogy with nucleation theory, presumably involve: (i) *Nucleation.* The coherent population of the global ground state in a small region of the finite system. The characteristic nucleation time is [275] $\tau_{\text{NUC}} = \hbar/k_B T_c$. For $(^4\text{He})_N$ clusters, $T_c \simeq T_\lambda$ and $\tau_{\text{NUC}} \simeq 3$ ps. Pursuing the analogy between the Bose–Einstein condensation and high-order phase transitions (Section III), τ_{NUC} corresponds to local, spontaneous symmetry-breaking, which results in the local evolution of the order parameter. (ii) *Growth of the small nucleus of the condensate.* All the states with low momentum states, with $p < \hbar/\lambda_{\text{DB}}$, where λ_{DB} is the de Broglie wavelength for energy $k_B T_c$, will cooperate in the coherent ground state of the sample. For a quantum cluster, or finite system with an interparticle distance r_0, it appears that the characteristic time for growth of the Bose–Einstein condensate is [275] $\tau_{\text{COH}} \simeq \tau_{\text{NUC}}(\lambda_{\text{DB}}/r_0)^2$, so that $\tau_{\text{NUC}} < \tau_{\text{COH}}$. These estimates of τ_{NUC} and τ_{COH} imply that for $(^4\text{He})_N$ clusters "instantaneous" cooling (or "ultracooling") has to be conducted on the subpicosecond time scale. Of course, for ultracold boson gases, where $k_B T_c$ is considerably lower, the time scale for the dynamics of Bose–Einstein condensation, which is expected to be proportional to $\sim(k_B T_c)^{-1}$, will be considerably longer; for example, for $T_c \sim 10^{-7}$ K one expects the time-resolved dynamics of the Bose–Einstein condensation in a dilute gas to occur on the μsec–msec scale.

3. *Electron Dynamics.* The exploration of electron dynamics in large finite systems will stem from concurrent progress in theory and experiment, which will focus on analysis and control of various channels of "pure" electron dynamics processes, without the involvement of nuclear motion, bypassing the constraints imposed by the Franck–Condon principle [183]. Of considerable interest is electron tunneling from electron bubbles in superfluid and normal clusters, as the time-resolved electron dynamics can be controlled, in principle, over a broad range, due to the exponential dependence of the tunneling rate on the initial bubble distance d from the cluster surface (Section IV). The "ordinary" electron tunneling rates from the ground electronic state (1 s) of the electron bubble are governed by the exponential factor exp($-2\alpha D$), where $\alpha \propto V_0 - E_e$, with E_e being the energy of the electronic state from which electron tunneling occurs. For the ground electronic $1s$ state we estimated that $(V_0 - E_e) = 0.92$ eV and $2\alpha = 1\ \text{Å}^{-1}$ (Section IV). A marked enhancement of the electron tunneling rate will be manifested upon optical excitation to higher electronic states of the electron bubble. A theoretical treatment of these electronic excitations [277] showed that for a bubble in an $(^4\text{He})_N$ $(N = 2 \times 10^5)$ cluster the $1p$ and $2p$ vertically excited states are energetically stable, with $(V_0 - E_e) = 0.7$ eV and $2\alpha = 0.85\ \text{Å}^{-1}$ for the $1p$ state and $(V_0 - E_e) = 0.1$ eV and $2\alpha = 0.31\ \text{Å}^{-1}$ for the $2p$ state. The electron tunneling rate from the $1p$ state will be enhanced by a numerical factor of $\sim$400 relative to the rate from the $1s$ state. As τ_{TUN}^{1s} lies in the millisecond–second time domain for the relevant experimental conditions for tunneling from the ground $1s$ state, we estimate τ^{1p} to be in the microsecond–millisecond range for the $1p$ state. This time scale is too long to compete with radiationless relaxation of the $1p$ state. On the other hand, for the $2p$ electronic excited state we estimated that the tunneling rate will be larger by a numerical factor of 10^{12} relative to the rate from the $1s$ ground state. Accordingly, $\tau_{\text{TUN}}^{2\text{p}}$ lies in the picosecond–femtosecond time domain. Such ultrafast electron dynamics will presumably overwhelm radiationless transitions from the $2p$ state to lower electronic states. Optical excitations of the electron bubble provide an avenue for ultrafast electron dynamics in these large ultracold systems.

It is fairly obvious that more theoretical and experimental developments are called for in this fascinating research area, and some of them should be obvious from the text of this review. With a sigh of relief, we therefore end this review.

Acknowledgments

We are grateful to Professor Peter Toennies for most stimulating and inspiring discussions and correspondence, and we thank Dr. Laurence Pruvost for fruitful collaboration on the dynamics of optical molasses. This research was supported in part by the German–Israeli James Franck Program on Laser–Matter Interaction.

References

1. J. K. Messer and F. C. De Lucia, *Phys. Rev. Lett.* **53**, 2555 (1984).
2. M. W. Beaky, T. M. Goyette, and F. C. De Lucia, *J. Chem. Phys.* **105**, 3994 (1996).
3. R. Cote and A. Dalgarno, *Chem. Phys. Lett.* **279**, 50 (1997).
4. G. Meijer, *ChemPhysChem* **3**, 495 (2002).
5. J. T. Bahns, P. L. Gould, and W. C. Stwalley, *Adv. At. Mol. Opt. Physics* **42**, 172 (2000).
6. J. P. Toennies and A. F. Vilosov, *Ann. Phys. Rev. Chem.* **49**, 1 (1998).
7. J. P. Toennies, A. F. Vilesov, and K. W. Whaley, *Phys. Today* **2**, 31 (2002).
8. J. A. Northby, *J. Chem. Phys.* **115**, 10065 (2001).
9. K. Callegari, K. K. Lehmn, R. Schmied, and G. Scoles, *J. Chem. Phys.* **115**, 10090 (2001).
10. F. Stienkemeier and A. F. Vilesov, *J. Chem. Phys.* **115**, 10119 (2001).
11. J. P. Toennies and A. F. Vilesov, *Angew. Chem. Int. Ed.* **43**, 2622 (2004).
12. S. Chu, L. Hollenberg, J. Bjorkholm, A. Cable, and A. Ashkin, *Phys. Rev. Lett.* **55**, 48 (1985).
13. J. Mourachko, D. Comparat, F. de Tomasi, A. Floretti, T. Nosbaum, V. M. Akulin, and P. Pillet, *Phys. Rev. Lett.* **80**, 253 (1998).
14. F. Dalfovo, S. Giorgini, and L. P. Pitaevskii, *Rev. Mod. Phys.* **61**, 463 (1999).
15. P. Kapitza, *Nature* **141**, 74 (1938).
16. J. F. Allen and A. D. Meisner, *Nature* **114**, 75 (1938).
17. L. Landau, *J. Phys. USSR* **5**, 71 (1941).
18. R. P. Feynman, *Phys. Rev.* **90**, 116 (1963); **91**, 1291 (1953); **91**, 1301 (1953).
19. I. M. Khalatnikov, *Introduction to the Theory of Superfluidity*, Benjamin, New York, 1965.
20. J. Wilks and D. S. Betts, *An Introduction to Liquid Helium*, Oxford University Press, New York, 1987.
21. D. M. Ceperley and E. L. Pollock, *Phys. Rev. Lett.* **56**, 351 (1986).
22. E. L. Pollock and D. M. Ceperley, *Phys. Rev. B* **36**, 8343 (1987).
23. D. M. Ceperley, *Rev. Mod. Phys.* **67**, 279 (1995).
24. A. Griffin, D. W. Snoke, and S. Stringari, eds., *Bose–Einstein Condensation*, Cambridge University Press, Cambridge, 1996.
25. P. Nozieres, Ref. 24, p. 15.
26. P. Sokol, Ref. 24, p. 51.
27. C. Bradley, A. Sacker, J. J. Tollett, and R. G. Hulet, *Phys. Rev. Lett.* **75**, 1687 (1995).
28. Y. Castin, J. Dalibard, and C. Cohen-Tannoudji, Ref. 24, p. 173.
29. S. Chu, J. Bjonkholm, A. Atkins, and A. Cable, *Phys. Rev. Lett.* **57**, 314 (1986).
30. P. Gould, P. Lett, P. Julienne, W. D. Phillips, W. Thorsheim, and J. Weiner, *Phys. Rev. Lett.* **60**, 788 (1988).
31. T. Hansch and A. Schawlow, *Opt. Commun.* **13**, 68 (1975).
32. J. Dalibard and C. Cohen-Tannoudji, *J. Opt. Soc. Am. B* **6**, 2023 (1989).
33. C. Cohen-Tannoudji, in *Fundamental Systems in Quantum Optics, Proceedings of Session LIII of Les Houches Summer School*, J. Dalibard, J. M. Raimond, and J. U. Zinn-Justin, eds., North Holland, Amsterdam, 1992.
34. E. Timmermans, P. Tommasini, M. Hussein, and A. Kerman, *Phys. Rep.* **315**, 199 (1999).

35. E. A. Donley, N. R. Claussen, S. T. Thompson, and C. E. Wieman, *Nature (London)* **417**, 529 (2002).
36. C. A. Regal, C. Ticknor, J. L. Bohn, and D. S. Jin, *Nature (London)* **424**, 47 (2003).
37. J. Herbig, T. Kraemer, M. Mark, T. Weber, C. Chin, H.-C. Nägerl, and R. Grimm, *Science* **301**, 1510 (2003).
38. S. Dürr, T. Volz, A. Marte, and G. Rempe, *Phys. Rev. Lett.* **92**, 020406 (2004).
39. K. Xu, T. Mukaiyama, J. R. Abo-Shaeer, J. K. Chin, D. E. Miller, and W. Ketterle, *Phys. Rev. Lett.* **91**, 210402 (2003).
40. S. Jochim, M. Bartemstein, A. Altmeyer, G. Hendl, S. Riedl, C. Chin, J. Hecker Denschlag, and R. Grimm, *Science* **302**, 2101 (2003).
41. M. Greiner, C. A. Regal, and S. D. Jin, *Nature (London)* **426**, 537 (2003).
42. M. W. Zwierlein, C. A. Stan, C. H. Schunck, S. M. F. Raupach, S. Gupta, Z. Hadzibabic, and W. Ketterle, *Phys. Rev. Lett.* **91**, 250401 (2003).
43. N. Vanhaecke, W. De Souza Melo, B. L. Tolra, D. Comparat, and P. Pillet, *Phys. Rev. Lett.* **89**, 063001 (2002).
44. C. Chin, A. J. Kerman, V. Vuletić, and S. Chu, *Phys. Rev. Lett.* **90**, 033201 (2003).
45. F. Masnou-Seeuws and P. Pillet, *Adv. At. Mol. Opt. Phys.* **47**, 53 (2001).
46. C. P. Koch, J. P. Palao, R. Kosloff, and F. Masnou-Seeuws, *Phys. Rev. A* **70**, 013402 (2004).
47. M. Anderson, J. Ensher, M. Matthews, C. Wieman, and E. Comell, *Science* **269**, 198 (1995).
48. K. Davis, et al., *Phys. Rev. Lett.* **75**, 3969 (1995).
49. C. Bradley, C. Sackett, J. Tollet, and R. Hulet, *Phys. Rev. Lett.* **78**, 985 (1997).
50. K. B. Whaley and R. E. Miller, eds., *Helium Nanodroplets: A Novel Medium for Chemistry and Physics*, Special Issue, *J. Chem. Phys.* **15**, 22 (2001).
51. V. H. Pandharipade, J. G. Zabolinsky, S. C. Pieper, R. B. Wiringa, and U. Helmbrecht, *Phys. Rev. Lett.* **50**, 1676 (1983).
52. S. C. Pieper, R. B. Wiringa, and V. R. Pandharipande, *Phys. Rev. B* **32**, 3341 (1985).
53. V. R. Pandharipande, S. C. Pieper, R. B. Wiringa, *Phys. Rev. B* **34**, 4571 (1986).
54. D. S. Lewart, V. R. Pandharipande and S. C. Pieper, *Phy. Rev. B* **37**, 4950 (1988).
55. M. V. Rama Krishna and K. B. Whaley, *J. Chem. Phys.* **93**, 6738 (1990).
56. S. Stringari and J. Treiner, *J. Chem. Phys.* **87**, 5021 (1987).
57. D. M. Brink and S. Stringari, *Z. Phys. D* **15**, 257 (1990).
58. S. A. Chin and E. Krotscheck, *Phys. Rev. B* **45**, 852 (1992).
59. M. V. Rama Krishna and K. B. Whaley, *Mod. Phys. Lett. B* **4**, 895 (1990).
60. M. Casas, F. Dalfovo, A. Lastri, L. Serra, and S. Stringari, *Z. Phys. D.* **35**, 67 (1995).
61. M. Casas and S. Stringari, *J. Low Temp. Phys.* **79**, 135 (1990).
62. E. Cheng, M. A. McMahon, and K. B. Whaley, *J. Chem. Phys.* **104**, 2669 (1996).
63. M. A. McMahon and K. B. Whaley, *J. Chem. Phys.* **103**, 2561 (1995).
64. M. A. McMahon, R. N. Barnett, and K. B. Whaley, *Z. Phys. B* **98**, 421 (1995).
65. P. Sindzingre, M. L. Klein, and D. M. Ceperley, *Phys. Rev. Lett.* **63**, 1601 (1989).
66. D. M. Ceperley and E. Manousakis, *J. Chem. Phys.* **115**, 10111 (2001).
67. L. Pitaevskii and S. Stringari, *Z. Phys. D* **16,** 299 (1990).
68. F. Dalfovo and S. Stringari, *J. Chem. Phys.* **115**, 10078 (2001).
69. M. Hartmann, R. E. Miller, J. P. Toennies, and A. Vilesov, *Phys. Rev. Lett.* **75**, 1566 (1995).

70. M. Hartmann, F. Mielke, J. P. Toennies, A. F. Vilesov, and G. Benedek, *Phys. Rev. Lett.* **76**, 4560 (1996).
71. S. Grebenev, J. P. Toennies, and A. F. Vilesov, *Science* **279**, 2083 (1998).
72. S. Grebenev, M. Hartmann, M. Havenith, B. Sartakov, J. P. Toennies, and A. F. Vilesov, *J. Chem. Phys.* **112**, 4485 (2000).
73. K. Nauta and R. E. Miller, *Phys. Rev. Lett.* **82**, 4480 (1999).
74. C. Callegari, A. Conjusteau, I. Reinhard, K. K. Lehmann, G. Scoles, and F. Dalfovo, *Phys. Rev. Lett.* **83**, 5058 (1999).
75. Y. Kwon and K. B. Whaley, *Phys. Rev. Lett.* **83**, 4108 (1999).
76. Y. Kwon, P. Huang, M. Patel, D. Blume, and K. B. Whaley, *J. Chem. Phys.* **113**, 6469 (2000).
77. K. Nauta, D. T. Moore, and R. E. Miller, *Faraday Discuss.* **113**, 261 (1999).
78. M. Hartmann, A. Lindinger, J. P. Toennies, and A. F. Vilesov, *Chem. Phys.* **239**, 139 (1998).
79. L. Pruvost, T. Serre, H. T. Duong, and J. Jortner, *Phys. Rev. A* **61**, 053408 (2000).
80. S. Grossman and M. Holthaus, *Phys. Lett. A* **208**, 188 (1995).
81. E. W. Schlag, R. Weinkauf, and R. E. Miller, eds., *Molecular Clusters*, Special Issue, *Chem. Phys.* **239**, 1 (1998).
82. T. Kondow, K. Kaya, and A. Terasaki, eds., *Structure and Dynamics of Clusters*, University Press, Tokyo, 1996.
83. C. Yannouleas and U. Landman, eds., *Small Particles and Inorganic Clusters (ISSP 10)*, *Eur. J. Phys. D* **16**, 9 (2001).
84. J. Jortner, *Z. Phys. D* **24**, 247 (1992).
85. J. Jortner, *Z. Phys. Chem.* **184**, 283 (1994).
86. J. Jortner, *J. Chim. Phys.* **92**, 205 (1995).
87. P. Alivisatos, *Science* **271**, 933 (1996).
88. J. Jellinek, ed., *Theory of Atomic and Molecular Clusters*, Springer, Berlin, 1999.
89. *Clusters of Atoms and Molecules*, H. Haberland, ed., Springer, Berlin, 1994.
90. A. Heidenreich, I. Last, U. Even, and J. Jortner, *Phys. Chem. Chem. Phys.* **3**, 2325 (2001).
91. C. Bréchignac, Ph. Cahuzac, F. Carliez, and M. Frutos, *Phys. Rev. Lett.* **64**, 2893 (1990).
92. U. Näher, S. Bjornholm, S. Fraundorf, F. Gracis, and C. Guet, *Phys. Rep.* **285**, 245 (1997).
93. I. Last, I. Schek, and J. Jortner. *J. Chem. Phys.* **107**, 6685 (1997).
94. I. Last, Y. Levy, and J. Jortner, *Proc. Natl. Acad. Sci. USA* **99**, 8107 (2002).
95. I. Last and J. Jortner, *Phys. Rev. Lett.* **87**, 033401 (2001).
96. I. Last and J. Jortner, *J. Chem. Phys.* **121**, 3030 (2004).
97. I. Last and J. Jortner, *J. Chem. Phys.* **121**, 8329 (2004).
98. I. Last and J. Jortner, *Chem Phys Chem* **3**, 845 (2002).
99. M. Farnik, U. Henne, B. Samelin, and J. P. Toennies, *Phys. Rev. Lett.* **81**, 3892 (1998).
100. E. L. Knuth, S. Schaper, and J. P. Toennies, *J. Chem. Phys.* **120**, 235 (2004).
101. F. Buyvol-Kot, A. Kalinin, O. Kornilov, J. P. Toennies, and J. A. Becker, *Solid State Commun.* (2005), in press.
102. G. Tejeda, J. M. Fernandez, S. Montero, D. Blume, and J. P. Toennies, *Phys. Rev. Lett.* **92**, 223401 (2004).
103. S. Grebenev, B. Sartakov, J. P. Toennies, and A. F. Vilesov, *Science* **289**, 1532 (2000).
104. Y. Kwon and K. B. Whaley, *Phys. Rev. Lett.* **89**, 28340 (2002).

105. R. Bruhl, R. Guardiola, A. Kalinin, O. Kornilov, J. Navarro, T. Savas, and J. P. Toennies, *Phys. Rev. Lett.* **92**, 185301 (2004).

106. F. Dalfovo and S. Stringari, *J. Chem. Phys.* **115**, 10078 (2001).

107. R. E. Griesenti, W. Schollkopf, J. P. Toennies, G. C. Hegerfeld, T. Köhler, and M. Stoll, *Phys. Rev. Lett.* **85**, 2284 (2000).

108. A. Kalinin, O. Kornilov, L. Rusin, J. P. Toennies, and G. Vladimirov, *Phys. Rev. Lett.* **93**, 163402 (2004).

109. E. Braaten and H. W. Hammer, *Phys. Rev. A* **67**, 042706 (2003).

110. V. Efimov, *Phys. Lett* **33B**, 563 (1970).

111. T. K. Lim, S. K. Duffy, and W. C. Damert, *Phys. Rev. Lett.* **38**, 341 (1977).

112. J. Harms, J. P. Toennies, and P. Delfovo, *Phys. Rev. B* **58**, 3341 (1998).

113. S. Stringari and J. Treiner, *J. Chem. Phys.* **87**, 5021 (1987).

114. L . V. Hau, B. D. Busch, C. Liu, Z. Dutton, M. M. Burns, and J. A. Golovchenko, *Phys. Rev. A* **58**, R54 (1998).

115. D. J. Han, R. H. Wynar, P. H. Courteille, and D. J. Heinzen, *Phys. Rev. A* **57**, R4114 (1998).

116. U. Ernst, A. Marte, F. Schreck, J. Schuster, and G. Rempe, *Europhys. Lett.* **41**, 1 (1998).

117. L. D. Landau and E. M. Lifshitz, *Course of Theoretical Physics*, Vol. 5, *Statistical Mechanics*, Pergamon, London, 1959.

118. F. London, *Superfluids. II. Macroscopic Theory of Superfluid Helium*, Dover, New York, 1954.

119. R. K. Pathria, *Statistical Mechanics*, Pergamon, Oxford, 1985.

120. F. London, *Nature* **141**, 643 (1938).

121. E. Hadi, *Physica* **41**, 289 (1969).

122. H. A. Kierstead, *Phys. Rev.* **162**, 153 (1967).

123. S. M. Apenko, *Phys. Rev. B* **60**, 3052 (1999).

124. P. Gruter, D. M. Ceperley, and F. Laloe, *Phys. Rev. Lett.* **79**, 3549 (1997).

125. M. C. Gordillo and D.M. Ceperley, *Phys. Rev. Lett.* **85**, 4735 (2000).

126. S. Grossman and M. Holtnaus, *Zeit. Für Phys. B* **97**, 319 (1995).

127. W. Ketterle and N. J. van Druten, *Phys. Rev. A* **54**, 656 (1996).

128. R. Krishna and K. B. Whaley, *J. Chem. Phys.* **93**, 746 (1990).

129. S. A. Chin and E. Krotschek, *Phys. Rev. B* **52**, 10405 (1995).

130. Ph. Nozieres and D. Pines, *The Theory of Quantum Liquids*, Vol. II, Addison-Wesley, Reading, MA, 1990.

131. O. Penrose and L. Onsager, *Phys. Rev.* **104**, 576 (1956).

132. C. N. Yang, *Rev. Mod. Phys.* **34**, 694 (1962).

133. L. Landau and I. M. Lifshitz, *Statistical Physics*, Pergamon Press, London, 1958, Chapter 14.

134. V. L. Ginzburg and L. P. Pitaevskii, *Sov. Phys. JETP* **7**, 858 (1958).

135. V. L. Ginzburg and A. A. Sobyanin, *Sov. Phys. Uspekhi* **19**, 773 (1977).

136. D. N. Bogoliobov, *J. Phys. (Moscow)* **11**, 23 (1947).

137. G. Careri, ed., *Liquid Helium*, Proceedings of the International School of Physics "Enrico Fermi," Course 21, Academic Press, New York, 1963.

138. L. P. Pitaevski, *Sov. Phys. JETP* **4**, 439 (1957).

139. R. P. Feynman, *Phys. Rev.* **94**, 262 (1954).

140. A. Bijil, *Physica* **7**, 869 (1940).

141. N. H. March and M. Parrinello, in *Collective Effects in Solids and Liquids*, Adam Hilger, Bristol, 1982.
142. P. Bosi, F. Dupre, F. Menzinger, F. Sacchetti, and M. C. Spinelli, *Lett. Novo Cimento* **71**, 436 (1978).
143. O. W. Dietrich et al., *Phys. Rev. A* **5**, 1377 (1972).
144. W. J. Buyers, V. F. Sears, P. A. Lonngi, and D. A. Lonngi, *Phys. Rev. A* **11**, 697 (1975).
145. J. Carneiro, N. Nielsen, and J. McTauge, *Phys. Rev. Lett.* **30**, 481 (1973).
146. T. D. Lee and C. N. Yang, *Phys. Rev.* **105**, 1119 (1957).
147. M. Cohen and R. P. Feynan, *Phys. Rev.* **107**, 13 (1957).
148. R. S. Berry, Ref. 88, p. 1.
149. V. L. Ginzburg and A. A. Sobyanin, *JETP Lett.* **15**, 242 (1972).
150. A. Aculichev and V. A. Bulanov, *JETP Lett.* **38**, 329 (1974).
151. P. Sindzingre, D. M. Ceperley, and M. L. Klein, *Phys. Rev. Lett.* **67**, 1871 (1991).
152. M. Gordillo and D. M. Ceperley, *Phys. Rev. Lett.* **90**, 3010 (1997).
153. H. J. Maris, G. M. Siedel, and F. I. B. Williams, *Phys. Rev. B* **36**, 6799 (1987).
154. F. Iachello, Ref. 24, p. 418
155. E. L. Pollock and K. J. Runger, *Phys. Rev. B* **46**, 3535 (1992).
156. E. G. Syskakis, F. Pobell and H. Ullmaier, *Phys. Rev. Lett.* **55**, 2964 (1985).
157. T. P. Chen and F. M. Gasparini, *Phys. Rev. Lett.* **40**, 331 (1978).
158. W. Huhn and V. Dohm, *Phys. Rev. Lett.* **61**, 1368 (1988).
159. J. Yoon and M. H. W. Chan, *Phys. Rev. Lett.* **78**, 4801 (1997).
160. G. M. Zassenhaus and J. D. Reppy, *Phys. Rev. Lett.* **83**, 4800 (1999).
161. F. Gasparini, G. Agnelot and J. D. Reppy, *Phys. Rev. B* **29**, 138 (1984).
162. I. Rhee, F. M. Gasparini, and D. J. Bishop, *Phys. Rev. Lett.* **63**, 410 (1989).
163. M. Bixon and J. Jortner, *J. Chem. Phys.* **91**, 1631 (1989).
164. R. S. Berry, *Israel J. Chem.* **44**, 211 (2004).
165. M. Schmidt, R. Kusche, T. Hippler, J. Donges, W. Krönmüllere, B. Von Issendorf, and H. Haberland, *Phys. Rev. Lett.* **86**,1191 (2001).
166. R. M. Ziff, G. E. Uhlenbeck, and M. Kac, *Phys. Rep. Phys. Lett.* **32C**, 169 (1977).
167. N. Angelescu, J. G. Brankov, and A. Verbeune, *J. Phys. A* **29**, 3341 (1996).
168. M. Gajda and K. Rzazewski, *Phys. Rev. Lett.* **78**, 2686 (1997).
169. P. Navez, D. Bitouk, M. Gajda, Z. Idziaszek, and K. Rzazewski, *Phys. Rev. Lett.* **79**, 1789 (1997).
170. M. Wilkens and C. Weiss, *J. Mod. Opt.* **44**, 1801 (1997).
171. R. Busani, M. Folkers, and O. Cheshnovsky, *Phys. Rev. Lett.* **81**, 3836 (1998).
172. O. C. Thomas, W. Zeng, and K. H. Bowen, *Phys. Rev. Lett.* **89**, 213403 (2002).
173. B. von Issendorff and O. Cheshnovsky, *Annu. Rev. Phys. Chem.* **56**, 549 (2005).
174. I. Becker, G. Markovich, and O. Cheshnovsky, *Phys. Rev. Lett.* **79**, 3391 (1997).
175. I. Becker and O. Cheshnovsky, *J. Chem. Phys.* **110**, 6288 (1999).
176. D. Scharf, U. Landman, and J. Jortner, *Phys. Rev. Lett.* **54**, 1860 (1985).
177. V. Bonacic-Koutecky and R. Mitric, *Chem. Rev.* **105**, 11 (2005).
178. M. Rosenblit and J. Jortner, *J. Chem. Phys.* **101**, 3029 (1994).

179. M. Rosenblit and J. Jortner, *Phys. Rev. B* **52**, 17461 (1995).
180. M. Rosenblit and J. Jortner, *J. Chem. Phys.* **101**, 9982 (1994).
181. A. H. Zewail, *Femtochemistry: Ultrafast Dynamics of the Chemical Bond*, World Scientific, Singapore, 1994.
182. T. M. Bernhard, J. Hagen, L. D. Socasiu, R. Mitric, A. Heidenreich, J. LeRoux, D. Popolan, M. Vaida, L. Wöste, V. Bonacic-Koutecky, and J. Jortner, *ChemPhysChem.* **6**, 243 (2005).
183. J. Jortner, *Philos. Trans. R. Soc. (London)* **356**, 477 (1997).
184. N. Del Fatti, C. Flytzanis, and F. Vallee, *Appl. Phys. B* **68**, 433 (1999).
185. N. Del Fatti, C. Viosin, F. Chevy, F. Vallee, and C. Flytzanis, *J. Chem. Phys.* **110**, 11484 (1999).
186. I. Last and J. Jortner, *J. Chem. Phys.* **120**, 1348 (2004).
187. M. Rosenblit and J. Jortner, *Phys. Rev. Lett.* **75**, 4079 (1995).
188. M. Rosenblit and J. Jortner, *J. Phys. Chem. A* **101**, 751 (1997).
189. Lord Rayleigh, *Philos. Mag.* **14**, 184 (1882).
190. N. Bohr and J. A. Wheeler, *Phys. Rev.* **56**, 426 (1939).
191. I. Last and J. Jortner, *J. Chem. Phys.* **120**, 1336 (2004).
192. F. M. Gasparini, T. P. Chen, and B. Bhattacharyya, *Phys. Rev. B* **23**, 5797 (1981).
193. M. E. Fisher, in *Critical Phenomena, Proceedings of the International School of Physics "Enrico Fermi," Course 51*, M. S. Green, ed., Academic Press, New York, 1971.
194. V. Privman, P. C. Hohenberg and A. Aharony, *Phase Transitions,* Vol. 14, Academic Press, New York, 1991, p. 4.
195. M. N. Barber, in *Phase Transitions and Critical Phenomena*, Vol. 8, C. Domb and J. L. Lebowitz, eds., Academic, New York, 1983, p. 145.
196. *Finite Size Scaling and Numerical Simulations of Statistical Systems*, V. Privman, ed., World Scientific, Singapore, 1990.
197. E. L. Pollock, in *Phase Transitions and Critical Phenomena*, Vol. 8, Academic Press, London, 1983, p. 146.
198. V. Privman and M. E. Fisher, *J. Stat. Phys.* **33**, 385 (1983).
199. J. Jortner, *J. Chem. Phys.* **119**, 11335 (2003).
200. L. S. Goldner, N. Mulders, and G. Ahlres, *J. Low Temp. Phys.* **93**, 131 (1993).
201. B. D. Josephson, *Phys. Lett.* **21**, 608 (1966).
202. J. C. Le Guillou and J. Zinn-Justin, *Phys. Rev. B* **21**, 3976 (1980).
203. P.B. Weichman and M.E. Fisher, *Phys. Rev. B* **34**, 7652 (1985).
204. M. E. Fisher and V. Privman, *Phys. Rev. B* **32**, 447 (1985).
205. Y.-H. Li and S. Teitel, *Phys. Rev. B* **40**, 9122 (1989); W. Janke, *Phys. Lett. A* **148**, 306 (1990).
206. M.-C. Cha, M. P. A. Fisher, S. M. Girvin, M. Wallin, and A. P. Young, *Phys. Rev. B* **44**, 6883 (1991).
207. L. Meyer and F. Reif, *Phys. Rev.* **119**, 1164 (1960).
208. T. Jiang, C. Kim, and J.A. Northby, *Phys. Rev. Lett.* **71**, 700 (1993).
209. J. A. Northby, C. Kim, and T. Jian, *Physica (Amsterdam)* **197B**, 426 (1994).
210. N. Kestner, J. Jortner, M. H. Cohen, and S. A. Rice, *Phys. Rev.* **140**, A56 (1965).
211. M. V. Rama Krishna and K. B. Whaley, *Phys. Rev. B* **38**, 11839 (1988).
212. B. R. Springett, J. Jortner, M.H. Cohen, *J. Chem. Phys.* **48**, 2720 (1968).
213. J. Jortner, N. Kestner, M. H. Cohen, and S. A. Rice, *J. Chem. Phys.* **43**, 2614 (1965).

214. K. Hiroiki, N. Kestner, S. A. Rice, J. Jortner, *J. Chem. Phys.* **43**, 2625 (1965).
215. E. Cheng, M. W. Cole, M. H. Cohen, *Phys. Rev. B* **50**,1136 (1994); *Phys. Rev. B* **50**,16134 (1994).
216. B. Space, D. Coker, Z. Liu, B. J. Berne, and G. Martyna, *J. Chem. Phys.* **97**, 2002 (1992).
217. W. T. Sommer, *Phys. Rev. Lett.* **12**, 271 (1964).
218. M. A. Woolf, G. W. Rayfield, *Phys. Rev. Lett.* **15**, 235 (1965).
219. J. R. Broomall, W. D. Johnson, D. G. Onn, *Phys. Rev. B* **14**, 2919 (1976).
220. B. Plenkiewicz, P. Plenkiewicz, J. P. Jay-Garin, *Chem. Phys. Lett.* **178**, 542 (1989).
221. K. Martini, J. P. Toennies, C. Winkler, *Chem. Phys. Lett.* **178**, 429 (1991).
222. N. Schwenter, E. E. Koch, J. Jortner, *Electronic Excitation in Condensed Rare Gases*, Springer, Berlin, 1985.
223. M. W. Cole and M. H. Cohen, *Phys. Rev. Lett.* **23,** 1238 (1969).
224. M. W. Cole, *Phys. Rev. B* **2**, 4239 (1970).
225. M. W. Cole, *Phys. Rev. B* **3**, 4418 (1971).
226. V. B. Shikin, *Sov. Phys. JETP* **31**, 936 (1970).
227. M. W. Cole, *Rev. Mod. Phys.* **46**, 451 (1974).
228. V. S. Edelnam, *Sov. Phys. Usp.* **23**, 227 (1980).
229. J. P. Hernandez, *Rev. Mod. Phys.* **63**, 675 (1991).
230. P. Leiderer, *J. Low Temp.* **87**, 247 (1992).
231. F. G. Saville, J. M. Goodkind, and P. M. Platzman, *Phys. Rev. Lett.* **70**, 1517 (1993).
232. A. L. Fetter, in *The Physics of Liquid and Solid Helium*, Part 1, K. H. Benneman and J. B. Ketterson, eds. Wiley, New York, 1976, p. 207.
233. B. R. Springett, J. Jortner, and M. H. Cohen, *Phys. Rev.* **159**, 183 (1967).
234. J. A. Northby and T. M. Sanders, *Phys. Rev. Lett.* **18**, 1184 (1967).
235. T. Miyakawa and D. L. Dexter, *Phys. Rev. A* **1**, 513 (1970).
236. C. C. Grimes and G. Adams, *Phys. Rev. B* **41**, 6366 (1990); **45**, 2305 (1992).
237. A. Ya. Parshin and V. Pereverzev, *Sov. Phys. JETP* **74**, 68 (1992).
238. W. B. Fowler and D. L. Dexter, *Phys. Rev.* **176**, 337 (1976).
239. W. Schoepe and G. W. Raydfield, *Phys. Rev. A* **7**, 2111 (1973).
240. M. H. Cohen and J. Jortner, *Phys. Rev.* **180**, 238 (1969).
241. J. Gspann, *Physica* **B169**, 519 (1991).
242. K. Martini, J. P. Toennies, and C. Winkler, *Chem. Phys. Lett.* **178**, 429 (1991).
243. U. Henne and J. P. Toennies, *J. Chem. Phys.* **108**, 9327 (1998).
244. M. Farnik, B. Samelin, and J. P. Toennies, *J. Chem. Phys.* **110**, 9195 (1999).
245. M. Farnik and J. P. Toennies, *J. Chem. Phys.* **118**, 4176 (2003).
246. S. Grebenev, J. P. Toennies, and A. F. Vilesov, *Science* **279**, 2083 (1998).
247. C. Ebner and W. F. Saam, *Phys. Rev. B* **12**, 923 (1975).
248. J. Wilks, *Liquid and Solid Helium*, Oxford University Press, Oxford, 1967.
249. L. Onsager, in *Modern Quantum Chemistry, Istanbul Lectures*, Vol. II, O. Sinanoglu, ed., Academic Press, New York, 1965, p. 123.
250. J. Jortner, S. A. Rice, and N. R. Kestner, in *Modern Quantum Chemistry, Istanbul Lectures*, Vol. II, O. Sinanoglu, ed., Academic Press, New York, 1965, p. 129.
251. M. Rosenblit and J. Jortner, Unpublished results, 1997, quoted in Ref. 245, and to be published.

252. D. G. Onn and M. Silver, *Phys. Rev.* **183**, 295 (1969).
253. M. Bixon and J. Jortner, *Adv. Chem. Phys.* **107**, 35 (1999).
254. F. Ancilotto and F. Toigo, *Phys. Rev. B* **50**, 12820 (1994).
255. J. L. Bohn, B. D. Esry, and C. H. Greene, *Phys. Rev. A* **58**, 584 (1998).
256. D. S. Jin, J. R. Ensher, M. R. Matthews, C. E. Wieman, and E. A. Cornell, *Phys. Rev. Lett.* **77**, 420 (1996).
257. L. M. Delvs, *Nuclear Phys.* **9**, 391 (1959); **20**, 268 (1962).
258. J. L. Ballot and M. Fabre de la Riepelle, *Ann. Phys. (N.Y.)* **127**, 62 (1980).
259. U. Fano, *Rep. Prog. Phys.* **46**, 97 (1983).
260. J. Avery, *Hyperspherical Harmonics: Applications in Quantum Theory*, Kluwer, Dordrecht, 1989.
261. M. Born and K. Huang, *Dynamical Theory of Crystal Lattices*, Oxford at the Calderon Press, Oxford, 1956, Appendix VII, p. 402.
262. Y. F. Smirnov and K. V. Shitikova, *Sov. J. Par. Nucl.* **8**, 344 (1977).
263. E. R. I. Abraham, W. I. McAlexander, C. A. Sackett, and R. G. Hulet, *Phys. Rev. Lett.* **74**, 1315 (1995).
264. M. Hubiers and H. T. C. Stoof, *Phys. Rev. A* **54**, 5055 (1996).
265. H. T. C. Stoof, *J. Stat. Phys.* **87**, 1353 (1997).
266. M. Ueda and A. J. Leggett, *Phys. Rev. Lett.* **80**, 1576 (1998).
267. J. Jortner and M. Bixon, in *Femtochemistry and Femtobiology: Ultrafast Reaction Dynamics on Atomic-Scale Resolution*, V. Sundström, ed., Imperial College Press, 1977, p. 349.
268. L. Pruvost, private communication.
269. S. A. Rice and J. Jortner, in *Physics of Solids at High Pressures*, C. T. Tomizuka and R. M. Emrich, eds., Academic Press, New York, 1965, pp. 163–170.
270. A. F. Andreev and I. M. Lifshitz, *Sov. Phys. JETP* **29**, 1107 (1969).
271. A. J. Legget, *Phys. Rev. Lett.* **25**, 1543 (1970).
272. G. Chester, *Phys. Rev. A* **2**, 256 (1970).
273. B. G. Levy, *Phys. Today*, April 2004, p. 21.
274. E.-S. Kim and M. H. W. Chan, *Nature* **247**, 225 (2004).
275. H. T. C. Stoof, Ref. 24, p. 226.
276. Yu. Kagan, Ref. 24, p. 202.
277. M. Rosenblit and J. Jortner (to be published).

AUTHOR INDEX

Adventures in Chemical Physics: A Special Volume in Advances in Chemical Physics, Volume 132, edited by R. Stephen Berry and Joshua Jortner. Series editor Stuart A. Rice

SUBJECT INDEX

Adventures in Chemical Physics: A Special Volume in Advances in Chemical Physics, Volume 132, edited by R. Stephen Berry and Joshua Jortner. Series editor Stuart A. Rice